DIGITAL
LOGIC
FUNDAMENTALS

Thomas L. Floyd

Mayland Technical Institute

CHARLES E. MERRILL PUBLISHING COMPANY
A Bell & Howell Company
Columbus   Toronto   London   Sydney

Published by
CHARLES E. MERRILL PUBLISHING COMPANY
A Bell & Howell Company
Columbus, Ohio 43216

This book was set in Times New Roman.
The production editors were James Leeke and Frances Margolin.
The cover was prepared by Will Chenoweth.

International Standard Book Number: 0-675-08495-4

Library of Congress Catalog Card Number: 77-70236

Printed in the United States of America

6 7 8   81 80

To
Jean, Debbie, and Cindy

Preface

Because of the increasing importance of digital techniques in a wide range of applications, it is imperative that the electronics technician and technologist have a thorough understanding of digital logic and its applications in the field of digital electronics.

This book provides a thorough introduction to digital logic and related topics for the electronics or electrical engineering technology student. It also provides an excellent review or reference for the more advanced reader. The level of the material and the technical content make this book ideal for the community college, the technical school, or the first digital course in a four-year program. The reader should have a background in basic electric circuitry and some fundamental exposure to semiconductor circuit theory in order to derive the maximum benefit from the study of this text.

Throughout the text, functional and logical operation and basic concepts are emphasized. Discussions of basic circuitry are used to reinforce and round out the reader's understanding of the functional or logical operation of a given device and to familiarize the reader with current circuit technologies. This approach is demanded by the advent of the integrated circuit (IC) in increasingly complex forms where many individual circuits are constructed on a single small chip of semiconductor material and encapsulated in a single package. Because of this technology, the understanding of the functional and operational characteristics is generally of greater importance from an applications viewpoint than is the actual circuit design. Also, circuit technology tends to change rapidly, but the basic logic functions remain unchanged. For example, an AND gate is an AND gate regardless of whether it is implemented with TTL circuitry or CMOS circuitry.

The material can be covered satisfactorily in a one-semester course or in two quarters. By omitting some topics and utilizing a selective presentation of those remaining, one can use the text for a one-quarter course. A large selection of problems appears at the end of each chapter to reinforce the material covered and to provide experience in practical problem solving. Answers to the odd-numbered problems are given at the back of the book, along with a glossary of important terms used throughout the text.

This book is the result of the efforts of many people. In particular, I would like to express my thanks to Sam Oppenheimer and Roger Everett, Broward Community College, and Richard Sandige, Texas A & M University, who reviewed the manuscript and provided many valuable comments and suggestions. Also, my appreciation goes to those who contributed moral or material support during the writing of this text. I sincerely hope that this book meets the needs of those who use it.

Contents

CHAPTER 4   Logic Gates   67

CHAPTER 5   Boolean Algebra   107

CHAPTER 6   Combinational Logic   135

CHAPTER 13   **Miscellaneous Digital Circuits and Associated Topics**   **409**

CHAPTER 14   **Digital Integrated Circuits**   **447**

CHAPTER 1

Introduction to Digital Logic

1–1 Digital Electronics

The beginning of digital electronics dates back to 1946 with the development of the first *electronic* digital computer implemented with vacuum-tube circuitry, the Electronic Numerical Integrator and Computer (ENIAC). Actually, the concept of a digital computer can be traced back to Charles Babbage, who, in the 1830's, developed the concept of a mechanical digital computer but never completed a working machine. The first digital computer was not built until around 1944, when a Harvard University professor completed a computer that was essentially electro-mechanical, not electronic, in design.

The term *digital* was derived from the basic method by which these computers perform their operations by counting numbers (digits).

For years the area of digital electronics was confined to applications in computer systems. In recent years, however, digital techniques have found wide application in many other areas of electronics such as telephone systems, telemetry and communication data processing, radar systems, missile guidance, navigation systems, television, automotive applications, and many military and consumer products.

The state of technology has progressed from systems using vacuum-tube circuitry through discrete semiconductor circuitry to integrated circuits in increasingly complex forms.

Digital electronics typically involves circuits and systems (a system is a combination of circuits connected in some way to perform a specified function) in which only two states are utilized. These two states are normally represented within the circuitry by two different voltage levels, one greater in value than the other. Other circuit conditions can also represent the two states. Examples are current values, open or closed switch, and on or off light. In digital systems, combinations of the two states are used to represent symbols, characters, numbers, and other types of information. The binary number system can readily be accommo-

1

dated to digital electronics because it is composed of only two digits, a 0 and a 1, which are called *bits*. The word *bit* is a contraction of the term *binary digit*. The two digits of the binary number system can be represented by the two voltage levels inherent in a digital circuit; one level can represent a 1 and the other a 0. The combination of states produced by several digital circuits together can represent combinations of binary digits that form complete numbers or represent other information.

1–2 The Transistor Switch

The transistor used as a switch (on-off device) is the basic element in a digital circuit. There are two types of transistors most commonly used in discrete and integrated digital circuits: the bipolar junction transistor and the metal oxide semiconductor field-effect transistor, normally called the MOSFET. The bipolar junction transistor is used in both discrete and integrated digital circuits. A *discrete* circuit is one constructed of individual components, and an *integrated* circuit (IC) is one where the entire circuit is fabricated on a single chip of semiconductor material. The MOSFET is in wide use in digital integrated circuits because of characteristics allowing construction of circuits requiring extremely low power consumption and very high packaging densities (large numbers of circuits in extremely small areas on a single chip).

Our purpose in this section is to review some of the basic switching characteristics of these two types of transistors. The theory of transistor operation is beyond the scope and purpose of this book.

First, we will discuss the bipolar transistor as a saturated switching device. The transistor is constructed of a semiconductor material, either silicon or germanium (most transistors used today for switching applications are silicon), and is basically a current-controlled device formed with three regions: the base, the emitter, and the collector. External connections are made to these three points via wire leads. The two types of bipolar junction transistors are the NPN and the PNP, for which the circuit symbols are shown in Figure 1–1.

The bipolar transistor can be operated in any one of three modes: active, saturation, and cutoff. For applications in digital circuits, the saturation and cutoff modes are of the greatest importance. The transistor is in *cutoff* when the base

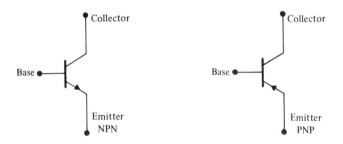

Figure 1–1. **Bipolar Transistor Circuit Symbols:**

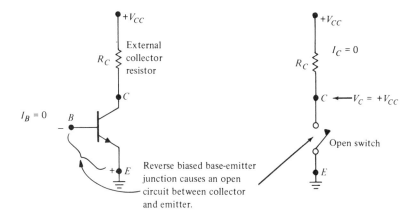

Figure 1–2.   Switch Analogy for an Off Transistor (Common Emitter).

current is zero ($I_B = 0$), resulting in a collector current that is quite small and that can usually be neglected for silicon transistors. The cutoff condition is caused by reverse biasing the base-emitter junction; this means that the base cannot be more than a few tenths of a volt more positive than the emitter for an NPN and a few tenths of volt more negative than the emitter for a PNP. For a silicon transistor this base-to-emitter voltage must be less than about 0.5 V in magnitude. When the transistor is cut off, it is approximately equivalent to an open switch between the collector and the emitter, and therefore essentially no collector current flows. Figure 1–2 illustrates the switch analogy of a transistor in cutoff.

The transistor *saturates* when the base-emitter junction and the base-collector junction are forward biased *and* sufficient base current is provided. To forward bias the base-emitter junction, the base must be made sufficiently more positive than the emitter for an NPN type and sufficiently more negative for a PNP type. Once the transistor turns on, the base is held to within 0.7 V of the emitter; that is, the base-emitter voltage does not exceed 0.7 V when the junction is forward biased (for silicon). Sufficient base current was mentioned as one condition for saturation once the transistor is turned on. The amount of base current required is dependent on the saturation value of collector current ($I_{Csat} \cong V_{CC}/R_C$) and the current gain ($\beta$) of the transistor as expressed by the equation

$$I_B = \frac{I_{Csat}}{\beta_{min}} \tag{1-1}$$

Let us assume the transistor is saturated and examine the characteristics in that state. When enough base current is provided to cause the transistor to saturate, it is approximately equivalent to a closed switch between the collector and the emitter, and therefore the *maximum* collector current flows and is determined by the limiting resistor, R_C. Figure 1–3 illustrates the switch analogy for a saturated transistor.

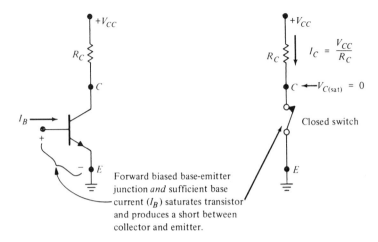

Figure 1–3. Switch Analogy for an Ideal Saturated Transistor.

In an actual bipolar transistor there is not an ideal short between collector and emitter when it is saturated, but rather a low value resistance appears between the two terminals, creating a small voltage drop. This voltage drop, called the collector saturation voltage (V_{CEsat}), is normally only a few tenths of a volt in most silicon transistors. Figure 1–4 illustrates a nonideal representation of a saturated transistor.

The bipolar transistor is used in logic circuits as both an inverting and a noninverting switch. Figure 1–5 shows an inverting transistor switch being driven between cutoff and saturation. This is a *common-emitter* configuration and the output is taken off the collector. Notice that the output is the inverse of the input.

Figure 1–6 shows a noninverting transistor switch known as a *common-collector* or *emitter-follower* circuit. Here the output is taken off the emitter and is noninverted; that is, it is a high level when the input is a high level and it is a low level when the input is a low level. A transistor in the common-collector configuration cannot saturate. Saturation can occur only in the common-emitter circuit.

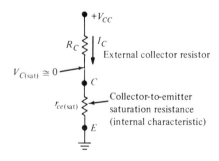

Figure 1–4. Nonideal Analogy for a Saturated Transistor.

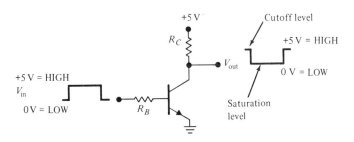

Figure 1-5.   Bipolar Transistor Inverter.

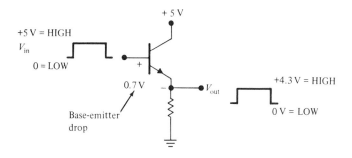

Figure 1-6.   Emitter-Follower Circuit.

If a bipolar transistor is operated in saturation, the speed with which it can switch to the off state is limited by a time delay due to internal charge storage. For higher-speed (faster-switching) applications, the transistor can be prevented from saturating by several methods. Bipolar digital circuits can therefore be classified as *saturating* and *nonsaturating* logic. We will discuss both types in Chapter 14 on integrated circuits.

The second type of transistor used a great deal in digital circuits is the MOSFET mentioned earlier. Because the enhancement MOSFET is the most common MOSFET in digital integrated circuits, we will restrict our discussion to this particular type, which is shown in Figure 1-7. The enhancement MOSFET is a *normally off* device; that is, with no voltage on the gate, there is no current flow between the drain and the source. For an N-channel device, a certain positive voltage on the gate "induces" a channel from the drain to the source and thereby creates a current path. For a P-channel device, a negative gate voltage is required to create a drain-to-source channel. In essence, the proper gate voltage will "turn on" the normally off MOSFET. This basic operation is illustrated in Figure 1-7.

A common source circuit with an N-channel enhancement MOSFET is

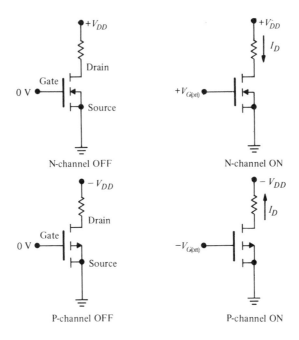

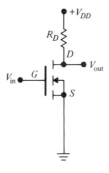

Figure 1–7. Enhancement MOSFET Operation.

shown in Figure 1–8. It is analogous to the common-emitter bipolar circuit (inverter). With no gate voltage, the MOSFET is off and the drain voltage is therefore equal to the supply voltage, V_{DD}. When a sufficiently positive voltage is applied to the gate, the MOSFET is on and the drain voltage is near ground. The common source MOSFET circuit is an inverter.

A common drain (source-follower) circuit is shown in Figure 1–9. It is analogous to the bipolar emitter-follower circuit. With zero gate voltage the source is 0 volts, and with a positive gate voltage the source is at a positive voltage. Therefore, the source follower is a noninverting circuit.

Figure 1–8. MOSFET Inverter.

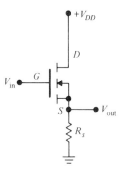

Figure 1-9.   MOSFET Source Follower.

1-3   Logic Levels and Pulse Waveforms

In digital systems two voltage levels represent the two binary digits, 1 and 0. If the higher of the two voltages represents a 1 and the lower voltage represents a 0, this is called *positive logic*. On the other hand, if the lower of the two voltages represents a 1 and the higher voltage represents a 0, we have what is called a *negative logic* system. To illustrate, suppose that we have +5 V and 0 V as our logic-level voltages. We will designate the +5 V as the HIGH level and the 0 V as the LOW level, so positive and negative logic can be defined as follows:

Positive Logic	Negative Logic
HIGH = 1	HIGH = 0
LOW = 0	LOW = 1

Both positive and negative logic are used in digital systems, but positive logic is the more common. For this reason we will use only positive logic in this text.

Pulses are very important in the operation of digital circuits and systems because the voltage levels are normally changing back and forth between the HIGH state and the LOW state. Figure 1-10(a) shows that a single *positive* pulse is generated when the voltage (or current) goes from its *normally* LOW level to its HIGH level and then back to its LOW level. The *negative* pulse in Figure 1-10(b) is generated when the voltage goes from its *normally* HIGH level to its LOW level and back to its HIGH level.

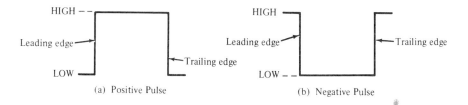

(a) Positive Pulse   (b) Negative Pulse

Figure 1-10.   Ideal Pulses.

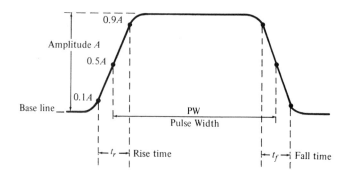

Figure 1–11. **Nonideal Pulse Characteristics.**

As indicated in Figure 1–10, the pulse is composed of two edges: a *leading* edge and a *trailing* edge. For a positive pulse, the leading edge is a *positive-going* transition, or *rising* edge, and the trailing edge is a *negative-going* transition, or *falling* edge. The pulses in Figure 1–10 are ideal because the rising and falling edges change in zero time (instantaneously). Actually, these transitions take time and never occur instantaneously, although for all practical purposes, for most digital logic work we can assume ideal pulses. Figure 1–11 shows a nonideal pulse with finite rising and falling edges. The time required for the pulse to go from its LOW level to its HIGH level is called the *rise time* (t_r), and the time required for the transition from the HIGH level to the LOW level is called the *fall time* (t_f). In actual practice it is common to measure rise time from 10 percent of the pulse amplitude to 90 percent of the pulse amplitude, and to measure the fall time from 90 percent to 10 percent of the pulse amplitude. This is due to nonlinearities that commonly occur near the bottom and the top of the pulse, as indicated in Figure 1–11.

The pulse width (PW) is a measure of the duration of the pulse and is typically defined as the time between the 50 percent points on the rising and falling edges, as indicated in Figure 1–11.

Most waveforms encountered in digital systems are composed of series of pulses and can be classified as *periodic* or *nonperiodic*. A periodic pulse waveform

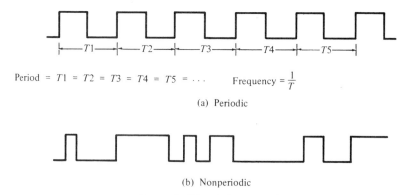

Period = $T1$ = $T2$ = $T3$ = $T4$ = $T5$ = $\ldots$ Frequency = $\dfrac{1}{T}$

(a) Periodic

(b) Nonperiodic

Figure 1–12. **Pulse Waveforms.**

is one that repeats itself at a fixed interval called the period. The frequency is the rate at which it repeats itself and is measured in pulses per second (pps) or Hertz (Hz). A nonperiodic pulse waveform, of course, does not repeat itself at fixed intervals and is composed of pulses of differing pulse widths and/or differing time intervals between the pulses. An example of each type is shown in Figure 1–12.

1–4   Elements of Digital Logic

In its basic form, logic is that realm of human reasoning that tells us a certain proposition is true if certain conditions or premises are true. "The light is on" is an example of a proposition (declarative statement) that can be true or false. "The light bulb is not burned out" and "The light switch is on" are other examples of propositions that can be classified as true or false.

Several propositions, when combined, form propositional or *logic* functions. For example, the propositional statement *"The light is on"* will be true if *"The light bulb is not burned out"* and if *"The light switch is on."* Therefore, the logical statement can be made as follows:

"The light is on if and only if the light bulb is not burned out and the light switch is on."

In this example the first statement is true only if the last two statements are true. The first statement ("The light is on") is then the basic proposition and the other two statements are the conditions or premises upon which the proposition depends.

Many situations, problems, and processes that we encounter in our daily lives can be expressed in the form of propositional or logic functions and can be readily automated. Since these functions are true/false or yes/no statements, it seems that digital circuits with their two-state characteristics are extremely applicable to this area of human endeavor.

A mathematical system for formulating logical statements with symbols, so problems can be written and solved in a manner similar to ordinary algebra, was developed by the Irish logician and mathematician George Boole in the 1850s. *Boolean algebra,* as it is known today, finds wide application in the design and analysis of digital logic circuits and systems.

The term *logic* is applied to digital circuits used to implement logical functions. Several basic digital circuits are the basic *elements* forming the building blocks for complex digital systems such as the computer. We will now look at these elements and discuss their functions in a very general way to give you an overall view of the primary ingredients of digital electronics. Later chapters will cover these circuits in full detail.

The first basic type of logic element is the NOT circuit. The primary function of this circuit is to change one logic level to the opposite logic level. If the higher level is applied to the input of the NOT circuit, the lower level appears on the output. If the lower level is applied to the input, the higher level appears on the output. This circuit is commonly called an *inverter*.

The second type of basic logic element is the AND gate. The primary function of this circuit is to produce a true condition on its output if *all* of its input conditions are true. For instance, let us assume that the higher voltage level (HIGH) represents a true condition and that the lower voltage level (LOW) represents a not-true (false) condition. Let us also use the example statement "The light is on" as our logic function. The AND gate can be used to tell us if this statement is true or false by applying the conditional functions "The light switch is on" and "The light bulb is not burned out" to the gate inputs. If both of these conditional statements are true, then both inputs to the AND gate are HIGH and, as a result, the output is also HIGH. If either condition is false as represented by a LOW, then the output is LOW. In other words, the AND gate "tells" us that *the light is on if and only if the light switch is on AND the light bulb is not burned out.*

The third basic type of logic element is the OR gate. The primary function of this circuit is to produce a true indication on its output when *one or more* of its input conditions are true. Again, let us assume that the higher voltage level represents a true condition and that the lower voltage level represents a false condition. Let us take as an example a door that will open automatically and that can be controlled from a remote transmitter or from a wall switch. The higher voltage level is required to energize the motor in order to open the door. Our propositional statement for this example is "The door opens" and the conditional statements are "The transmitter is on" and "The wall switch is on." The OR gate can be used to "tell" us when the propositional statement is true or false by applying the conditional functions to its inputs. If either one or both of these conditional statements are true, then the output of the OR gate will be HIGH. If neither conditional statement is true, the output of the gate is LOW. For this example, the OR gate "tells" us that *the door opens if the transmitter is on, OR the wall switch is on, OR both are on.* The gate output can be used to effect the opening of the door by properly energizing the activating motor.

The fourth basic type of logic element, which belongs to a class of circuits called multivibrators, is the bistable multivibrator, or *flip-flop*. The primary function of this circuit is to store or "memorize" a binary digit for an indefinite period of time. This device is distinguished from those previously discussed by its ability to retain either logic level after input conditions have been removed. Actually, the flip-flop can be constructed from combinations of basic gates, but it is treated as a distinct logic element because of its great importance in digital systems.

There are several other basic categories of digital logic elements that are normally used in supportive or secondary applications. These are mentioned here and are also given more detailed treatment in later chapters. Two additional types of multivibrators are the one-shot (monostable) and the free-running oscillator (astable). The one-shot is used to generate single pulses that can be used for delay and timing applications. The free-running multivibrator can be used as a source of periodic pulse waveforms for timing purposes in certain applications. The Schmitt trigger is a type of circuit used for wave shaping and noise elimination. There are also various types of storage devices in addition to the flip-flop. These include several types of magnetic devices and capacitive elements. Figure 1-13 shows the pyramid of "building blocks" in a digital system. The basic elements

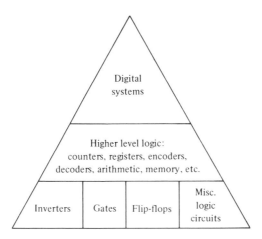

Figure 1–13.   **Building Blocks of a Digital System.**

such as the gates, inverter, flip-flop, and miscellaneous circuits are used to construct higher-level logic circuits such as counters, registers, decoders, memories,
and many others. These more complex logic functions are then combined to form
complete digital systems designed to perform specified tasks.

1–5   Digital Logic Operations

The inverter, the basic gates, and the flip-flop are used in a great number of ways
in digital systems to construct higher-level logic circuits that perform many logic
operations, several of which duplicate the basic functions of the human mind.
Some of the more common logic operations are comparison, addition, subtraction, multiplication, division, counting, memorizing, decoding, encoding, and
multiplexing. We will see what these functions mean in terms of their "block" operations; many are already quite familiar to you because they are common operations.

The *comparison* operation is performed by a logic circuit called a *comparator*.
Its function is to compare two quantities and to indicate if they are equal or not
equal. For example, suppose we have two numbers and we wish to know if they
are equal or not equal, and, if unequal, which is greater. Figure 1–14 is a block

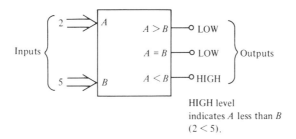

Figure 1–14.   **The General Comparator.**

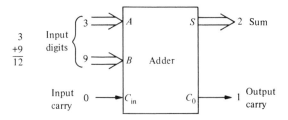

Figure 1–15.   The General Adder.

diagram of a comparator. One number is applied to input *A* and the other number is applied to input *B*. The outputs indicate the relation of the two numbers by producing a HIGH level on the proper output line. Suppose a binary representation of the number 2 is applied to input *A* and a binary representation of the number 5 is applied to input *B* (we will discuss the representation of numbers and symbols in detail in Chapters 2 and 3). A HIGH level will appear on the "*A* < *B*" output, indicating the relationship between the two numbers.

The *addition* operation is performed by a logic circuit called an *adder*. Its function is to add two numbers and generate a *sum* and an *output carry*. The block diagram in Figure 1–15 indicates the addition of the digit 3 and the digit 9. We all know that the sum is 12, and the adder indicates this result by producing the digit 2 on the *sum* output and the digit 1 on the *carry* output. Note that we assume the input carry in this example to be 0. In later chapters we will learn in detail how numbers can be represented by the logic levels of a digital circuit.

Subtraction is the second arithmetic function that can be performed by digital logic circuits. A *subtractor* block diagram is shown in Figure 1–16 indicating three inputs: the two numbers to be subtracted and an input borrow. The two outputs are the *difference* and the *output borrow*. When, for instance, 5 is subtracted from 8 with no input borrow, the difference is 3 with no output borrow. We will see later how subtraction can actually be performed by a basic adder circuit with certain modifications, because subtraction is simply a special case of addition.

The third arithmetic operation that can be performed by logic circuits is *multiplication*. A block diagram illustrating the general *multiplier* is shown in Figure 1–17. Since numbers are always multiplied two at a time, two inputs are required. The output of the multiplier, the *product*, can be one decimal digit larger than either input number (as illustrated in the figure). Since multiplication is sim-

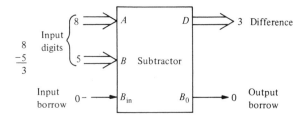

Figure 1–16.   The General Subtractor.

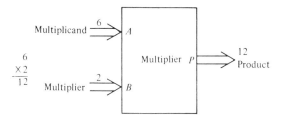

Figure 1–17.   The Multiplication Function.

ply a series of additions with shifts in the positions of the partial products, we will see in a later chapter that multiplication can be performed by a modified adder.

Division, the fourth type of arithmetic operation, turns out to be a series of subtractions, comparisons, and shifts and so can also be performed by a modified adder, as we shall see later. Two inputs to the *divider* are required, and the outputs generated are called the *quotient* and the *remainder*. An example is illustrated in Figure 1–18.

The *encoding* function is one that is probably less familiar than comparison and arithmetic. Basically, the *encoder* converts information such as a decimal number or an alphabetic character into some coded form. For example, a certain type of encoder converts each of the decimal digits, 0 through 9, to a binary code as shown in Figure 1–19. A HIGH level on a given input corresponding to a specific decimal digit produces the proper four-bit code on the output lines.

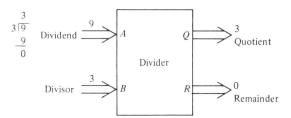

Figure 1–18.   The Division Function.

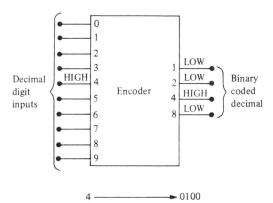

Figure 1–19.   Example of an Encoder.

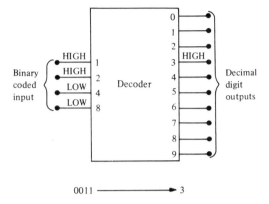

Figure 1–20.   Example of a Decoder.

The *decoder* performs the inverse function of the encoder and converts coded information, such as binary, into a recognizable form such as decimal. For example, a particular type of decoder converts a four-bit binary code into the appropriate decimal digit, as illustrated in Figure 1–20.

The *counting* operation is very important in digital systems, and the device that performs this operation is called, suprisingly enough, a *counter*. There are many types of digital counters, but the basic function is to count events represented by changing levels or pulses or to generate a particular sequence of numbers. In order to count, the counter must "remember" the present number so that it can go to the next proper number in sequence. Therefore, storage or memory capability is an important characteristic of all counters and, generally, flip-flops are used to implement these devices. Another important counter application is frequency division. Figure 1–21 illustrates two simple counter operations.

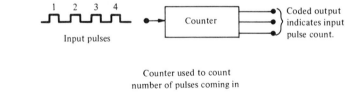

Counter used to count
number of pulses coming in

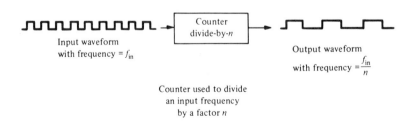

Counter used to divide
an input frequency
by a factor n

Figure 1–21.   Two Examples of Counter Application.

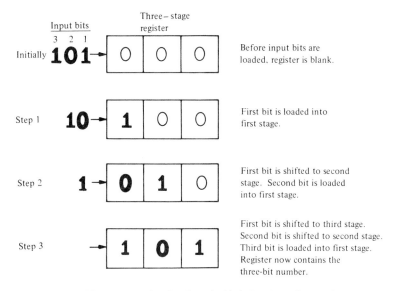

Figure 1–22.   Example of a Simple Shift Register Operation.

Registers are digital circuits used for the temporary storage and shifting of information. For instance, a number in binary form can be stored in a register, and then its position within the register can be changed by shifting it one way or the other. Figure 1–22 illustrates a simple form of register operation.

Multiplexing is an interesting operation performed with digital logic circuits called *multiplexers*. Multiplexing allows information to be switched from several lines onto a single line in some sequence. A simple multiplexer can be represented by a switch operation that sequentially contacts each of the input lines with the output, as illustrated in Figure 1–23. Assume that we have logic levels as indicated on the three inputs (a multiplexer can have any number of inputs). During time interval T1, input *A* is connected to the output; during interval T2, input *B* is connected to the output; and during interval T3, input *C* is connected to the output. As a result of this *multiplexing action*, we have the three logic levels (information) on the inputs appearing in sequence on the output line. The switching action, of course, is accomplished with logic circuits, as we will learn later. It might be well to mention at this point that the inverse of the multiplexing function is called *demultiplexing*. Here, logic data from a single input line are sequentially switched onto several output lines, as shown in Figure 1–24.

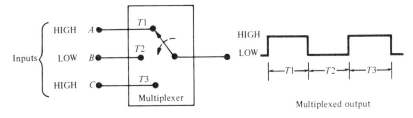

Figure 1–23.   Example of Simple Multiplexer Operation.

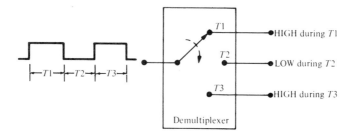

Figure 1–24.   Example of Simple Demultiplexer Operation.

Digital logic circuits can be used to perform a variety of tasks limited only by the imagination. Here, in order to give you a general picture of some of the basic aspects of digital logic, we have touched on only a few of the more basic and common functions that can be used as building blocks for increasingly complex systems. In the following chapters we will study in detail the fundamentals of this fascinating field.

PROBLEMS

1–1   Define the term *bit*. How many types of bits are in a binary number system?

1–2   What is the HIGH output voltage level for the transistor inverter shown in Figure - 1–25? For what input level does this occur?

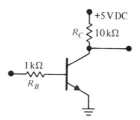

Figure 1–25.

1–3   For the circuit of Problem 1–2, determine the maximum collector current that flows when the transistor is saturated.

1–4   If the transistor in the circuit of Problem 1–2 has a current gain (β) of 50, what is the minimum base current required to saturate it?

1–5   For the pulse shown in Figure 1–26, determine the following :
        (a) rise time   (b) fall time   (c) pulse width   (d) amplitude

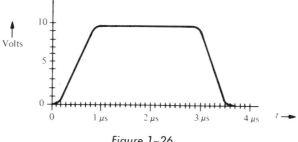

Figure 1–26.

1–6 Determine the period for the pulse waveform graphed in Figure 1–27.

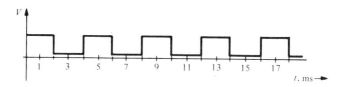

Figure 1–27.

1–7 What is the frequency for the waveform in Problem 1–6?

1–8 Explain the difference between an AND gate and an OR gate.

1–9 Define one characteristic that distinguishes a flip-flop from a gate.

1–10 A basic logic element requires HIGHs on all of the inputs to make the output HIGH. What type of logic circuit is this?

1–11 A basic logic element has a HIGH on one input, a LOW on the other input, and the output is LOW. Identify the element.

1–12 A basic logic element has a HIGH on one input, a LOW on the other input, and the output is HIGH. What is the logic element?

CHAPTER 2

Number Systems

2-1 Decimal Numbers

We will begin our discussion of number systems with the one with which we are all familiar—the decimal system. Because human anatomy is characterized by four fingers and a thumb on each hand, it was only natural that our method of counting involve the use of ten digits, that is, a system with a base of ten. Each of the ten decimal digits, 0 through 9, represents a certain quantity. The ten symbols (digits) do not limit us to expressing only ten different quantities because we use the various digits in appropriate positions within a number to indicate the magnitude of the quantity. We can express quantities up through nine before we run out of digits; if we wish to express a quantity greater than nine we use two or more digits, and the position of each digit within the number tells us the magnitude it represents. If, for instance, we wish to express the quanity twenty-three, we use (by their respective positions in the number) the digit 2 to represent the quantity twenty and the digit 3 to represent the quantity three. Therefore, the position of the digit in the decimal number indicates the magnitude of the quantity represented and can be assigned a "weight." A decimal number is the sum of the products of the digits and their respective weights. The following examples will illustrate:

Example 2-1

$$23 = 2 \times 10 + 3 \times 1 = 20 + 3$$

The digit 2 has a weight of 10, as indicated by its position, and the digit 3 has a weight of 1, as indicated by its position.

Example 2-2

$$568 = 5 \times 100 + 6 \times 10 + 8 \times 1 = 500 + 60 + 8$$

The digit 5 has a weight of 100, the digit 6 has a weight of 10, and the digit 8 has a weight of 1.

2-2   9's and 10's Complements

The 9's and 10's complements of decimal numbers will be defined in this section. The 9's complement of a decimal number is found by subtracting each digit in the number from 9. The 9's complement of each of the decimal digits is as follows:

Digit	9's Complement
0	9
1	8
2	7
3	6
4	5
5	4
6	3
7	2
8	1
9	0

Several examples will illustrate how to determine the 9's complement of a decimal number.

Example 2-3   Find the 9's complement of each of the following decimal numbers: 12, 28, 56, 99, 115, 562, 3497.

Solutions:
To get the 9's complement of a number, we subtract *each* digit in the number from 9.

$$\begin{array}{r} 99 \\ -12 \\ \hline 87 \end{array}$$ 9's complement of 12 $$\begin{array}{r} 99 \\ -28 \\ \hline 71 \end{array}$$ 9's complement of 28

$$\begin{array}{r} 99 \\ -56 \\ \hline 43 \end{array}$$ 9's complement of 56 $$\begin{array}{r} 999 \\ -115 \\ \hline 884 \end{array}$$ 9's complement of 115

$$\begin{array}{r} 999 \\ -562 \\ \hline 437 \end{array}$$ 9's complement of 562 $$\begin{array}{r} 9999 \\ -3497 \\ \hline 6502 \end{array}$$ 9's complement of 3497

The usefulness of the 9's complement stems from the fact that subtraction of a smaller decimal number from a larger one can be accomplished by adding the 9's complement of the number to be subtracted (in this case the smaller number) to the other number and then adding the carry to the result. When subtracting a larger number from a smaller one, there is no carry, and the result is in 9's complement form and negative. This procedure is a distinct advantage in implementing certain types of arithmetic logic, as we will learn in a later chapter. A few examples will demonstrate decimal subtraction using the 9's complement method.

Example 2-4 Perform the following subtractions using the 9's complement method:

Regular Subtraction	9's Complement Subtraction	
8	8	
−3	+6	9's complement of 3
5	① 4	
	+1	add carry to result
	5	
13	13	
−7	+92	9's complement of 07
6	① 05	
	+1	add carry to result
	6	
54	54	
−21	+78	9's complement of 21
33	① 32	
	+1	add carry to result
	33	
15	15	
−28	+71	9's complement of 28
−13	86	9's complement of result
	↓	
	−13	

The 10's complement of a decimal number is equal to the 9's complement plus 1. This is illustrated in the following example.

Example 2-5 Convert the following decimal numbers to their 10's complement form: 8, 17, 52, 428.

Solutions:
First we find the 9's complement, and then we add 1.

9			99		
−8			−17		
1	9's complement of 8		82	9's complement of 17	
+1	add 1		+1	add 1	
2	10's complement of 8		83	10's complement of 17	
99			999		
−52			−428		
47	9's complement of 52		571	9's complement of 428	
+1	add 1		+1	add 1	
48	10's complement of 52		572	10's complement of 428	

The 10's complement, like the 9's complement, can be used to perform subtraction by adding one number to the 10's complement of the number to be subtracted and *dropping* the carry. This is illustrated in the following example:

Example 2-6 Perform the following subtractions using the 10's complement method:

Regular Subtraction	10's Complement Subtraction	
8	8	
− 3	+7	10's complement of 3
5	⥿5	drop carry
13	13	
−7	+93	10's complement of 7
6	⥿06	drop carry
54	54	
−21	+79	10's complement of 21
33	⥿33	drop carry
196	196	
−155	+845	10's complement of 155
41	⥿041	drop carry

Note that the 10's complement has an advantage over the 9's complement in that we do not have to add the carry to the result. However, the 10's complement is not as easy to generate as the 9's complement.

2-3   Binary Numbers

The binary number system is simply another way to count. It is less complicated than the decimal system because it is composed of only *two* digits. It may seem more difficult at first because it is unfamiliar to you.

Just as the decimal system with its ten digits is a base-ten system, the binary system with its two digits is a base-two system. The two binary digits (bits) are 1 and 0. The position of the 1 or 0 in a binary number indicates its "weight" or value within the number, just as the position of a decimal digit determines the magnitude of that digit. The weight of each successively higher position in a binary number is an increasing power of two.

To learn to count in binary, let us first look at how we count in decimal. We start at 0 and count up to 9 before we run out of digits. We then start another digit position and continue counting 10 through 99. At this point we have exhausted all two-digit combinations, so a third digit is needed in order to count from 100 through 999.

A comparable situation occurs when counting in binary, except that we have only two digits. We begin counting—0, 1; at this point we have used both digits, so we include another digit position and continue—10, 11. We have now exhausted all combinations of two digits, so a third is required. With three digits we can count 100, 101, 110, and 111. Now we need a fourth digit to continue, and so on. A binary count of 0 through 31 is shown in Table 2–1.

TABLE 2–1

Count	Decimal Number	Binary Number
zero	0	00000
one	1	00001
two	2	00010
three	3	00011
four	4	00100
five	5	00101
six	6	00110
seven	7	00111
eight	8	01000
nine	9	01001
ten	10	01010
eleven	11	01011
twelve	12	01100
thirteen	13	01101
fourteen	14	01110
fifteen	15	01111
sixteen	16	10000
seventeen	17	10001
eighteen	18	10010
nineteen	19	10011
twenty	20	10100
twenty-one	21	10101
twenty-two	22	10110
twenty-three	23	10111
twenty-four	24	11000
twenty-five	25	11001
twenty-six	26	11010
twenty-seven	27	11011
twenty-eight	28	11100
twenty-nine	29	11101
thirty	30	11110
thirty-one	31	11111

An easy way to remember how to write a binary sequence such as in Table 2–1 is as follows:

1. The right-most column in the binary number begins with a 0 and alternates each bit.
2. The next column begins with two 0s and alternates every two bits.
3. The next column begins with four 0s and alternates every four bits.
4. The next column begins with eight 0s and alternates every eight bits.
5. The left-most column begins with sixteen 0s and alternates every sixteen bits.

As you have seen, it takes five binary digits (bits) to count from 0 to 31. The following formula tells us how high we can count, beginning with zero, with *n* bits:

$$\text{Highest number} = 2^n - 1 \qquad\qquad (2\text{–}1)$$

For instance, with two bits we can count from 0 through 3.

$$2^2 - 1 = 4 - 1 = 3$$

With four bits, we can count from 0 through 15.

$$2^4 - 1 = 16 - 1 = 15$$

A table of powers of two (2^n) is shown in Appendix A.

A binary number is a weighted number, as mentioned previously. The value of a given binary number in terms of its decimal equivalent can be determined by adding the product of each digit and its weight. The right-most digit is the *least significant* digit in a binary number and has a weight of $2^0 = 1$. The weights increase by a power of two for each digit from right to left. The method of evaluating a binary number is illustrated by the following example:

Example 2–7

Binary weight:	2^6	2^5	2^4	2^3	2^2	2^1	2^0
Weight value:	64	32	16	8	4	2	1
Binary number:	1	1	0	1	1	0	1

$$
\begin{aligned}
&1 \times 64 + 1 \times 32 + 0 \times 16 + 1 \times 8 + 1 \times 4 + 0 \times 2 + 1 \times 1 \\
=\ &64\ +\ 32\ +\ 0\ +\ 8\ +\ 4\ +\ 0\ +\ 1 \\
=\ &109
\end{aligned}
$$

This is the equivalent decimal value of the binary number.

The binary numbers we have seen so far have been whole numbers. Fractional numbers can also be represented in binary by placing digits to the right of the binary point just as decimal digits are placed to the right of the decimal point.

The general form of a binary number can be expressed as

$$2^n \cdots 2^3 2^2 2^1 2^0 . 2^{-1} 2^{-2} 2^{-3} \cdots 2^{-n}$$
$$\underset{\text{binary point}}{\Big\uparrow}$$

This indicates that all the digits to the left of the binary point have weights that are positive powers of two, as we have previously discussed. All digits to the right of the binary point have weights that are negative powers of two, or fractional weights as illustrated in the following example:

Example 2–8   Determine the value of the binary fractional number 0.1011.

Solution:

First, we determine the weight of each digit and then sum the weights times the digits.

Binary weight:	2^{-1}	2^{-2}	2^{-3}	2^{-4}
Weight value:	0.5	0.25	0.125	0.0625
Binary number:	0.1	0	1	1

$$1 \times 0.5 + 0 \times 0.25 + 1 \times 0.125 + 1 \times 0.0625$$
$$= \quad 0.5 \quad + \quad 0 \quad + \quad 0.125 \quad + \quad 0.0625$$
$$= \quad 0.6875$$

It should be pointed out here that to determine the decimal value of a binary number, fractional or whole, we simply *add the weights of each 1 and ignore each 0* because the product of a 0 and its weight is 0. Another method of evaluating a binary fraction is to determine the whole-number value of the bits and divide by the total possible combinations of the number of bits appearing in the fraction. For instance, for the binary fraction in Example 2–8:

If we neglect the binary point, the value of 1011 is 11 in decimal.

With four bits, there are 16 possible combinations.

Dividing, we obtain $11/16 = 0.6875$.

The following examples will illustrate the evaluation of binary numbers in terms of their equivalent decimal values:

Example 2–9 Determine the decimal value of 11101.011.

Solution:

$$(16 + 8 + 4 + 1).(0.25 + 0.125) = 29.375$$

Using the alternate method to evaluate the fraction, we have $3/8 = 0.375$ (same result).

Example 2–10 Determine the decimal value of 110101.11.

Solution:

$$(32 + 16 + 4 + 1).(0.5 + 0.25) = 53.75$$

Using the alternate method to evaluate the fraction gives us $3/4 = 0.75$ (same result).

2–4   Binary Addition

There are four basic rules for adding binary digits. They are as follows:

$0 + 0 = 0$	zero plus 0 equals 0
$0 + 1 = 1$	zero plus 1 equals 1
$1 + 0 = 1$	one plus 0 equals 1
$1 + 1 = 10$	one plus 1 equals 0 with a carry of 1 (binary two)

Notice that three of the addition rules result in a single binary digit, and that the addition of two 1s yields a binary two (10). When adding binary numbers, the latter condition creates a sum of 0 in a given column and a carry of 1 over to the next higher column, as illustrated at the top of the next page.

$$
\begin{array}{r}
11 \\
+01 \\
\hline
100
\end{array}
$$

In the right column, $1 + 1 = 0$ with a carry of 1 to the next column.
In the next column, $1 + 1 + 0 = 0$ with a carry of 1 to the next column.
In the left column, $1 + 0 + 0 = 1$.

When there is a carry, we have a situation where three bits are being added (a bit in each of the two numbers and a carry bit). The rules for this are as follows:

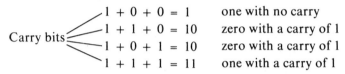

Carry bits
$$
\begin{array}{ll}
1 + 0 + 0 = 1 & \text{one with no carry} \\
1 + 1 + 0 = 10 & \text{zero with a carry of 1} \\
1 + 0 + 1 = 10 & \text{zero with a carry of 1} \\
1 + 1 + 1 = 11 & \text{one with a carry of 1}
\end{array}
$$

The following examples will illustrate binary addition:

Example 2–11

$$
\begin{array}{cccc}
\begin{array}{r} 3 \\ +3 \\ \hline 6 \end{array} &
\begin{array}{r} 11 \\ +11 \\ \hline 110 \end{array} &
\begin{array}{r} 4 \\ +2 \\ \hline 6 \end{array} &
\begin{array}{r} 100 \\ +10 \\ \hline 110 \end{array}
\end{array}
$$

$$
\begin{array}{cccc}
\begin{array}{r} 7 \\ +3 \\ \hline 10 \end{array} &
\begin{array}{r} 111 \\ +11 \\ \hline 1010 \end{array} &
\begin{array}{r} 6 \\ +4 \\ \hline 10 \end{array} &
\begin{array}{r} 110 \\ +100 \\ \hline 1010 \end{array}
\end{array}
$$

$$
\begin{array}{cccc}
\begin{array}{r} 15 \\ +12 \\ \hline 27 \end{array} &
\begin{array}{r} 1111 \\ +1100 \\ \hline 11011 \end{array} &
\begin{array}{r} 28 \\ +19 \\ \hline 47 \end{array} &
\begin{array}{r} 11100 \\ +10011 \\ \hline 101111 \end{array}
\end{array}
$$

2–5 Binary Subtraction

There are four basic rules for subtracting binary digits, which are as follows:

$$
\begin{array}{ll}
0 - 0 = 0 & \text{zero minus 0 equals 0} \\
1 - 1 = 0 & \text{one minus 1 equals 0} \\
1 - 0 = 1 & \text{one minus 0 equals 1} \\
10 - 1 = 1 & \text{two minus 1 equals 1}
\end{array}
$$

When subtracting numbers, we sometimes have to borrow from the next higher column. The only time a borrow is required is when we try to subtract a 1 from a 0. In this case, when a 1 is borrowed from the next higher column, a 10 is created in the column being subtracted, and the last of the four basic rules listed above must be applied. The following examples illustrate binary subtraction:

Example 2–12

$$
\begin{array}{cccc}
\begin{array}{r} 11 \\ -01 \\ \hline 10 \end{array} &
\begin{array}{r} 3 \\ -1 \\ \hline 2 \end{array} &
\begin{array}{r} 11 \\ -10 \\ \hline 1 \end{array} &
\begin{array}{r} 3 \\ -2 \\ \hline 1 \end{array}
\end{array}
$$

No borrows were required in this example.

Example 2-13

$$\begin{array}{rr} 101 & 5 \\ -011 & -3 \\ \hline 010 & 2 \end{array}$$

Let us examine exactly what was done to subtract the two binary numbers.

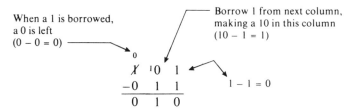

Example 2-14

$$\begin{array}{rr} 1001 & 9 \\ -0110 & -6 \\ \hline 0011 & 3 \end{array}$$

Let us look at the subtraction process for these two numbers.

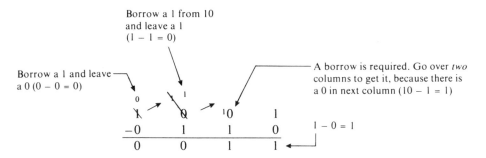

2-6 Binary Multiplication

The following are the four basic rules for multiplying binary digits:

$$\begin{array}{ccccc} 0 & \times & 0 & = & 0 \\ 0 & \times & 1 & = & 0 \\ 1 & \times & 0 & = & 0 \\ 1 & \times & 1 & = & 1 \end{array}$$

Multiplication is performed in binary in the same manner as with decimal numbers. It involves forming the partial products, shifting each successive partial product left one place, and then adding all the partial products. A few examples will illustrate the procedure and the equivalent decimal multiplication.

Example 2-15

11	3	11	3
× 1	× 1	× 11	× 3
11	3	11	9
		11	
		1001	

111	7	1011	11
× 101	× 5	× 1001	× 9
111	35	1011	99
000		0000	
111		0000	
100011		1011	
		1100011	

2-7   Binary Division

Division in binary follows the same procedure as division in decimal.

Example 2-16

$$
\begin{array}{r} 10 \\ 11\overline{)110} \\ 11 \\ \hline 00 \end{array}
\qquad
\begin{array}{r} 2 \\ 3\overline{)6} \\ 6 \\ \hline 0 \end{array}
\qquad
\begin{array}{r} 11 \\ 10\overline{)110} \\ 10 \\ \hline 10 \\ 10 \\ \hline 00 \end{array}
\qquad
\begin{array}{r} 3 \\ 2\overline{)6} \\ 6 \\ \hline 0 \end{array}
$$

$$
\begin{array}{r} 11 \\ 100\overline{)1100} \\ 100 \\ \hline 100 \\ 100 \\ \hline 000 \end{array}
\qquad
\begin{array}{r} 3 \\ 4\overline{)12} \\ 12 \\ \hline 0 \end{array}
\qquad
\begin{array}{r} 10.1 \\ 110\overline{)1111.0} \\ 110 \\ \hline 11\,0 \\ 11\,0 \\ \hline 00\,0 \end{array}
\qquad
\begin{array}{r} 2.5 \\ 6\overline{)15.0} \\ 12 \\ \hline 3\,0 \\ 3\,0 \\ \hline 0 \end{array}
$$

2-8   1's Complement Method of Subtraction

The 1's complement of a binary number is found by simply changing all 1s to 0s and all 0s to 1s, as illustrated by a few examples.

Binary Number	1's Complement
10101	01010
10111	01000
111100	000011
11011011	00100100

Subtraction of binary numbers can be accomplished by the direct method described in Section 2-5 or by using the 1's complement method, which allows us to subtract using only addition.

When subtracting a smaller number from a larger number, the 1's complement method is as follows:

1. Determine the 1's complement of the smaller number.
2. Add the 1's complement to the larger number.
3. Remove the carry and add it to the result. This is called "end-around" carry.

Example 2–17   Subtract 10011 from 11001 using the 1's complement method. Show direct subtraction for comparison.

Solution:

Direct Subtraction	1's Complement Method	
11001	11001	
− 10011	+01100	1's complement of 10011
00110	(1)00101	
	+1	add end-around carry
	00110	final answer

When subtracting a larger number from a smaller, the 1's complement method is as follows:

1. Determine the 1's complement of the larger number.
2. Add the 1's complement to the smaller number.
3. The answer is negative and the 1's complement of the result. There is no carry.

Example 2–18   Subtract 1101 from 1001 using the 1's complement method. Show direct subtraction for comparison.

Solution:

Direct Subtraction	1's Complement Method	
1001	1001	
− 1101	+0010	1's complement of 1101
− 0100	1011	answer in 1's complement form
	→ − 0100 final answer	

The 1's complement method is particularly useful in arithmetic logic circuits because subtraction can be accomplished with an adder. Also, the 1's complement of a number is easily obtained by inverting each bit in the number.

2–9   2's Complement Method of Subtraction

The 2's complement of a binary number is found by adding 1 to the 1's complement. A few examples will show how this is done.

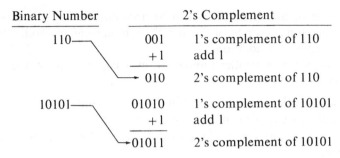

An alternate method of obtaining the 2's complement is demonstrated for the binary number 1101011101000.

1. Start at the right and write the bits *as they are* up to and including the first 1.
2. Take the 1's complement of the remaining bits

First 1 going right to left

$$\underbrace{001010001}_{\substack{\text{1's Complements of} \\ \text{original bits}}}\ \underbrace{1000}_{\text{These bits remain as they were}}$$

Subtraction can be done using the 2's complement method as indicated below.

When subtracting a smaller number from a larger, the 2's complement method is applied as follows:

1. Determine the 2's complement of the smaller number.
2. Add the 2's complement to the larger number.
3. Discard the carry (there is always a carry in this case).

Example 2-19   Subtract 1011 from 1100 using the 2's complement method. Show direct subtraction for comparison.

Solution:

Direct Subtraction	2's Complement Method	
1100	1100	
-1011	$+0101$	2's complement of 1011
0001	✶ 0001	discard carry
	⮡ 0001	final answer

When subtracting a larger number from a smaller one, the 2's complement method is as follows:

1. Determine the 2's complement of the larger number.
2. Add the 2's complement to the smaller number.
3. There is no carry. The result is in 2's complement form and is negative.
4. To get an answer in true form, take the 2's complement and change the sign.

Example 2–20   Subtract 11100 from 10011 using the 2's complement method.

Solution:

Direct Subtraction	2's Complement Method	
10011	10011	
−11100	+00100	2's complement of 11100
−01001	10111	no carry; 2's complement of answer
		⟶ −01001 final answer

At this point, both the 1's and 2's complement methods of subtraction may seem excessively complex compared with direct subtraction. However, as was mentioned before, they both have distinct advantages when implemented with logic circuits because they allow subtraction to be done using only addition. Both the 1's and the 2's complements of a binary number are relatively easy to accomplish with logic circuits; and the 2's complement has an advantage over the 1's complement in that an end-around carry operation does not have to be performed. We will see later how arithmetic operations are implemented with logic circuits using both methods.

2–10   Decimal to Binary Conversion

In Section 2.3 we discussed how to determine the equivalent decimal value of a binary number. Now, you will learn two ways of converting from a decimal to a binary number.

One way to find the binary number equivalent to a given decimal number is to determine the set binary weight values whose sum is equal to the decimal number. For instance, the decimal number 9 can be expressed as the sum of binary weights as follows:

$$9 = 8 + 1 = 2^3 + 2^0$$

By placing a 1 in the appropriate weight positions, 2^3 and 2^0, and a 0 in the other positions, we have the binary number for decimal 9.

$$2^3 \quad 2^2 \quad 2^1 \quad 2^0$$
$$1 \quad \ \ 0 \quad \ \ 0 \quad \ \ 1 \qquad \text{binary nine}$$

Example 2–21   Convert the decimal numbers 12, 25, 58, 82 to binary.

Solutions:

$$12 = 8 + 4 = 2^3 + 2^2 \longrightarrow 1100$$
$$25 = 16 + 8 + 1 = 2^4 + 2^3 + 2^0 \longrightarrow 11001$$
$$58 = 32 + 16 + 8 + 2 = 2^5 + 2^4 + 2^3 + 2^1 \rightarrow 111010$$
$$82 = 64 + 16 + 2 = 2^6 + 2^4 + 2^1 \longrightarrow 1010010$$

A more systematic method of converting from decimal to binary is the *successive division by two* process. For example, to convert the decimal number 12 to binary we begin by dividing 12 by 2, and then we divide each resulting quotient by

2 until we have a 0 quotient. The remainder generated by each division forms the binary number. The first remainder to be produced is the least significant bit in the binary number.

Step 1.

$$2 \overline{)12} \quad \underline{6} \quad \underline{\text{Remainder}}$$
$$\underline{12}$$
$$0 \longrightarrow 0$$

Step 2.

$$2 \overline{)6} \quad 3$$
$$\underline{6}$$
$$0 \longrightarrow 0$$

Step 3.

$$2 \overline{)3} \quad 1$$
$$\underline{2}$$
$$1 \longrightarrow 1$$

Step 4.

$$2 \overline{)1} \quad 0$$
$$\underline{0}$$
$$1 \longrightarrow 1$$

$$\longrightarrow \boxed{1100}$$

Example 2–22   Convert the following decimal numbers to binary: 19, 45.

Solutions:

$$2 \overline{)19} \quad 9 \quad \underline{\text{Remainder}}$$
$$\underline{18}$$
$$1 \longrightarrow 1$$

$$2 \overline{)9} \quad 4$$
$$\underline{8}$$
$$1 \longrightarrow 1$$

$$2 \overline{)4} \quad 2$$
$$\underline{4}$$
$$0 \longrightarrow 0$$

$$2 \overline{)2} \quad 1$$
$$\underline{2}$$
$$0 \longrightarrow 0$$

$$2 \overline{)45} \quad 22 \quad \underline{\text{Remainder}}$$
$$\underline{44}$$
$$1 \longrightarrow 1$$

$$2 \overline{)22} \quad 11$$
$$\underline{22}$$
$$0 \longrightarrow 0$$

$$2 \overline{)11} \quad 5$$
$$\underline{10}$$
$$1 \longrightarrow 1$$

$$2 \overline{)5} \quad 2$$
$$\underline{4}$$
$$1 \longrightarrow 1$$

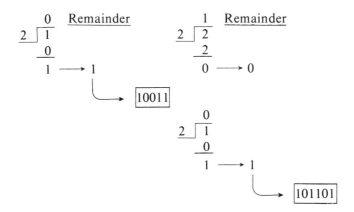

2–11   Octal Numbers

The octal number system is composed of eight digits, which are

0, 1, 2, 3, 4, 5, 6, 7

To count above 7, we begin another column and start over.

10, 11, 12, 13, 14, 15, 16, 17, 20, 21, etc.

Counting in octal is the same as counting decimal, except any number with an 8 or 9 is omitted. Table 2–2 shows the equivalent octal numbers for decimal numbers through 31.

TABLE 2–2

Decimal	Octal
0	0
1	1
2	2
3	3
4	4
5	5
6	6
7	7
8	10
9	11
10	12
11	13
12	14
13	15
14	16
15	17
16	20
17	21
18	22

TABLE 2–2, cont.

Decimal	Octal
19	23
20	24
21	25
22	26
23	27
24	30
25	31
26	32
27	33
28	34
29	35
30	36
31	37

To distinguish octal numbers from decimal numbers, we will use the subscript 8 to indicate an octal number. For instance, 15_8 is equivalent to 13_{10}.

Since the octal number system has a base of eight, each successive digit position is an increasing power of eight, beginning in the right-most column with 8^0. The evaluation of an octal number in terms of its decimal equivalent is accomplished by multiplying each digit by its weight and summing the products, as illustrated below for octal 2374.

$$
\begin{array}{llcccc}
\text{Weight:} & & 8^3 & 8^2 & 8^1 & 8^0 \\
\text{Value:} & & 512 & 64 & 8 & 1 \\
\text{Octal number:} & & 2 & 3 & 7 & 4
\end{array}
$$

$$
\begin{aligned}
2374_8 &= 2 \times 8^3 + 3 \times 8^2 + 7 \times 8^1 + 4 \times 8^0 \\
&= 2 \times 512 + 3 \times 64 + 7 \times 8 + 4 \times 1 \\
&= 1024 + 192 + 56 + 4 = 1276_{10}
\end{aligned}
$$

A method of converting a decimal number into an octal number is the *successive division by eight* method, which is similar to the method used in conversion of decimal to binary. To show how it works, we will convert the decimal number 359 to octal. Each successive division by 8 yields a remainder which is a digit in the equivalent octal number. The first remainder generated is the least significant digit.

$$
\begin{array}{ll}
& \quad\quad \underline{44} \quad\quad \text{Remainder} \\
\text{Step 1.} & 8\,\overline{)\,359} \\
& \quad\quad \underline{32} \\
& \quad\quad\, 39 \\
& \quad\quad\, \underline{32} \\
& \quad\quad\quad 7 \longrightarrow 7 \\
\text{Step 2.} & \quad\quad\, 5 \\
& 8\,\overline{)\,44} \\
& \quad\, \underline{40} \\
& \quad\quad 4 \longrightarrow 4
\end{array}
$$

Step 3.

$$8 \overline{)5} \quad \frac{0}{} \quad \text{Remainder}$$

$$\frac{0}{5} \longrightarrow \quad 5$$

Therefore, $359_{10} = 547_8$

The octal numbers that we have seen to this point have been whole numbers. Fractional octal numbers are represented by digits to the right of the octal point.

The general form of an octal number can be expressed as

$$8^n \cdots 8^3 8^2 8^1 8^0 . 8^{-1} 8^{-2} 8^{-3} \cdots 8^{-n}$$

This shows that all digits to the left of the octal point have weights that are positive powers of eight, as we have seen previously. All digits to the right of the octal point have fractional weights, or negative powers of eight as illustrated in the following example.

Example 2–23   Determine the value of the octal fraction 0.325.

Solution:

First, we determine the weights of each digit and then sum the weight times the digit.

Octal weight:	8^{-1}	8^{-2}	8^{-3}
Value:	0.125	0.01562	0.00195
Octal number:	0.3	2	5

$$0.325_8 = 3(0.125) + 2(0.01562) + 5(0.00195)$$
$$= 0.375 + 0.03124 + 0.00975 = 0.41599_{10}$$

2–12   Conversions Between Octal and Binary Systems

Because all three-bit binary numbers are required to represent the eight octal digits, it is very easy to convert from octal to binary and from binary to octal. For this reason, the octal number system is used in some digital systems especially for input/output applications.

Each octal digit can be represented by three binary digits (bits) as indicated.

Octal Digit	Binary
0	000
1	001
2	010
3	011
4	100
5	101
6	110
7	111

To convert an octal number to a binary number, simply replace each octal digit by the appropriate three binary digits. This is illustrated in the following example:

Example 2–24 Convert each of the following octal numbers to binary: 13, 25, 47, 170, 752, 5276.

Solutions:

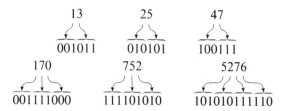

Conversion of a binary number to an octal number is also a straightforward process. Beginning at the right, simply break the binary number into groups of three bits and convert each group into the appropriate octal digit.

Example 2–25 Convert each of the following binary numbers into its octal equivalent: 110101, 101111001, 1011100110.

Solutions:

$$
\begin{array}{ccc}
\underline{110}\,\underline{101} & \underline{101}\,\underline{111}\,\underline{001} & \text{00}\,\underline{1011}\,100\,110 \\
\downarrow\quad\downarrow & \downarrow\quad\downarrow\quad\downarrow & \downarrow\quad\downarrow\quad\downarrow\quad\downarrow \\
6\quad 5 & 5\quad 7\quad 1 & 1\quad 3\quad 4\quad 6
\end{array}
$$

2–13 Hexadecimal Numbers

The hexadecimal is a base-sixteen system; that is, it is composed of 16 digits and characters. Some digital systems process binary data in four-bit groups, making the hexadecimal number very convenient because each hexadecimal digit represents a four-bit binary number (as listed in Table 2–3). Other symbols can be used, but ten digits and six alphabetic characters are most prevalent.

TABLE 2–3

Decimal	Hexadecimal	Binary
0	0	0000
1	1	0001
2	2	0010
3	3	0011
4	4	0100
5	5	0101
6	6	0110
7	7	0111
8	8	1000
9	9	1001
10	A	1010
11	B	1011
12	C	1100
13	D	1101
14	E	1110
15	F	1111

Converting a binary number to hexadecimal is a straightforward procedure. Simply break the binary number into four-bit groups starting at the right and replace each group by the equivalent hexadecimal symbol.

Example 2–26 Convert the following binary numbers to hexadecimal: 1100101001010111, 111111000101101001.

Solutions:

To go from a hexadecimal number to a binary number, we reverse the process and replace each hexadecimal symbol with the appropriate four binary digits.

Example 2–27 Determine the binary numbers for the hexadecimal numbers 10A4, CF83, and 9742.

Solutions:

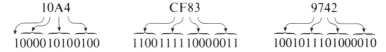

The use of letters to represent quantities may seem strange at first, but keep in mind that *any* number system is only a set of symbols. If we understand what these symbols mean in terms of quantities represented, then the form of the symbols themselves is unimportant once we get accustomed to using them.

It should be clear that it is much easier to write the hexadecimal number than the equivalent binary number, and since conversion is so easy, the hexadecimal system is a "natural" for communicating with digital systems utilizing four-bit groupings of binary numbers.

Finally, how do we count in hexadecimal once we get to F? Simply start over with another column and continue as follows:

10, 11, 12, 13, 14, 15, 16, 17, 18, 19, 1A, 1B, 1C, 1D, 1E, 1F,
20, 21, 22, 23, 24, 25, 26, 27, 28, 29, 2A, 2B, 2C, 2D, 2E, 2F,
30, 31, etc.

PROBLEMS

2–1 Determine the 9's complement of each decimal number.

(a) 3	(b) 5	(c) 8	(d) 12
(e) 17	(f) 25	(g) 49	(h) 86
(i) 127	(j) 381	(k) 690	(l) 1354

2–2 Perform the following subtractions using the 9's complement method:

(a) 6 − 2	(b) 15 − 7	(c) 23 − 14	(d) 48 − 33
(e) 69 − 68	(f) 91 − 70	(g) 98 − 59	(h) 100 − 82

2–3 Perform the following subtractions using the 9's complement method:

(a) 115 − 92	(b) 159 − 125	(c) 298 − 200
(d) 561 − 443	(e) 846 − 709	(f) 1024 − 837

2-4   Determine the 10's complement of each decimal number.
   (a) 7   (b) 19   (c) 36
   (d) 52   (e) 84   (f) 90

2-5   Convert the following binary numbers to decimal:
   (a) 11   (b) 100   (c) 111   (d) 1000
   (e) 1001   (f) 1100   (g) 1011   (h) 1111

2-6   Convert the following binary numbers to decimal:
   (a) 1110   (b) 1010   (c) 11100   (d) 10000
   (e) 10101   (f) 11101   (g) 10111   (h) 11111

2-7   Convert each binary number to decimal.
   (a) 110011.11   (b) 101010.01   (c) 1000001.111
   (d) 1111000.101   (e) 1011100.10101   (f) 1110001.0001
   (g) 1011010.1010   (h) 1111111.11111

2-8   What is the highest decimal number that can be represented by the following number of binary digits (bits)?
   (a) 2   (b) 3   (c) 4   (d) 5
   (e) 6   (f) 7   (g) 8   (h) 9
   (i) 10   (j) 11

2-9   How many bits are required to represent the following decimal numbers?
   (a) 17   (b) 35   (c) 49   (d) 68
   (e) 81   (f) 114   (g) 132   (h) 205
   (i) 271

2-10   Generate the following binary count sequences:
   (a) 0 through 7   (b) 8 through 15   (c) 16 through 31
   (d) 32 through 63   (e) 64 through 75

2-11   Add the binary numbers.
   (a) 11 + 1   (b) 10 + 10   (c) 101 + 11
   (d) 111 + 110   (e) 1001 + 101   (f) 1101 + 1011

2-12   Use direct subtraction on the following binary numbers:
   (a) 11 − 1   (b) 101 − 100   (c) 110 − 101
   (d) 1110 − 11   (e) 1100 − 1001   (f) 11010 − 10111

2-13   Perform the following binary multiplications:
   (a) 11 × 11   (b) 100 × 10   (c) 111 × 101
   (d) 1001 × 110   (e) 1101 × 1101   (f) 1110 × 1101

2-14   Divide the binary numbers as indicated.
   (a) 100 ÷ 10   (b) 1001 ÷ 11   (c) 1100 ÷ 100

2-15   Determine the 1's complement of each binary number.
   (a) 101   (b) 110   (c) 1010
   (d) 11010111   (e) 1110101   (f) 00001

2-16   Perform the following subtractions using the 1's complement method:
   (a) 11 − 10   (b) 100 − 11   (c) 1010 − 111
   (d) 1101 − 1010   (e) 11100 − 1101   (f) 100001 − 1010
   (g) 1001 − 1110   (h) 10111 − 11111

2-17   Determine the 2's complement of each binary number.
   (a) 10   (b) 111   (c) 1001
   (d) 1101   (e) 11100   (f) 10011

2-18   Perform the following subtractions using the 2's complement method:
     (a) $10 - 01$ (b) $111 - 110$ (c) $1101 - 1001$
     (d) $1111 - 1101$ (e) $10111 - 10011$ (f) $10001 - 11100$
     (g) $10101 - 10111$ (h) $1111000 - 1111111$

2-19   Convert each decimal number to binary using the sum of weights method.
     (a)   10   (b)   17   (c)   24   (d)   48
     (e)   61   (f)   93   (g)   125   (h)   186
     (i)   298

2-20   Convert each decimal number to binary by successive division by two.
     (a)   15   (b)   21   (c)   28   (d)   34
     (e)   40   (f)   59   (g)   65   (h)   73
     (i)   99

2-21   Convert each octal number to decimal.
     (a) 12_8 (b) 27_8 (c) 56_8
     (d) 64_8 (e) 103_8 (f) 557_8
     (g) 163_8 (h) 1024_8 (i) 7765_8

2-22   Convert each decimal number to octal by successive division by eight.
     (a)   15   (b)   27   (c)   46   (d)   70
     (e)   100   (f)   142   (g)   219   (h)   435
     (i)   791

2-23   Convert each octal number to binary.
     (a) 13_8 (b) 57_8 (c) 101_8
     (d) 321_8 (e) 540_8 (f) 4653_8
     (g) 13271_8 (h) 45600_8 (i) 100213_8

2-24   Convert each binary number to octal.
     (a)   111   (b)   10   (c)   110111
     (d)   101010   (e)   1100   (f)   1011110
     (g)   101100011001   (h)   10110000011   (i)   111111101111000

2-25   Convert each hexadecimal number to binary.
     (a) 38_{16} (b) 59_{16} (c) $A14_{16}$
     (d) $5C8_{16}$ (e) 4100_{16} (f) $FB17_{16}$

2-26   Convert each binary number to hexadecimal.
     (a)   1110   (b)   10   (c)   10111
     (d)   10100110   (e)   1111110000   (f)   100110000010

CHAPTER 3

Digital Codes and Operations

As you learned in the last chapter, decimal, octal, and hexadecimal numbers can be represented by binary digits. In this chapter we will see that not only numbers, but letters and other symbols, can be represented by 1s and 0s. In fact, *any* entity expressible as numbers, letters, or other symbols can be represented by binary digits, and therefore can be processed by digital logic circuits.

Combinations of binary digits that represent numbers, letters, or symbols are *digital codes*. In many applications, special codes are used for such auxiliary functions as error detection.

3–1 The 8421 Code

The 8421 code is a type of *binary coded decimal* (BCD) code and is composed of four binary digits representing the decimal digits 0 through 9. The designation "8421" indicates the binary weights of the four bits (2^3, 2^2, 2^1, 2^0). The ease of conversion between 8421 code numbers and the familiar decimal numbers is the main advantage of this type of code. The 8421 code is the predominant BCD code, and when we refer to BCD we always mean the 8421 code unless otherwise stated.

"Binary coded decimal" means that *each* decimal digit is represented by a binary code, so all you have to remember are the ten binary combinations that represent the ten decimal digits (as shown in Table 3–1).

You should realize that with four bits, sixteen numbers (2^4) can be represented, and that in the 8421 code only ten of these are utilized. The six code combinations that are not used—1010, 1011, 1100, 1101, 1110, and 1111—are *unallowed* or *invalid* in the 8421 BCD code.

To express any decimal number in 8421 code, simply replace each decimal digit by the appropriate four-bit code, as shown by the following example:

41

TABLE 3–1
The 8421 BCD Code.

8421 (BCD)	Decimal
0000	0
0001	1
0010	2
0011	3
0100	4
0101	5
0110	6
0111	7
1000	8
1001	9

Example 3–1   Convert each of the following decimal numbers into the 8421 code: 3, 9, 18, 34, 65, 92, 150, 321, 1472.

Solutions:

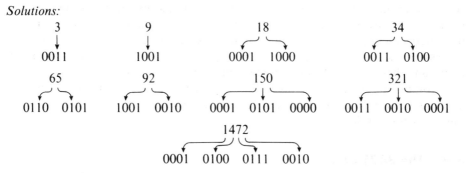

It is equally easy to determine a decimal number from an 8421 code number. Start at the right and break the code into groups of four bits, and then write the decimal digit represented by each four-bit group. An example will illustrate.

Example 3–2   Find the decimal numbers represented by the following 8421 BCD codes: 10000110, 00110001, 1010011, 100101110100, and 1100001100000.

Solutions:

```
1000   0110      0011   0001      101   0011
  ↓      ↓          ↓      ↓        ↓      ↓
  8      6          3      1        5      3

1001   0111   0100      1   1000   0110   0000
  ↓      ↓      ↓        ↓     ↓      ↓      ↓
  9      7      4        1     8      6      0
```

Note: If there are not four bits in the left-most group, zeros are implied.

3–2   BCD (8421) Addition

The 8421 is a numerical code, and many applications require that arithmetic operations be performed. Addition is the most important operation because the other three operations (subtraction, multiplication, and division) can be accomplished using addition. Here is how to add two 8421 BCD numbers:

1.   Add the two numbers, using the rules for binary addition in Section 2–4.
2.   If a four-bit group (sum) is equal to or less than 9, it is a *valid* 8421 BCD number.
3.   If a four-bit group (sum) is greater than 9, or if a carry is generated from the group, it is an *invalid* result. Add 6 (0110) to the four-bit group in order to skip the six invalid states and return the code to 8421. If a carry results when 6 is added, simply add it to the next four-bit group.

Several examples will illustrate 8421 BCD addition for the case where the sum of any four-bit column does not exceed 9.

Example 3–3   Add the following 8421 BCD numbers (the equivalent decimal addition is shown for comparison):

0011		3		0110		6	
+0100		+4		+0010		+2	
0111		7		1000		8	
0010	0011	23		1000	0110	86	
+0001	0101	+15		+0001	0011	+13	
0011	1000	38		1001	1001	99	
0100	0101	0000	450	1000	0111	0011	873
+0100	0001	0111	+417	+0001	0001	0010	+112
1000	0110	0111	867	1001	1000	0101	985

Note that in each case the sum in any four-bit column does not exceed 9, and the results are valid 8421 BCD numbers.

Next, we will deal with the case of an invalid sum (greater than 9 or a carry) by illustrating the procedure with several examples.

Example 3–4   Add the following 8421 BCD numbers:

9	1001	
+4	+0100	
13	1101	invalid BCD number
	+0110	add 6
0001	0011	valid BCD number
↓	↓	
1	3	

```
  9        1001
 +9       +1001
 18      1  0010        invalid because of carry
            +0110       add 6
      0001  1000        valid BCD number
        ↓     ↓
        1     8

 16      0001  0110
+15     +0001  0101
 31      0010  1011     right group is invalid, left group valid
               +0110    add 6 to invalid code (add carry to next group)
         0011  0001     valid BCD number
           ↓     ↓
           3     1

 67      0110  0111
+53     +0101  0011
120      1011  1010     both groups are invalid
         +0110 +0110    add 6 to both groups
   0001  0010  0000     valid BCD number
     ↓     ↓     ↓
     1     2     0
```

3–3 The Excess-3 Code

The Excess-3 is an important BCD code that is derived by adding 3 to *each* decimal digit and then converting the result to four-bit binary. Since no definite weights can be assigned to the four digit positions, Excess-3 is an unweighted code. For instance, the Excess-3 code for the decimal 2 is

$$
\begin{array}{r}
2 \\
+3 \\
\hline
5
\end{array} \rightarrow 0101
$$

The Excess-3 code for the decimal 9 is

$$
\begin{array}{r}
9 \\
+3 \\
\hline
12
\end{array} \rightarrow 1100
$$

The Excess-3 code for each decimal digit is found by the same procedure, and the entire code is shown in Table 3–2.

Notice that ten of a possible 16 code combinations are used in the Excess-3 code. The six *invalid* combinations are 0000, 0001, 0010, 1101, 1110, and 1111.

The key feature of the Excess-3 code is that it is *self-complementing*. This means that the 1's complement of an Excess-3 number is the 9's complement of the corresponding decimal number. For example, the Excess-3 code for the deci-

TABLE 3–2
Excess-3 Code.

Decimal	Excess-3
0	0011
1	0100
2	0101
3	0110
4	0111
5	1000
6	1001
7	1010
8	1011
9	1100

mal 4 is 0111. If we take the 1's complement of this we have 1000, which is the Excess-3 code for the decimal 5 (and 5 is the 9's complement of 4).

It should be noted that the 1's complement is easily produced with digital logic circuits by simply inverting each bit. The self-complementing property makes the Excess-3 code extremely useful in arithmetic operations, because subtraction can be performed using the 9's complement method.

Example 3–5 Convert each of the following decimal numbers to Excess-3 code: 13, 35, 87, 159, 430.

Solutions:

First, add 3 to *each* digit in the decimal number, and then convert each sum to its equivalent binary code.

```
  1     3                        1     5     9
 +3    +3                       +3    +3    +3
 ‾‾    ‾‾                       ‾‾    ‾‾    ‾‾
  4     6                        4     8    12
  ↓     ↓                        ↓     ↓     ↓
0100  0110   Excess-3 for 13   0100  1000  1100   Excess-3 for 159

  3     5                        4     3     0
 +3    +3                       +3    +3    +3
 ‾‾    ‾‾                       ‾‾    ‾‾    ‾‾
  6     8                        7     6     3
  ↓     ↓                        ↓     ↓     ↓
0110  1000   Excess-3 for 35   0111  0110  0011   Excess-3 for 430

  8     7
 +3    +3
 ‾‾    ‾‾
 11    10
  ↓     ↓
1011  1010   Excess-3 for 87
```

3–4   Excess-3 Addition

In this section we will cover addition using Excess-3 numbers. The three rules for Excess-3 addition are

1. Add the Excess-3 numbers using the rules for binary addition (Section 2-4).
2. If there is *no* carry from a four-bit group, *subtract* 3 (0011) from that group to get the Excess-3 code for the digit.
3. If there is a carry from a four-bit group, *add* 3 (0011) to that group to get the Excess-3 code for the digit, and add 3 to any *new* column (digit) generated by the last carry.

No carry from a particular four-bit group indicates that the sum is Excess-6 because we have added two Excess-3 numbers. Therefore, we have to subtract 3 to get back to Excess-3. A carry indicates an invalid result that requires the addition of 3 in order to skip the invalid states and return the digit to Excess-3 form.

Rule 2 is illustrated in Example 3–6, and rule 3 in Example 3–7.

Example 3–6   Convert each of the decimal numbers to Excess-3 and add as indicated.

```
     8      1011        Excess-3 for 8
    +1     +0100        Excess-3 for 1
     9      1111        no carry
            -11         subtract 3
            1100        Excess-3 for 9


    15     0100  1000   Excess-3 for 15
   +12    +0100  0101   Excess-3 for 12
    27     1000  1101   no carries
           -11   -11    subtract 3 from each group
           0101  1010   Excess-3 for 27


    35     0110  1000   Excess-3 for 35
   +24    +0101  0111   Excess-3 for 24
    59     1011  1111   no carries
           -11   -11    subtract 3 from each group
           1000  1100   Excess-3 for 59


   273    0101  1010  0110   Excess-3 for 273
  +126   +0100  0101  1001   Excess-3 for 126
   399    1001  1111  1111   no carries
          -11   -11   -11    subtract 3 from each group
          0110  1100  1100   Excess-3 for 399
```

Example 3–7   Convert each decimal number to Excess-3 and add as indicated.

7	1010	Excess-3 for 7
+6	+1001	Excess-3 for 6
13	1 0011	there is a carry
	+11 +11	add 3 to both groups
	0100 0110	Excess-3 for 13

15	0100 1000	Excess-3 for 15
+15	+0100 1000	
30	1001 0000	carry out of right-most group only
	−11 +11	subtract 3 from left; add 3 to right
	0110 0011	Excess-3 for 30

98	1100 1011	Excess-3 for 98
+86	+1011 1001	Excess-3 for 86
184	1 1000 0100	carry out of both groups
	+11 +11 +11	add 3 to each group
	0100 1011 0111	Excess-3 for 184

In Chapter 12 on arithmetic logic, we will extend the basic rules of addition presented in this chapter and examine how arithmetic operations are performed by digital logic circuits. We will look at all of the special cases of addition, which include subtraction. Multiplication and division will also be considered.

3–5   The Gray Code

Another important code is the Gray code. The Gray code is an unweighted code that exhibits only a *single-bit change* from one code number to the next. This property is important in many applications where data are transmitted from one portion of a system to another and error susceptibility increases with the number of bit changes between adjacent numbers in a sequence. The Gray code is not an arithmetic code.

Table 3–3 is a listing of four-bit Gray code numbers for decimal numbers 0 through 15. Notice the single-bit change between any two Gray code numbers. Binary numbers are shown for reference. For instance, in going from decimal 4 to 5, the Gray code changes from 0110 to 0111. The only bit change is in the right-most bit; the others remain the same.

By representing the ten decimal digits with a four-bit Gray code, we have another form of BCD code. The Gray code, however, can be extended to any number of bits, and conversion between binary code and Gray code is sometimes useful. First, we will discuss how to convert from a binary number to a Gray code number. The following rules apply:

1. The most significant digit (left-most) in the Gray code is the same as the corresponding digit in the binary number.

TABLE 3–3
Four-Bit Gray Code.

Decimal	Gray	Binary
0	0000	0000
1	0001	0001
2	0011	0010
3	0010	0011
4	0110	0100
5	0111	0101
6	0101	0110
7	0100	0111
8	1100	1000
9	1101	1001
10	1111	1010
11	1110	1011
12	1010	1100
13	1011	1101
14	1001	1110
15	1000	1111

2. Going from left to right, add each adjacent pair of binary digits to get the next Gray code digit. Disregard carries.

For example, let us convert the binary number 10110 to Gray code.

Step 1. The left-most Gray digit is the same as the left-most binary digit.

$$1 \quad 0 \quad 1 \quad 1 \quad 0 \qquad \text{binary}$$
$$\downarrow$$
$$1 \qquad\qquad\qquad\qquad \text{Gray}$$

Step 2. Add the left-most binary digit to the adjacent one.

$$+$$
$$1 \quad 0 \quad 1 \quad 1 \quad 0 \qquad \text{binary}$$
$$\downarrow$$
$$1 \quad 1 \qquad\qquad\qquad \text{Gray}$$

Step 3. Add the next adjacent pair.

$$+$$
$$1 \quad 0 \quad 1 \quad 1 \quad 0 \qquad \text{binary}$$
$$\downarrow$$
$$1 \quad 1 \quad 1 \qquad\qquad \text{Gray}$$

Step 4. Add the next adjacent pair and discard carry.

$$+$$
$$1 \quad 0 \quad 1 \quad 1 \quad 0 \qquad \text{binary}$$
$$\downarrow$$
$$1 \quad 1 \quad 1 \quad 0 \qquad\quad \text{Gray}$$

Step 5.   Add the last adjacent pair.

$$+$$

1 0 1 1 0   binary

1 1 1 0 1   Gray

The conversion is now complete and the Gray code is 11101.

To convert from Gray code to binary, a similar method is used, but there are some differences. The following rules apply:

1. The most significant digit (left-most) in the binary code is the same as the corresponding digit in the Gray code.
2. Add each binary digit generated to the Gray digit in the next adjacent position. Disregard carries.

For example, the conversion of the Gray code number 11011 to binary is as follows:

Step 1.   The left-most digits are the same.

1 1 0 1 1   Gray

1   binary

Step 2.   Add the last binary digit just generated to the Gray digit in the next position. Discard carry.

1 1 0 1 1   Gray
+↗↓
1 0   binary

Step 3.   Add the last binary digit generated to the next Gray digit.

1 1 0 1 1   Gray
   +↗↓
1 0 0   binary

Step 4.   Add the last binary digit generated to the next Gray digit.

1 1 0 1 1   Gray
      +↗↓
1 0 0 1   binary

Step 5.   Add the last binary digit generated to the next Gray digit. Discard carry.

1 1 0 1 1   Gray
         +↗↓
1 0 0 1 0   binary

This completes the conversion. The final binary number is 10010.

3-6   Alphanumeric Codes

In order to communicate, we need not only numbers, but also letters and other symbols. In the strictest sense, codes that represent numbers and alphabetic characters (letters) are called *alphanumeric* codes. Most of these codes, however, also represent symbols and various instructions necessary for conveying intelligible information.

At a minimum, the alphanumeric code must represent ten decimal digits and 26 letters of the alphabet, for a total of 36 items. This requires six bits in each code combination because five bits are insufficient ($2^5 = 32$). There are 64 total combinations of six bits, so we therefore have 28 unused code combinations. Obviously, in many applications symbols other than just numbers and letters are necessary to communicate completely. We need spaces to separate words, periods to mark the end of sentences or for decimal points, instructions to tell the receiving system what to do with the information, and more. So, with codes that are six bits long, we can handle decimal numbers, the alphabet, and 28 other symbols. This should give you an idea of the requirements for a basic alphanumeric code.

One standardized alphanumeric code, called the *American Standard Code for Information Interchange* (ASCII), is perhaps the most widely used type. This is a seven-bit code where the decimal digits are represented by the 8421 BCD code preceded by 011. The letters of the alphabet and other symbols and instructions are represented by other code combinations as shown in Table 3-4. For instance, the letter *A* is represented by 1000001, the letter *B* by 1000010, the comma by 0101100, and EOM (end of message) by 0000011.

Another alphanumeric code also frequently encountered is called the *Extended Binary Coded Decimal Interchange Code* (EBCDIC). This is an eight-bit code in which the decimal digits are represented by the 8421 BCD code preceded by 1111. Both lowercase and uppercase letters are represented, in addition to numerous other symbols and commands, as shown in Table 3-5. For example, uppercase *A* is 11000001 and lowercase *a* is 10000001.

3-7   Parity and Error Detection

Errors can occur as digital codes are being transferred from one point to another within a digital system or while codes are being transmitted from one system to another. The errors take the form of changes in the bits that make up the coded information; that is, a 1 can change to a 0, or a 0 to 1, due to component malfunctions or electrical noise (which we will discuss in the next chapter). In most digital systems the probability that even a single-bit error will occur is very small and the likelihood of more than one occurring is even smaller. Many systems, however, employ a *parity bit* as a means of detecting a bit error. Binary information is normally handled by a digital system in groups of bits sometimes called *words*. A word always contains either an even or an odd number of 1s. A parity bit is attached to the group of information bits in order to make the *total* number of 1s

TABLE 3-4. American Standard Code for Information Interchange.

	000	001	010	011	100	101	110	111
0000	NULL	DC$_0$	b	0	@	P		
0001	SOM	DC$_1$	!	1	A	Q		
0010	EOA	DC$_2$	"	2	B	R		
0011	EOM	DC$_3$	#	3	C	S		
0100	EOT	DC$_4$	$	4	D	T		
0101	WRU	ERR	%	5	E	U		
0110	RU	SYNC	&	6	F	V		
0111	BELL	LEM	,	7	G	W		
1000	FE$_0$	S$_0$	(	8	H	X		
1001	HT / SK	S$_1$	)	9	I	Y		
1010	LF	S$_2$	*	:	J	Z		
1011	V$_{tab}$	S$_3$	+	;	K	[		
1100	FF	S$_4$	,	<	L	\		ACK
1101	CR	S$_5$	−	=	M	]		②
1110	SO	S$_6$	★	>	N	↑		ESC
1111	SI	S$_7$	/	?	O	←		DEL

Definitions of control abbreviations:

ACK	Acknowledge	LEM	Logical end of media
BELL	Audible signal	LF	Line feed
CR	Carriage return	RU	"Are you...?"
DC$_0$–DC$_4$	Device control	SK	Skip
DEL	Delete idle	SI	Shift in
EOA	End of address	SO	Shift out
EOM	End of message	S$_0$–S$_7$	Separator (space)
EOT	End of transmission	SOM	Start of message
ERR	Error	V$_{tab}$	Vertical tabulation
ESC	Escape	WRU	"Who are you?"
FE	Format effector	②	Unassigned control
FF	Form feed	SYNC	Synchronous idle
HT	Horizontal tabulation		

Example of code format:

$$B_7 \quad B_1$$

$\underbrace{100}_{\text{three-bit group}}\underbrace{0100}_{\text{four-bit group}}$ is the code for D

TABLE 3–5.   Partial EBCDIC Table.

Bit positions 4,5,6,7 ↓ \ Bit positions 2,3 → \ Bit positions 0,1 →	00				01				10				11			
	00	01	10	11	00	01	10	11	00	01	10	11	00	01	10	11
0000	NUL		DS		SP	&	–									0
0001			SOS				/		a	j			A	J		1
0010			FS						b	k	s		B	K	S	2
0011		TM							c	l	t		C	L	T	3
0100	PF	RES	BYP	PN					d	m	u		D	M	U	4
0101	HT	NL	LF	RS					e	n	v		E	N	V	5
0110	LC	BS	EOB	UC					f	o	w		F	O	W	6
0111	DL	IL	PRE	EOT					g	p	x		G	P	X	7
1000									h	q	y		H	Q	Y	8
1001									i	r	z		I	R	Z	9
1010		CC	SM		¢	!		:								
1011						$	,	#								
1100					<	*	%	@								
1101					(	)	–	'								
1110					+	;	>	=								
1111	CU1	CU2	CU3				?	"								

Definitions of control abbreviations:

BS	Backspace	LC	Lowercase
BYP	Bypass	LF	Line feed
CC	Cursor control	NL	New line
CU1	Customer use	PF	Punch off
CU2	Customer use	PN	Punch on
CU3	Customer use	PRE	Prefix
DL	Delete	RES	Restore
DS	Digit select	RS	Reader stop
EOB	End of block	SM	Set mode
EOT	End of transmission	SP	Space
FS	Field separator	TM	Tape mark
HT	Horizontal tab	UC	Uppercase
IL	Idle		

Meanings of unfamiliar symbols:

|	Vertical bar: logical OR
¬	Logical NOT
-	Hyphen or minus sign
_	Underscore (01101101)

Example of code format:

01234567	Bit positions
11000110	is F

52

always even or *always odd.* An even parity bit makes the total number of 1s even, and an odd parity bit makes the total odd.

A given system operates with even or odd parity, but not both. For instance, if a system operates with even parity, a check is made on each group of bits received to make sure the total number of 1s in that group is even. If there is an odd number of 1s, an error has occurred.

As an illustration of how parity bits are attached to a code word, Table 3-6 lists the parity bits for each 8421 BCD code number for both even and odd parity. The parity bit for each BCD number is in the P column.

TABLE 3-6. 8421 BCD
Code with Parity Bits.

Even Parity	Odd Parity
8421P	8421P
00000	00001
00011	00010
00101	00100
00110	00111
01001	01000
01010	01011
01100	01101
01111	01110
10001	10000
10010	10011

The parity bit can be attached to the code group at either the beginning or the end, depending on system design. Notice that the total number of 1s, *including the parity bit,* is even for even parity and odd for odd parity.

A parity bit provides for the detection of a *single* error (or any odd number of errors, which is very unlikely) but cannot check for two errors. For instance, let us assume that we wish to transmit the BCD code 0101 (parity can be used with any number of bits; we are using four for illustration). The total code transmitted, including the parity bit, is

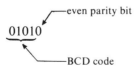

Now, let us assume an error occurs in the second bit from the left (the 1 becomes a 0), as follows:

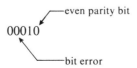

When this code is received, the parity check circuitry determines that there is only a single 1 (odd number), when there should be an even number of 1s. Because an even number of 1s does not appear in the code when it is received, an error is indicated.

Let us now consider what happens if two bit errors occur as follows:

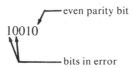

When a check is made, an even number of 1s appears, and although there are two errors, the parity check indicates a correct code.

An odd parity bit also provides in a similar manner for the detection of a single error in a given group of bits.

Example 3–8 Assign the proper even parity bit to the following code words: 1010, 111000, 101101, 100011100101, 101101011111.

Solutions:

Make the parity bits either 1 or 0 to make the total number of 1s even. The parity will be the right-most bit.

$$10100$$
$$1110001$$
$$1011010$$
$$1000111001010$$
$$1011010111111$$

Example 3–9 Assign the proper odd parity bit to the following code words: 1011, 101010, 1110001, 110011100.

Solutions:

Make the parity bits either 1 or 0 in order to make the total number of 1s odd. The parity will be the right-most bit.

$$10110$$
$$1010100$$
$$11100011$$
$$1100111000$$

Example 3–10 An odd parity system receives the following code words: 10110, 11010, 110011, 110101110100, 1100010101010. Determine which ones, if any, are in error.

Solutions:

Since odd parity is required, any code with an even number of 1s is incorrect. Errors are 110011 and 1100010101010.

Several specific codes also provide inherent error detection; we will discuss a few of the most important.

The *2-out-of-5* code is sometimes used in communications work. It utilizes five bits to represent the ten decimal digits, so it is a form of BCD code. Each code word has exactly two 1s, which facilitates decoding and provides for better error detection than the single parity bit method. If other than two 1s appear, an error is indicated.

The *63210 BCD* code is also characterized by having exactly two 1s in each of the five-bit groups. Like the 2-out-of-5 code, it provides reliable error detection and is used in some applications.

The *biquinary* code (two-five) is used in certain counters and is composed of a two-bit group and a five-bit group, each with a single 1. Its weights are 50 43210. The two-bit group, having weights 50, indicates whether the number represented is less than, equal to, or greater than 5. The five bit-group indicates the count above or below 5.

The *ring counter* code has ten bits, one for each decimal digit, and a single 1 makes error detection possible. It is easy to decode, but wastes bits and requires more circuitry to implement than the four- or five-bit codes. The name is derived from the fact that the code is generated by a certain type of shift register, or "ring counter." It is also known as the 9876543210 code because of its weights.

Each of these codes is listed in Table 3–7. You should realize that this is not an exhaustive coverage of all codes, but simply an introduction to some of them.

TABLE 3–7.   Some Codes with Error Detection Properties.

Decimal	2-out-of-5	63210	5043210	9876543210
0	00011	00110	01 00001	0000000001
1	00101	00011	01 00010	0000000010
2	00110	00101	01 00100	0000000100
3	01001	01001	01 01000	0000001000
4	01010	01010	01 10000	0000010000
5	01100	01100	10 00001	0000100000
6	10001	10001	10 00010	0001000000
7	10010	10010	10 00100	0010000000
8	10100	10100	10 01000	0100000000
9	11000	11000	10 10000	1000000000

3–8   Error Correction

This section discusses a method, generally known as the Hamming code, which not only provides for the detection of a bit error, but also identifies which bit is in error so it can be corrected. The code uses a number of parity bits (dependent on the number of information bits) located at certain positions in the code group.

The Hamming code is constructed as follows for *single-error* correction:

If the number of information bits is designated m, then the number of parity bits, p, is determined by the following relationship:

$$2^p \geq m + p + 1 \tag{3–1}$$

For example, if we have four information bits, then p is found by trial and error using Equation (3–1). Let $p = 2$. Then,

$$2^p = 2^2 = 4$$

and $$m + p + 1 = 4 + 2 + 1 = 7$$

Since 2^p has to be equal to or greater than $m + p + 1$, the relationship in Equation (3–1) is *not* satisfied. We have to try again. Let $p = 3$. Then,

$$2^p = 2^3 = 8$$

and $$m + p + 1 = 4 + 3 + 1 = 8$$

This value of p satisfies the relationship of Equation (3–1), and therefore three parity bits are required to provide single-error correction for four information bits. It should be noted here that error detection and correction are provided for *all* bits, both parity and information, in a code group.

Now that we have found the number of parity bits required in our particular example, we arrange the bits properly in the code. At this point you should realize that, in this example, the code is composed of the four information bits and the three parity bits. The left-most bit is designated *bit 1,* the next bit is *bit 2,* and so on, as shown below:

bit 1, bit 2, bit 3, bit 4, bit 5, bit 6, bit 7

The parity bits are located in the positions that are numbered corresponding to ascending powers of two: 1, 2, 4, 8, etc., as indicated:

$$P_1, P_2, M_1, P_3, M_2, M_3, M_4$$

The symbol P_n designates a particular parity bit, and M_n designates a particular information bit.

Finally, we properly assign a 1 or 0 value to each parity bit. Since each parity bit provides a check on certain other bits in the total code, we have to know the value of these others in order to assign the parity-bit value. To do this, first number each bit position in binary; that is, write the binary number for each decimal position number (as shown in the second two rows of Table 3–8). Next, indicate the parity and information bit locations, as shown in the first row of Table 3–8. Notice that the binary position number of parity bit P_1 has a 1 for its right-most digit. This parity bit checks all bit positions, including itself, that have 1s in the same location in the binary position numbers. Parity bit P_1 checks bit positions 1, 3, 5, and 7.

TABLE 3–8.   Bit Position Table for a Seven-Bit Error Correcting Code.

Bit Designation	P_1	P_2	M_1	P_3	M_2	M_3	M_4
Bit Position	1	2	3	4	5	6	7
Binary Position Number	001	010	011	100	101	110	111
Information Bits (M_n)							
Parity Bits (P_n)							

Parity bit P_2 has a 1 for its second-from-right digit. It checks all bit positions, including itself, that have 1s in this same position. Parity bit P_2 checks bit positions 2, 3, 6, and 7.

Parity bit P_3 has a 1 for its third-from-right digit. It checks all bit positions, including itself, that have 1s in this same position. Parity bit P_3 checks bit positions 4, 5, 6, and 7.

In each case the parity bit is assigned a value to make the quantity of 1s in the set of bits that it checks odd or even, depending on which is specified. The following examples should make this procedure clear:

Example 3–11   Determine the single-error correcting code for the BCD number 1001 (information bits) using even parity.

Solution:

Step 1.   Find the number of parity bits required. Let $p = 3$. Then,

$$2^p = 2^3 = 8$$
$$m + p + 1 = 4 + 3 + 1 = 8$$

Three parity bits are sufficient.
Total code bits $= 4 + 3 = 7$.

Step 2.   Construct a bit position table.

Bit Designation	P_1	P_2	M_1	P_3	M_2	M_3	M_4
Bit Position	1	2	3	4	5	6	7
Binary Position Number	001	010	011	100	101	110	111
Information Bits			1		0	0	1
Parity Bits	0	0		1			

Parity bits are determined in the following steps:

Step 3.   Determine the parity bits as follows:

P_1 checks bit positions 1, 3, 5, and 7 and must be a 0 in order to have an even number of 1s (2) in this group.
P_2 checks bit positions 2, 3, 6, and 7 and must be a 0 in order to have an even number of 1s (2) in this group.
P_3 checks bit positions 4, 5, 6, and 7 and must be a 1 in order to have an even number of 1s (2) in this group.

Step 4.   These parity bits are entered into the table, and the resulting combined code is 0011001.

Example 3–12   Determine the single-error correcting code for the information code 10110 for odd parity.

Solution:

Step 1. Determine the number of parity bits required. In this case the number of information bits, m, is five.

From the previous example we know that $p = 3$ will not work. Try $p = 4$.

$$2^p = 2^4 = 16$$
$$m + p + 1 = 5 + 4 + 1 = 10$$

Four parity bits are sufficient.

Total code bits $= 5 + 4 = 9$.

Step 2. Construct a bit position table.

Bit Designation	P_1	P_2	M_1	P_3	M_2	M_3	M_4	P_4	M_5
Bit Position	1	2	3	4	5	6	7	8	9
Binary Position Number	0001	0010	0011	0100	0101	0110	0111	1000	1001
Information Bits			1		0	1	1		0
Parity Bits	1	0		1				1	

Parity bits are determined in the following steps:

Step 3. Determine the parity bits as follows:

P_1 checks bit positions 1, 3, 5, 7, and 9 and must be a 1 to have an odd number of 1s (3) in this group.

P_2 checks bit positions 2, 3, 6, and 7 and must be a 0 to have an odd number of 1s (3) in this group.

P_3 checks bit positions 4, 5, 6, and 7 and must be a 1 to have an odd number of 1s (3) in this group.

P_4 checks bit positions 8 and 9 and must be a 1 to have an odd number of 1s (1) in this group.

Step 4. These parity bits are entered into the table, and the resulting combined code is 101101110.

Now that a method for constructing an error correcting code has been covered, how do we use it to locate and correct an error? Each parity bit, along with its corresponding group of bits, must be checked for the proper parity. If there are three parity bits in a code word, then three parity checks are made. If there are four parity bits, four checks have to be made, and so on. Each parity check will yield a good or a bad result. The total result of all the parity checks indicates the bit, if any, that is in error, as follows:

Step 1. Start with the group corresponding to P_1.

Step 2. Check the group for proper parity. A 0 represents a good parity check and a 1 represents a bad check.

Step 3. Repeat Step 2 for each parity group.

Step 4.   The binary number formed by the results of each parity check designates the position of the code bit that is in error. The first parity
check generates the least significant bit (LSB). If all checks are good,
there is no error.

Example 3–13   Assume that the code word in Example 3–11 (0011001) is transmitted, and that 0010001 is received. The receiver does not "know" what was
transmitted and must look for proper parities to determine if the code is correct.
Designate any error that has occurred in transmission if even parity is used.

Solution:
First, make up a bit position table.

Bit Designation	P_1	P_2	M_1	P_3	M_2	M_3	M_4
Bit Position	1	2	3	4	5	6	7
Binary Position Number	001	010	011	100	101	110	111
Received Code	0	0	1	0	0	0	1

First parity check:
   P_1 checks positions 1, 3, 5, and 7.
   There are two 1s in this group.
   Parity check is good. ───────────────────────────────────→ 0
Second parity check:
   P_2 checks positions 2, 3, 6, and 7.
   There are two 1s in this group.
   Parity check is good. ───────────────────────────────────→ 0
Third parity check:
   P_3 checks positions 4, 5, 6, and 7.
   There is one 1 in this group.
   Parity check is bad.───────────────────────────────────→ 1
Result:
   The binary number generated by the good and bad checks is 100 (binary
   4). This says that the bit in the number 4 position is in error. It is a 0 and
   should be a 1. The corrected code is 0011001, which agrees with the
   transmitted code.

Example 3–14   The code 101101010 is received. Correct any errors. There are
four parity bits and odd parity is used.

Solution:
First, make a bit position table.

Bit Designation	P_1	P_2	M_1	P_3	M_2	M_3	M_4	P_4	M_5
Bit Position	1	2	3	4	5	6	7	8	9
Binary Position Number	0001	0010	0011	0100	0101	0110	0111	1000	1001
Received Code	1	0	1	1	0	1	0	1	0

First parity check:
    P_1 checks positions 1, 3, 5, 7, and 9.
    There are two 1s in this group.
    Parity check is bad.————————————————————————————→ 1
Second parity check:
    P_2 checks positions 2, 3, 6, and 7.
    There are two 1s in this group.
    Parity check is bad.————————————————————————————→ 1
Third parity check:
    P_3 checks positions 4, 5, 6, and 7.
    There are two 1s in this group.
    Parity check is bad.————————————————————————————→ 1
Fourth parity check:
    P_4 checks positions 8 and 9.
    There is one 1 in this group.
    Parity check is good. ————————————————————————————→ 0
Result:
    The binary number generated by the good and bad checks is 0111
    (binary 7). This says that the bit in the number 7 position is in error.
    The corrected code is therefore 101101110.

In this section we have discussed the Hamming code for single-error correction. It should be pointed out that this method can be extended to multiple-error
correction, which is beyond the scope of this book.

3–9   Serial and Parallel Data Formats

Data or information in binary form can be transferred from one location to
another within a digital system by one of two basic methods, serial or parallel, or
by a combination of both. These two methods are based on the relationship of the
bits as they are being moved from place to place.

Serial data means that the bits follow one another so that only one bit at a
time is transferred on a single line, as illustrated in Figure 3–1.

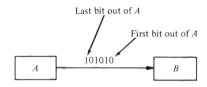

Figure 3–1.   Serial Transfer of Bits from A to B.

Parallel data means that all bits in a given group are transferred simultaneously on separate lines, as illustrated in Figure 3–2.

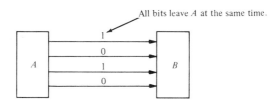

Figure 3–2.   Parallel Transfer of Bits from *A* to *B*.

3–10   Asynchronous and Synchronous Waveforms

Two waveforms are *asynchronous* if there is no fixed time relationship between them, as shown in Figure 3–3(a). Notice that changes in waveform *B* do not occur at any definite time with respect to waveform *A*.

Two waveforms are *synchronous* if there is a definite time relationship between them, as illustrated in Figure 3–3(b). In this example, changes in waveform *B* occur only on the positive-going edges of waveform *A*, creating a definite, fixed relationship.

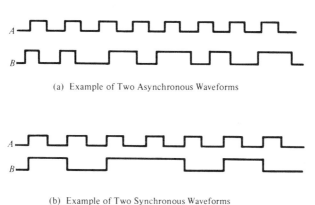

(a) Example of Two Asynchronous Waveforms

(b) Example of Two Synchronous Waveforms

Figure 3–3.   Waveform Relationships.

3–11   Some Ways to Represent Digital Codes

Three important ways in which digital data can be represented for purposes of transmission or recording are called *return-to-zero* (RZ), *non-return-to-zero* (NRZ), and *biphase*. These waveform representations are separated into *bit times,* the intervals during which the level of the waveform indicates a 1 or 0 bit. These bit times are definable by their relation to a basic system timing signal or *clock*.

Figure 3–4 shows an example of a return-to-zero (RZ) waveform. In this case a fixed-width pulse occurring during a bit time represents a 1, and no pulse during a bit time is a 0. There is always a return to the 0 level after a 1 occurs. The period of the clock waveform determines the bit time interval.

Figure 3–4. An RZ Waveform Representing 101011000111.

Figure 3–5 illustrates a non-return-to-zero (NRZ) waveform. In this case a 1 or 0 level remains during the entire bit time. If two or more 1s occur in succession, the waveform does not return to the 0 level until a 0 occurs.

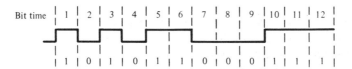

Figure 3–5. An NRZ Waveform Representing 101011000111.

Figure 3–6 is an illustration of a biphase waveform. In this type, a 1 is a HIGH level for the first half of a bit time and a LOW level for the second half, so a high-to-low transition occurring in the middle of a bit time is interpreted as a 1. A 0 is represented by a LOW level during the first half of a bit time followed by a HIGH level during the second half, so a low-to-high transition in the middle of a bit time is interpreted as a 0.

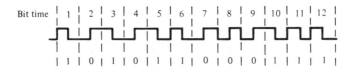

Figure 3–6. A Biphase Waveform Representing 101011000111.

PROBLEMS

3–1 Convert each of the following decimal numbers to 8421 BCD:

(a)	10	(b)	13	(c)	18
(d)	21	(e)	25	(f)	36
(g)	44	(h)	57	(i)	69
(j)	98	(k)	125	(l)	156

3–2 Convert each of the decimal numbers in Problem 3–1 to straight binary (refer to Chapter 2), and compare the number of bits required with that required in BCD.

3–3 Convert the following decimal numbers to BCD:

(a)	104	(b)	128	(c)	132
(d)	150	(e)	186	(f)	210
(g)	359	(h)	547	(i)	1051
(j)	2563				

3-4 Convert each of the 8421 BCD code numbers to decimal.
  (a) 0001   (b) 0110   (c) 1001
  (d) 00011000   (e) 11001   (f) 00110010
  (g) 1000101   (h) 10011000   (i) 100001110000
  (j) 011000011001

3-5 Convert each of the BCD code numbers to decimal.
  (a) 10000000   (b) 1000110111
  (c) 1101000110   (d) 1000010001
  (e) 11101010100   (f) 100000000000
  (g) 100101111000   (h) 1011010000011
  (i) 1001000000011000   (j) 0110011001100111

3-6 Add the following BCD numbers:
  (a) 0010 + 0001   (b) 0101 + 0011
  (c) 0111 + 0010   (d) 1000 + 0001
  (e) 00011000 + 00010001   (f) 01100100 + 00110011
  (g) 01000000 + 01000111   (h) 10000101 + 00010011

3-7 Add the following BCD numbers:
  (a) 1000 + 0110   (b) 0111 + 0101
  (c) 1001 + 1000   (d) 1001 + 0111
  (e) 00100101 + 00100111   (f) 01010001 + 01011000
  (g) 10011000 + 10010111   (h) 010101100001 + 0011100001000

3-8 Convert each pair of decimal numbers to BCD and add as indicated.
  (a) 4 + 3   (b) 5 + 2   (c) 6 + 4
  (d) 17 + 12   (e) 28 + 23   (f) 65 + 58
  (g) 113 + 101   (h) 295 + 157

3-9 Convert each of the following decimal numbers to Excess-3 code.
  (a) 1   (b) 3   (c) 6
  (d) 10   (e) 18   (f) 29
  (g) 56   (h) 75   (i) 107
  (j) 149   (k) 231   (l) 500
  (m) 1251   (n) 2379   (o) 6841

3-10 Convert each Excess-3 code number to decimal.
  (a) 0011   (b) 1001
  (c) 0111   (d) 01000110
  (e) 01111100   (f) 10000101
  (g) 10010101011   (h) 110000110110
  (i) 101001011000 (j) 101101000111

3-11 Perform each of the following Excess-3 additions:
  (a) 0011 + 0011   (b) 0101 + 0100
  (c) 0111 + 0101   (d) 1001 + 0110
  (e) 1011 + 1000   (f) 01010101 + 00110110
  (g) 01011011 + 01001100   (h) 01110011 + 01100101
  (i) 10011010 + 10011100   (j) 11001011 + 10110011

3-12 Convert each of the following *binary* numbers to a Gray code number:
  (a) 0100   (b) 1011   (c) 11010
  (d) 10111   (e) 100001   (f) 111111
  (g) 1010101   (h) 11010110   (i) 1100001011
  (j) 101110101001

3-13   Convert each of the following Gray code numbers to binary:

    (a) 1101   (b) 1010

    (c) 110110   (d) 1111000

    (e) 100011000   (f) 1100110011

    (g) 1111111111   (h) 101111101111

    (i) 100000010000   (j) 10101010101010

3-14   Decode the following ASCII coded message. Bit 7 of the first character is the left-most bit in the first row.

    100100110101000011000100100010000011010011001100 0

    100001010001011000101100111000110001010011000001

    100100110001000011000101010010010001000000110101 00

    001100010001111000101100111010010011010101101001 1

    001100010010011010011001100001110010111001001100 0

    010010100110001010000100010110100101010010011101 0000

    100100110100101000001101010010010011001111100111 0

    001100010000011001110100010000110000110010011000

    010010100110001001001100111010100111010000100100 1

    101001010000011010100100100100110011111100111000 00011

3-15   Convert your name and address to ASCII code.

3-16   Convert your name and address to EBCDIC.

3-17   For each of the codes, assign an even parity bit.

    (a) 1101   (b) 1001

    (c) 101011   (d) 111000

    (e) 1010111   (f) 10111001101

    (g) 10111001100011   (h) 10111110101010

    (i) 10111010111111   (j) 111011011001000

3-18   Repeat Problem 3-17 for odd parity.

3-19   Each of the following code groups contains an even parity bit. Determine the code groups that are in error.

    (a) 11011   (b) 10100

    (c) 1010101   (d) 10110101

    (e) 11011101   (f) 101011111

    (g) 111101000   (h) 100001000

3-20   Each of the following code groups contains an odd parity bit. Determine the code groups that are in error.

    (a) 10110   (b) 10000

    (c) 01010   (d) 1011101

    (e) 1110001   (f) 1011001100000

    (g) 1111111011111 (h) 10111000111010

3-21   The following are information bits to be transmitted. Determine the total code that must be transmitted in each case for single-error correction.

    (a) 1010   (b) 0110   (c) 1110

    (d) 1000   (e) 10111   (f) 11100

    (g) 10001   (h) 110011   (i) 100010

    (j) 111000

3-22   What is the total transmitted code for decimal 1854 represented in straight binary for single-error correction?

3-23 Detect and correct any errors in the following code words. Each code word includes information bits and even parity bits.

      (a) 1101001   (b) 1000111   (c) 0110001
      (d) 1011110   (e) 0101010   (f) 1100110
      (g) 1011100   (h) 0010000   (i) 1110000
      (j) 0011001

3-24 Sketch an RZ waveform for each of the following groups of bits:

      (a) 101010101010   (b) 110111011110110
      (c) 11100001111001   (d) 11111110000100
      (e) 11010100011010

3-25 Repeat Problem 3-24 for an NRZ waveform.

3-26 Repeat Problem 3-24 for a biphase waveform.

CHAPTER 4

Logic Gates

In this chapter we will study the various types of logic gates that make up a typical digital system. The emphasis is on the logical operation of the circuits and the limitations and considerations involved in their operation. It is very important to know what the output of a gate is for various combinations of inputs and to understand how the electrical characteristics affect its operation so that we can predict and analyze how a circuit will perform in a given system.

Because of the growing use of integrated circuits, detailed knowledge of circuit design is becoming less important for those involved in the development, application, or maintenance of digital equipment. However, some basic circuit implementations for each type of gate are discussed to give you a better "feel" for what a gate is made of and to help satisfy your natural curiosity of how it works.

American National Standard Institute (ANSI) graphic symbols for logic diagrams are used in this chapter and throughout the book.

4–1 The NOT Circuit (Inverter)

The NOT circuit, which performs a basic logic function normally called *inversion* or *complementation*, is commonly referred to as an *inverter*.

The purpose of the inverter is to convert or change one logic level to the other logic level. If a HIGH is applied to its input, a LOW will appear on its out-

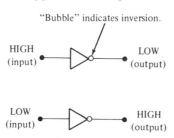

Figure 4–1. Inverter Operation Showing Standard Logic Symbol.

67

put. If a LOW is applied to its input, a HIGH will appear on its output. This is illustrated in Figure 4–1, where the inverter is represented by a standard logic symbol.

We can summarize this operation with Table 4–1, which shows the output for each possible input. This is called a *truth table.*

TABLE 4–1
Truth Table for
the Inverter.

Input	Output
LOW	HIGH
HIGH	LOW

Figure 4–2 shows the output of an inverter for a pulse input where t_1 and t_2 indicate the corresponding points on the input and output pulse waveforms. Note that when the input is LOW, the output is HIGH and when the input is HIGH, the output is LOW, thereby producing an inverted output pulse.

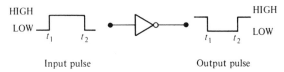

Input pulse   Output pulse

Figure 4–2.   Inverter with Pulse Input.

The bipolar transistor, because of its inherent switching characteristics, is the basic electronic device in many digital circuits. It is used in the basic inverter (NOT) circuit. Let us examine several methods of constructing a NOT circuit, beginning with resistor-transistor logic (RTL), so named because a resistor is used to couple the input to the transistor. Figure 4–3(a) shows an RTL inverter circuit. If a HIGH (say, +5 V) is applied to the input, the base-emitter junction is forward biased, allowing the transistor to turn on. A sufficient amount of base current is provided by the input source to saturate the transistor, and because the transistor acts approximately like a closed switch when it is saturated, the output (collector) is near ground potential, or approximately 0 V (which represents a LOW). So, for a HIGH input we get a LOW output, as illustrated in Figure 4–3(b).

If a LOW (0 V) is applied to the input, the transistor is cut off and the output is at +5 V, because the transistor acts approximately like an open switch when it is off. So, for a LOW input we get a HIGH output, as illustrated in Figure 4–3(c).

The diode-transistor logic (DTL) implementation of the NOT circuit uses a diode as the coupling device from the input to the base of the transistor. A typical DTL inverter is shown in Figure 4–4. If, for example, the input is +5 V (HIGH), then diode D_1 is reverse biased and the DC source (V_{CC}) supplies sufficient bias

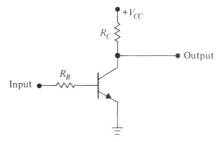

(a) Basic RTL Inverter

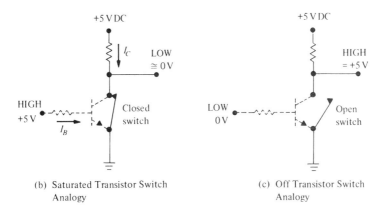

(b) Saturated Transistor Switch
Analogy

(c) Off Transistor Switch
Analogy

Figure 4–3.   RTL Inverter Operation.

voltage and base current through R_B and D_2 to saturate the transistor. This condition results in the output being at approximately 0 V (LOW), as illustrated in Figure 4–4(a).

If the input is at or near 0 V (LOW), the diode D_1 is forward biased. This reverse biases diode D_2 and the base-emitter junction of the transistor, and pulls current away from the base, keeping the transistor off. As a result, the output is at +5 V (HIGH). Figure 4–4(b) illustrates this condition.

Transistor-transistor logic (TTL or T^2L) is very similar in operation to DTL, except that a transistor is used as the input coupling device. As shown in Figure 4–5, the base-emitter junction diode acts as diode D_1 in the comparable DTL circuit, and the base-collector junction diode acts as diode D_2. The main reason for the development of TTL technology is that in integrated circuits where both transistor and diode functions are required, the procedures and economics of integrated circuit fabrication make it easier to use the base-emitter junction of a transistor for the diode function.

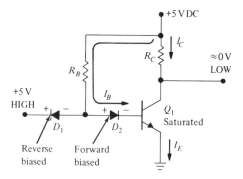

(a) Basic DTL Inverter in the ON Condition

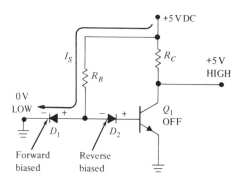

(b) Basic DTL Inverter in the OFF Condition

Figure 4–4.   Basic DTL Inverter.

The field-effect transistor (FET) is widely used in integrated digital circuits for reasons we will discuss in Chapter 14. The most prevalent type of FET used in digital circuits is the metal oxide semiconductor FET, or MOSFET for short. The "normally off" or enhancement MOSFET lends itself particularly well to applications in digital circuits because, without any gate-to-source bias, it stays off or nonconducting. Figure 4–6 shows a basic inverter implemented with an N-channel enhancement MOSFET. With 0 gate-to-source bias voltage, the device is not conducting; that is, it looks like an open switch. In this condition the output (drain) is HIGH. When a positive voltage is applied to the gate, the MOSFET is biased on and a LOW output results. One attraction of this device is that no gate current is required to turn the transistor on, only a voltage, so that the circuit can operate at very low current levels.

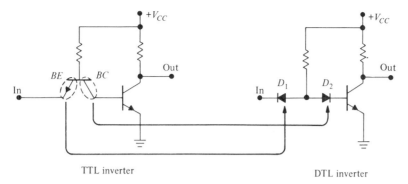

Figure 4–5.   TTL/DTL Comparison.

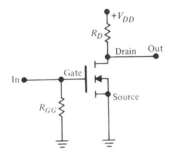

Figure 4–6.   Basic MOSFET Inverter.

Example 4–1   A pulse waveform is applied to an inverter in Figure 4–7(a). Determine the output waveform corresponding to the input.

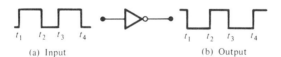

(a) Input   (b) Output

Figure 4–7.

Solution:
The output waveform is exactly opposite to the input (inverted) at each point, as shown in Figure 4–7(b).

4–2   The AND Gate and Its Function

The AND gate performs the basic operation of logic multiplication commonly known as the AND function. The mathematical aspects of this are discussed in Chapter 5 on Boolean algebra.

The AND gate is composed of two or more inputs and a single output, as indicated by the standard logic symbols shown in Figure 4–8. Gates with two, three,

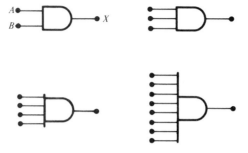

Figure 4–8. Standard AND Gate Logic Symbols with Various Numbers of Inputs.

four, and eight inputs are shown; however, an AND gate can have any number of inputs greater than one.

The operation of the AND gate is such that the output is HIGH only when *all* of the inputs are HIGH. If *any* of the inputs are LOW, the output is LOW. Therefore, the basic purpose of an AND gate is to determine when certain conditions are simultaneously true as indicated by all HIGH levels on its inputs and to produce a HIGH on its output indicating this condition. The inputs of the two-input AND gate in Figure 4–8 are labeled A and B and the output is labeled X. We can express the gate operation with the following description:

> If A **AND** B are HIGH, then X is HIGH.
> If A is LOW, or if B is LOW, or if both
> A and B are LOW, then X is LOW.

The HIGH level is the prime or *active* output level for the AND gate. Figure 4–9 illustrates a two-input AND gate with all four possibilities of input level combinations, and the resulting output for each.

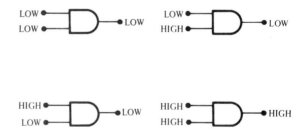

Figure 4–9. All Possible Logic Conditions for a Two-Input AND Gate.

We generally express the logical operation of a gate with a table that lists all input combinations and the corresponding outputs. This table of combinations is commonly called a *truth table,* and is illustrated in Table 4–2 for a two-input AND gate. Tables 4–3 and 4–4 are the truth tables for a three-input AND gate and a four-input AND gate, respectively. Notice that for each gate, regardless of the number of inputs, the output is HIGH *only* when *all* inputs are HIGH.

TABLE 4–2. Truth Table for a Two-Input AND Gate.

Inputs		Output
A	B	X
LOW	LOW	LOW
LOW	HIGH	LOW
HIGH	LOW	LOW
HIGH	HIGH	HIGH

The total number of possible combinations of binary inputs is determined by the following formula:

$$N = 2^n \qquad\qquad (4\text{–}1)$$

where N is the total possible combinations and n is the number of input variables. To illustrate, the following calculations are made using Equation (4–1):

For two input variables: $2^2 = 4$

For three input variables: $2^3 = 8$

For four input variables: $2^4 = 16$

This is how we determine the number of combinations for each of the truth tables.

In a great majority of applications, the inputs to a gate are not stationary levels, but are voltages that change frequently between two logic levels and that can be classified as pulse waveforms. We will now look at the operation of AND gates with pulsed input waveforms. Keep in mind that an AND gate obeys the truth table operation regardless of whether its inputs are constant levels or pulsed levels.

In examining the pulsed operation of the AND gate, we will look at the input levels with respect to each other in order to determine the output level at any given time. For example, in Figure 4–10 the inputs are both HIGH during the interval $T1$, making the output HIGH during this interval. During interval $T2$, input A is

TABLE 4–3. Truth Table for a Three-Input AND Gate.

Inputs			Output
A	B	C	X
LOW	LOW	LOW	LOW
LOW	LOW	HIGH	LOW
LOW	HIGH	LOW	LOW
LOW	HIGH	HIGH	LOW
HIGH	LOW	LOW	LOW
HIGH	LOW	HIGH	LOW
HIGH	HIGH	LOW	LOW
HIGH	HIGH	HIGH	HIGH

TABLE 4–4.   Truth Table for a
Four-Input AND Gate.

Inputs				Output
A	B	C	D	X
LOW	LOW	LOW	LOW	LOW
LOW	LOW	LOW	HIGH	LOW
LOW	LOW	HIGH	LOW	LOW
LOW	LOW	HIGH	HIGH	LOW
LOW	HIGH	LOW	LOW	LOW
LOW	HIGH	LOW	HIGH	LOW
LOW	HIGH	HIGH	LOW	LOW
LOW	HIGH	HIGH	HIGH	LOW
HIGH	LOW	LOW	LOW	LOW
HIGH	LOW	LOW	HIGH	LOW
HIGH	LOW	HIGH	LOW	LOW
HIGH	LOW	HIGH	HIGH	LOW
HIGH	HIGH	LOW	LOW	LOW
HIGH	HIGH	LOW	HIGH	LOW
HIGH	HIGH	HIGH	LOW	LOW
HIGH	HIGH	HIGH	HIGH	HIGH

LOW and input B is HIGH, so the output is LOW. During interval $T3$, both inputs are HIGH again, and therefore the output is HIGH. During interval $T4$, input A is HIGH and input B is LOW, resulting in a LOW output. Finally, during interval $T5$, input A is LOW, input B is LOW, and the output is therefore LOW.

It is very important, when analyzing the pulsed operation of logic gates, to pay very careful attention to the time relationships of all the inputs with respect to each other and with respect to the output.

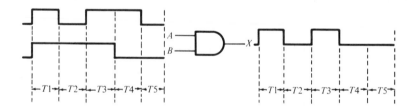

Figure 4–10.   Example of Pulsed AND Gate Operation.

Example 4–2   If the two waveforms are applied to the AND gate in Figure 4–11(a), what is the resulting output waveform?

Solution:
See Figure 4–11(b).

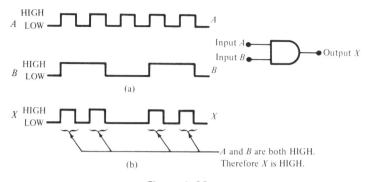

Figure 4-11.

Example 4-3   For the two input waveforms graphed in Figure 4-12(a), sketch the output waveform showing its proper relation to the inputs for a two-input AND gate.

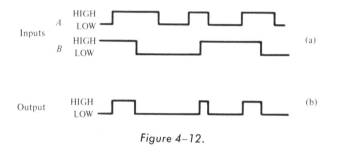

Figure 4-12.

Solution:
The output is HIGH only when both of the inputs are HIGH. See Figure 4-12(b).

Example 4-4   For the three-input AND gate in Figure 4-13(a), determine the output waveform in proper relation to the inputs.

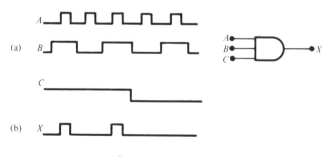

Figure 4-13.

Solution:
See Figure 4-13(b).

Now, let us look at some basic circuit configurations for the AND gate in order to get a "feel" for the inner workings of this logic element. Keep in mind that, regardless of the details of the circuit design, the logical function of an AND gate is always the same. However, each of the various types of circuit implementations has certain advantages and disadvantages, depending on its particular electrical characteristics. We will discuss gate characteristics in general later in this chapter and more specifically in Chapter 14 on integrated circuits.

Figure 4–14 represents one basic form of a DTL two-input AND gate. Let us begin with both inputs at the HIGH level. In this case both input diodes, D_1 and D_2, are reverse biased, and therefore act approximately as open switches. This allows the base of the transistor to be forward biased by V_{CC} through R_B. Sufficient base current is provided through R_B to keep the transistor on, resulting in a HIGH output at the emitter. This condition is illustrated in Figure 4–15(a).

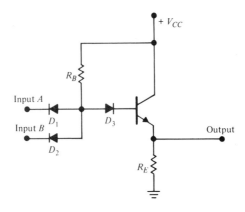

Figure 4–14.   A Basic Two-Input DTL AND Gate Circuit.

If a LOW (0 V) is applied to either input, then the associated diode is forward biased and approximates a closed switch. This holds point a to 0.7 V (for a silicon diode), and diode D_3 provides an additional 0.7-V margin to assure that the transistor is biased off. The resulting output is LOW, as illustrated in Figures 4–15(b) and (c).

If both inputs are LOW, the transistor is held off because of the clipping action of the input diodes, and the output is therefore LOW. This condition is shown in Figure 4–15(d).

Now we will examine a simple form of a TTL two-input AND gate, as shown in Figure 4–16. As we learned in the previous section, a TTL circuit is very similar to DTL in its basic operation. However, let us examine the basic operation in order to reinforce our understanding of this circuit.

As illustrated in Figure 4–17, when either input is LOW, the corresponding base-emitter junction of the multiple-emitter transistor, Q_1, is forward biased. If both inputs are LOW, then both base-emitter junctions of Q_1 are forward biased. In either case, current flows from V_{CC} through R_1 and the forward biased base-emitter junction(s) keeping the base-collector junction of Q_1 reverse biased. As a result, transistor Q_2 is off, and therefore the output (emitter of Q_2) is LOW.

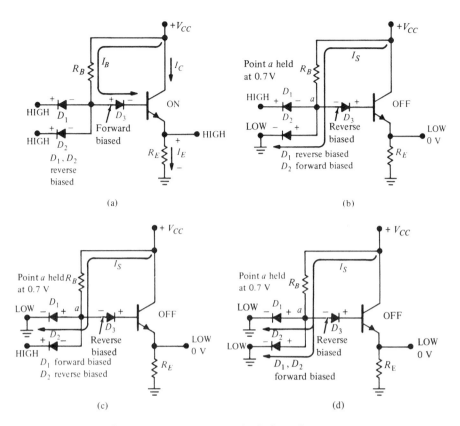

Figure 4–15.   Basic DTL AND Gate Operation.

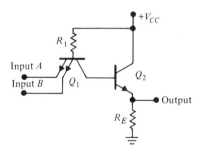

Figure 4–16.   A Basic Two-Input TTL AND Gate Circuit.

As shown in Figure 4–17(d), when both inputs are HIGH, the base-emitter junctions of the input transistor Q_1 are both reverse biased. Current now flows from V_{CC} through R_1 and the base-collector junction of Q_1 (which is now forward biased). As a result, Q_2 is turned on and a HIGH level voltage appears on the output.

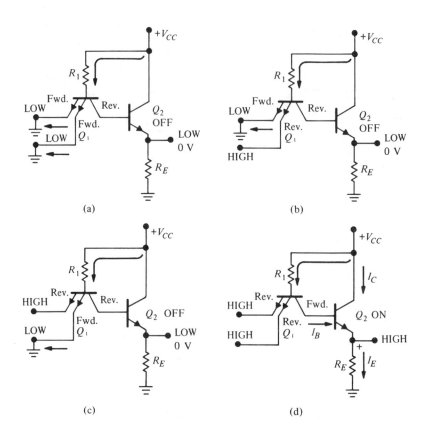

Figure 4-17.   Operation of Basic TTL AND Gate.

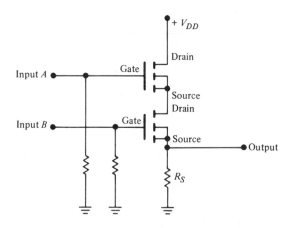

Figure 4-18.   A Basic MOSFET Two-Input AND Gate Circuit.

One basic form of MOSFET AND gate is shown in Figure 4–18. When either input or both inputs are LOW (0 V), the circuit is nonconducting; therefore a LOW appears on the output (source). Since the FETs are N-channel enhancement mode types, a positive voltage on the gate is required to turn them on. Therefore, if a HIGH is applied to each input, both MOSFETs are turned on and current flows through R_s, creating a HIGH output at the source.

4–3 The OR Gate and Its Function

The OR gate performs the basic operation of logic addition, which we usually refer to as the OR function. The mathematical aspects of the operation will be covered in Chapter 5 on Boolean algebra.

The OR gate has two or more inputs and one output, and is normally represented by a standard logic symbol, as shown in Figure 4–19, where OR gates with various numbers of inputs are illustrated. It should be noted, however, that an OR gate can have any number of inputs greater than one.

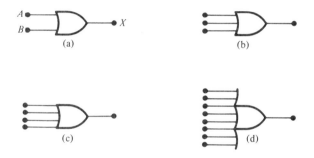

Figure 4–19. Standard OR Gate Symbols with Various Numbers of Inputs.

The operation of the OR gate is such that a HIGH on the output is produced when *any* of the inputs are HIGH. The output is LOW only when *all* of the inputs are LOW. Therefore, the purpose of an OR gate is to determine when one or more of its inputs are HIGH and to produce a HIGH on its output to indicate this condition. The inputs of the two-input OR gate in Figure 4–19(a) are labeled A and B, and the output is labeled X. We can express the operation of the gate as follows:

*If either A **OR** B **OR** both are HIGH, then X is*
HIGH. If both A and B are LOW, then X is LOW.

The HIGH level is the prime or *active* output level for the OR gate. Figure 4–20 illustrates the logic operation for a two-input OR gate for all four possible input level combinations.

The logical operation of the two-input OR gate can be described in the truth table form shown in Table 4–5. Truth tables for three-input OR gates and four-input OR gates are given in Tables 4–6 and 4–7, respectively. Notice that for each gate, regardless of the number of inputs, the output is HIGH when *any* of the inputs are HIGH.

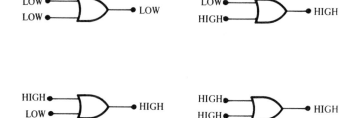

Figure 4–20. All Possible Logic Conditions for a Two-Input OR Gate.

TABLE 4–5.   Truth Table for
a Two-Input OR Gate.

Inputs		Output
A	*B*	*X*
LOW	LOW	LOW
LOW	HIGH	HIGH
HIGH	LOW	HIGH
HIGH	HIGH	HIGH

TABLE 4–6.   Truth Table for a
Three-Input OR Gate.

Inputs			Output
A	*B*	*C*	*X*
LOW	LOW	LOW	LOW
LOW	LOW	HIGH	HIGH
LOW	HIGH	LOW	HIGH
LOW	HIGH	HIGH	HIGH
HIGH	LOW	LOW	HIGH
HIGH	LOW	HIGH	HIGH
HIGH	HIGH	LOW	HIGH
HIGH	HIGH	HIGH	HIGH

Let us now turn our attention to the operation of an OR gate with pulsed inputs, keeping in mind what we have learned about its logical operation.

Again, the important thing in analysis of gate operation with pulsed waveforms is the relationship of all the waveforms involved. For example, in Figure 4–21, the inputs *A* and *B* are both HIGH during interval *T*1, making the output

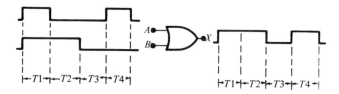

Figure 4–21. An Example of Pulsed OR Gate Operation.

TABLE 4–7.   Truth Table for a
Four-Input OR Gate.

Inputs				Output
A	B	C	D	X
LOW	LOW	LOW	LOW	LOW
LOW	LOW	LOW	HIGH	HIGH
LOW	LOW	HIGH	LOW	HIGH
LOW	LOW	HIGH	HIGH	HIGH
LOW	HIGH	LOW	LOW	HIGH
LOW	HIGH	LOW	HIGH	HIGH
LOW	HIGH	HIGH	LOW	HIGH
LOW	HIGH	HIGH	HIGH	HIGH
HIGH	LOW	LOW	LOW	HIGH
HIGH	LOW	LOW	HIGH	HIGH
HIGH	LOW	HIGH	LOW	HIGH
HIGH	LOW	HIGH	HIGH	HIGH
HIGH	HIGH	LOW	LOW	HIGH
HIGH	HIGH	LOW	HIGH	HIGH
HIGH	HIGH	HIGH	LOW	HIGH
HIGH	HIGH	HIGH	HIGH	HIGH

HIGH. During interval $T2$, input A is LOW, but because input B is HIGH, the output is HIGH. Both inputs are LOW during interval $T3$, and we have a LOW output during this time. During $T4$, the output is HIGH because input A is HIGH.

In this illustration, we have simply applied the truth table operation of the OR gate to each of the intervals during which the levels are nonchanging. A few examples will further illustrate OR gate operation with pulse waveforms on the inputs.

Example 4–5   If the two waveforms are applied to the OR gate in Figure 4–22(a), what is the resulting output waveform?

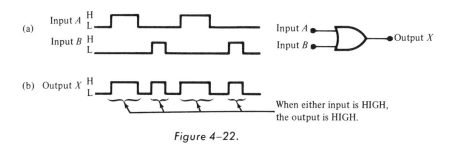

Figure 4–22.

Solution:
See Figure 4–22(b).

Example 4–6   For the two input waveforms in Figure 4–23(a), sketch the output waveform showing its proper relation to the inputs for a two-input OR gate.

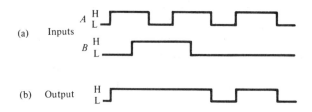

Figure 4–23.

Solution:

When either input or both inputs are HIGH, the output is HIGH. See Figure 4–23(b).

Example 4–7   For the three-input OR gate in Figure 4–24(a), determine the output waveform in proper relation to the inputs.

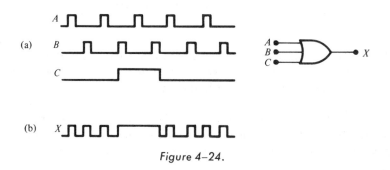

Figure 4–24.

Solution:
See Figure 4–24(b).

We will now look at some basic circuit implementations for the OR gate. Remember that an OR gate always performs the same logical function regardless of the circuitry with which it is implemented. Also keep in mind that the presentation of basic circuitry in this chapter is by no means exhaustive or complete, but intended only to give you a "feel" for gate operation at the circuit level.

Figure 4–25 illustrates one basic form of an RTL OR gate with two inputs. The figure can be extended to OR gates with any number of inputs. If the inputs are both LOW (0 V), transistors Q_1 and Q_2 are both off and the resulting output is LOW because no current flows through R_E. If one of the inputs is HIGH and the other LOW, the transistor with the HIGH input is turned on and a HIGH level voltage is produced at the output by emitter current through R_E. If both inputs are HIGH, a HIGH is produced on the output because current is flowing

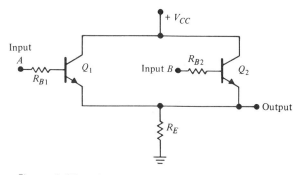

Figure 4–25.   A Basic RTL Two-Input OR Gate Circuit.

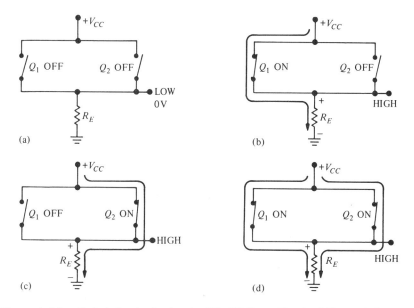

Figure 4–26.   Switch Analogies for the RTL OR Gate Circuit of Figure 4–25.

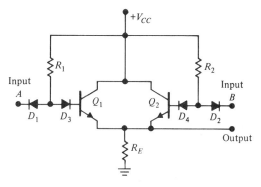

Figure 4–27.   A Basic DTL Two-Input OR Gate Circuit.

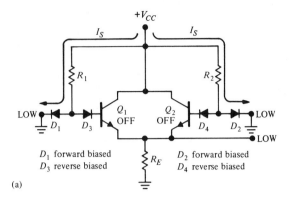

(a)

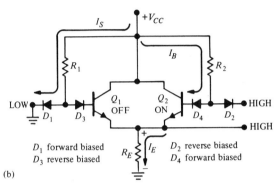

(b)

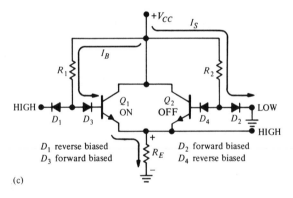

(c)

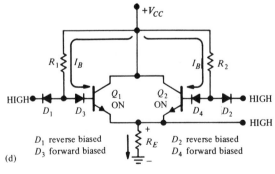

(d)

Figure 4–28.   DTL OR Gate Operation.

through R_E. Output voltage is limited to within a few tenths of a volt of the base voltage of each transistor because of the base-emitter diode junction in the transistors. Switch analogies for each of the input conditions are shown in Figure 4-26.

Figure 4-27 represents one basic form of a DTL two-input OR gate. In this circuit, if both inputs are LOW (0 V) then both input diodes, D_1 and D_2, are forward biased. This action holds Q_1 and Q_2 off, and the resulting output is LOW because there is no emitter current from either transistor. If a HIGH level voltage is applied to one of the inputs, with the other input still LOW, the corresponding transistor is turned on because the input diode is reverse biased. The resulting output is HIGH due to the emitter current through R_E. If both of the inputs are HIGH, both D_1 and D_2 are reverse biased and base current is supplied to both transistors. The output is therefore HIGH. Figure 4-28 (opposite) illustrates the operation of this circuit for the various input conditions.

A basic form of TTL OR gate is shown in Figure 4-29. As was pointed out before, the multiple-emitter transistors perform the function of the input diodes of

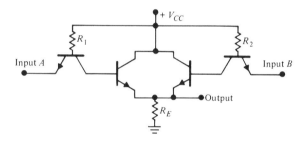

Figure 4-29.   A Basic Two-Input TTL OR Gate Circuit.

an equivalent DTL circuit, and therefore the operation is very similar. Figure 4-30 illustrates TTL OR gate operation for the four input conditions.

The OR function can be implemented using MOSFETs, and one way to do it is shown in Figure 4-31. In this circuit, if both inputs are LOW, the MOSFETs are both off, and the output taken off the source is therefore LOW. A HIGH on either input or both inputs will turn the corresponding FETs on and produce a HIGH on the output by current through R_S.

4-4   The NAND Gate

The term NAND is a contraction of NOT-AND, and implies an AND function with a complemented (inverted) output. A standard logic symbol for a two-input NAND gate and its equivalency to an AND gate followed by an inverter are shown in Figure 4-32.

The NAND gate is a very popular logic function because it is a "universal" function; that is, it can be used to construct an AND gate, an OR gate, an inverter, or any combination of these functions. In Chapter 6 we will examine this "universal" property of the NAND gate. In this chapter we are going to look at the logical operation of the NAND gate and some basic circuit implementations.

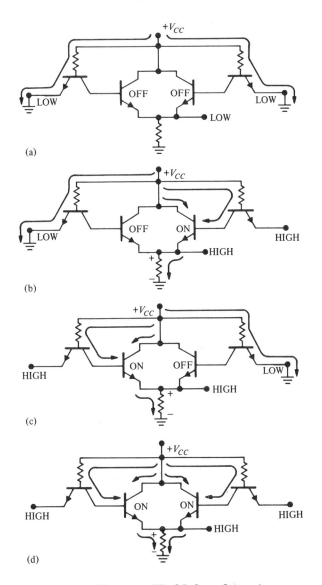

Figure 4–30. Basic TTL OR Gate Operation.

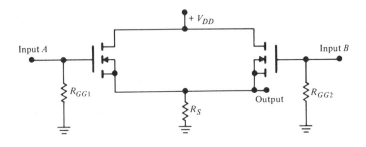

Figure 4–31. A Basic MOSFET Two-Input OR Gate Circuit.

Figure 4–32.   Standard NAND Gate Logic Symbol and Its NOT/AND Equivalent.

The logical operation of the NAND gate is such that a LOW output occurs only if *all* inputs are HIGH. If *any* of the inputs are LOW, the output will be HIGH. For the specific case of a two-input NAND gate as shown in Figure 4–32, with the inputs labeled *A* and *B* and the output labeled *X*, we can state the operation as follows:

> *If A* **AND** *B are HIGH, then X is LOW. If A is LOW, or B is LOW, or if both A and B are LOW, then X is HIGH.*

Note that this operation is the opposite of the AND as far as output is concerned. In NAND operation, the LOW level is the *active* output level. The circle or "bubble" on the output indicates that the output is *active LOW*. Figure 4–33 illustrates the logical operation of a two-input NAND gate for all four input level combinations.

Figure 4–33.   Logical Operation of a Two-Input NAND Gate.

The truth table summarizing the logical operation of the two-input NAND gate is shown in Table 4–8. The extensions of the truth table to three- and four-input NAND gates appear as problems at the end of this chapter.

TABLE 4–8.   Truth Table
for a Two-Input NAND
Gate.

Inputs		Output
A	*B*	*X*
LOW	LOW	HIGH
LOW	HIGH	HIGH
HIGH	LOW	HIGH
HIGH	HIGH	LOW

The NAND gate has another function inherent in its operation, as exhibited by the fact that a LOW level on any input produces a HIGH on the output. In reference to the two-input gate of Figure 4–34(a), this aspect of its operation is stated as follows:

If A is LOW OR B is LOW OR if both A and B are
LOW, then X is HIGH.

Here we have an OR operation that requires LOW inputs to produce a HIGH output and that is commonly referred to as *negative-OR*. This function of the NAND gate can be considered as a secondary aspect of its operation, and is represented by the standard logic symbol in Figure 4–34(b). The two symbols in Figure 4–34 represent the same gate, but they also serve to define its role in a particular application, as illustrated by the following examples.

(a) NAND   (b) Negative-OR

Figure 4–34.   Standard Symbols Representing the Dual Function of the Same Gate.

Example 4–8   The simultaneous occurrence of two HIGH level voltages must be detected and indicated by a LOW level output that is used to drive an indicator light. Sketch the operation.

Solution:

This application requires a NAND function, since the output has to be *active LOW* in order to produce an indication when two *HIGHs* occur on its inputs. The *NAND* symbol is therefore used to show the operation. See Figure 4–35.

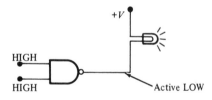

Figure 4–35.

Example 4–9   A gate is required to monitor two lines, to detect the occurrence of LOW level voltages on either or both lines, and to generate a HIGH level output used to activate an alarm light. Sketch the operation.

Solution:

This application requires a *negative-OR* function because the output has to be *active HIGH* in order to produce an indication of the occurrence of one or more *LOW* levels on its inputs. In this case, the gate functions in the *negative-OR* mode, and is represented by the appropriate symbol shown in Figure 4–36. A LOW on either input or both inputs causes an active-HIGH output to activate the lamp.

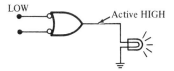

Figure 4–36.

We will now look at the pulsed operation of the NAND gate. Remember from the truth table that any time *all* of the inputs are HIGH, the output will be LOW, and this is the *only* time a LOW output occurs. A few examples will serve to illustrate pulsed operation.

Example 4–10   If the two waveforms shown in Figure 4–37(a) are applied to the NAND gate, determine the resulting output waveform.

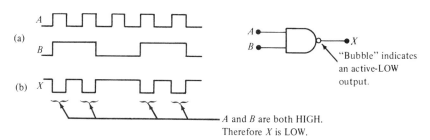

Figure 4–37.

Solution:
See Figure 4–37(b).

Example 4–11   Sketch the output waveform for the three-input NAND gate in Figure 4–38(a), showing its proper relationship to the inputs.

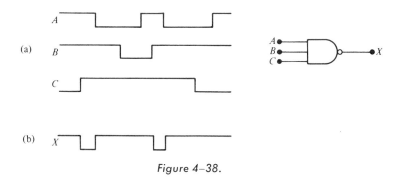

Figure 4–38.

Solution:
The output is LOW only when all three inputs are HIGH. See Figure 4–38(b).

Example 4–12 For the four-input NAND gate in Figure 4–39(a) operating in the *negative-OR* mode, determine the output with respect to the inputs.

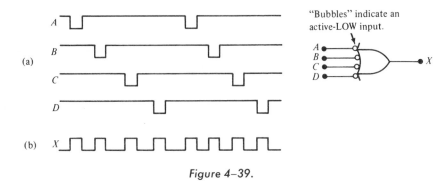

Figure 4–39.

Solution:

The output is HIGH any time an input is LOW. See Figure 4–39(b).

Figure 4–40 shows four basic circuit implementations for the NAND gate. The circuits are similar to the AND gate circuits we discussed earlier, except for

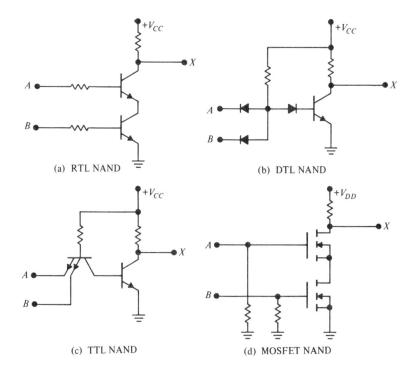

Figure 4–40. Some Basic Two-Input NAND Gate Configurations.

the *inverted* output. The inversion is accomplished by taking the output off the collector instead of the emitter in the cases of the bipolar transistor circuits, and off the drain rather than the source in the case of the FET circuits. Since the circuit operations are the same as those previously discussed, with the exception of the inverted output, they will not be described here.

4–5   The NOR Gate

The term NOR is a contraction of NOT-OR and implies an OR function with an inverted output. A standard logic symbol for a two-input NOR gate and its equivalent OR gate followed by an inverter are shown in Figure 4–41.

Figure 4–41   Standard NOR Gate Logic Symbol and Its NOT/OR Equivalent.

The NOR gate, like the NAND, is a very useful logic gate because it also is a universal type of function. We will examine the universal property of this gate in detail in Chapter 6.

The logical operation of the NOR gate is such that a LOW output occurs when *any* of its inputs are HIGH. Only when *all* of its inputs are LOW is the output HIGH. For the specific case of a two-input NOR gate, as shown in Figure 4–41 with the inputs labeled *A* and *B* and the output labeled *X*, we can state the operation as follows:

If A **OR** *B* **OR** *both are HIGH, then X is* **LOW**.
If both A and B are LOW, then X is HIGH.

Note that this operation results in an output opposite that of the OR gate. In NOR operation, the LOW output is the *active* output level. As was pointed out for the NAND gate, the "bubble" on the output tells us that the OR function is *active LOW*. Figure 4–42 illustrates the logical operation of a two-input NOR gate for all four possible input combinations.

The truth table for the two-input NOR gate is given in Table 4–9. The extensions of the table to three-input and four-input NOR gates appear as problems at the end of this chapter.

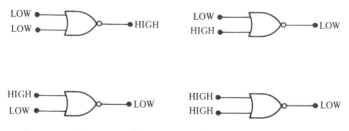

Figure 4–42.   Logical Operation of a Two-Input NOR Gate.

TABLE 4–9.   Truth Table
for a Two-Input NOR Gate.

Inputs		Output
A	B	X
LOW	LOW	HIGH
LOW	HIGH	LOW
HIGH	LOW	LOW
HIGH	HIGH	LOW

The NOR gate, like the NAND, also displays another mode of operation that is inherent in the way it logically functions. This can be seen from the fact that a HIGH is produced on the gate output only if *all* of the inputs are LOW. In reference to Figure 4–43(a), this aspect of NOR operation is stated as follows:

*If both A **AND** B are LOW, then X is HIGH*

We have essentially an AND operation that requires all *LOW inputs* to produce a HIGH output. This is commonly called *negative-AND*, and can be considered a secondary aspect of NOR gate operation. The standard symbol for the negative-AND function of the NOR gate is shown in Figure 4–43(b). It is important to remember that the two symbols in this figure represent the same gate and serve only to distinguish between the two facets of logical operation. The following two examples will illustrate this dual operation.

(a)   (b)

Figure 4–43   *Standard Symbols Representing the Dual Functions of the Same Gate.*

Example 4–13   A certain application requires that two lines be monitored for the occurrence of a HIGH level voltage on either or both. Upon detection of a HIGH level, the circuit must provide a LOW voltage to energize a particular indicating device. Sketch the operation.

Solution:
This application requires a NOR function, since the output has to be *active LOW* in order to give an indication of at least one HIGH on its inputs. The NOR symbol is therefore used to show the operation. See Figure 4–44.

HIGH

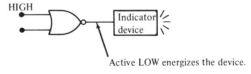
Active LOW energizes the device.

Figure 4–44.

Example 4–14   A device is needed to indicate when two LOW levels occur simultaneously on its inputs and to produce a HIGH output as an indication. Sketch the operation.

Solution:
Here, a negative-AND function is required, as shown in Figure 4–45.

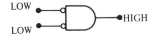

Figure 4–45.

The next three examples will illustrate the logical operation of the NOR gate with pulsed inputs. Again, as with the other types of gates, we will simply follow the truth table operation in order to determine the output waveforms.

Example 4–15   If the two waveforms shown in Figure 4–46(a) are applied to the NOR gate, what is the resulting output waveform?

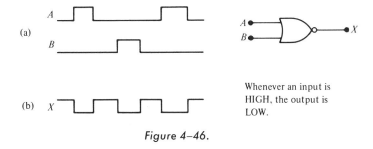

Figure 4–46.

Solution:
Whenever an input is HIGH, the output is LOW. See Figure 4–46(b).

Example 4–16   Sketch the output waveform for the three-input NOR gate in Figure 4–47(a), showing the proper relation to the inputs.

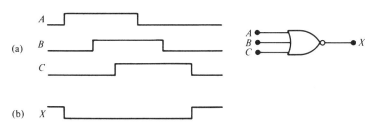

Figure 4–47.

Solution:
See Figure 4–47(b).

Example 4–17   For the four-input NOR gate operating in the *negative-AND* mode in Figure 4–48(a), determine the output relative to the inputs.

Solution:
Any time *all* of the inputs are LOW, the output is HIGH. See Figure 4–48(b).

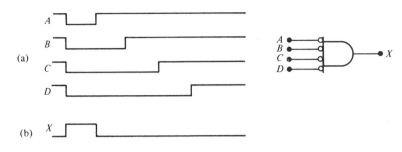

Figure 4–48.

Figure 4–49 shows four basic circuit configurations for the NOR gate. Except for the inverted output, these circuits are identical to those for the OR gate. The inversion is accomplished by taking the output off the collector of the bipolar transistors or off the drain of the MOSFETs. Since the circuit operations are the same as those previously described, with the exception of the inverted output, we will not repeat them here.

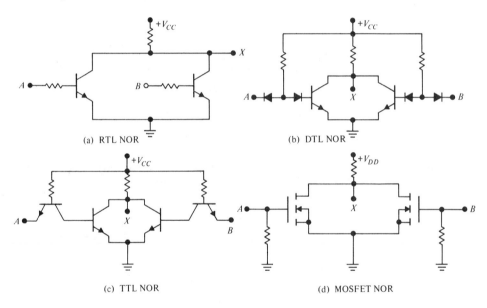

Figure 4–49.   Some Basic Two-Input NOR Gate Circuit Configurations.

4–6 Electrical Characteristics of Gates

Propagation Delays

Propagation delay is a very important characteristic of logic circuits because it limits the speed (frequency) at which they can operate. The terms *low speed* and *high speed*, when applied to logic circuits, refer to the propagation delays; the shorter the propagation delay, the higher the speed of the circuit.

A propagation delay of a gate is basically the *time interval between the application of an input pulse and the occurrence of the resulting output pulse.* There are two propagation delays associated with a logic gate: the delay time from the positive-going edge of the input pulse to the negative-going edge of the output pulse is called the turn-on delay, and the delay time from the negative-going edge of the input pulse to the positive-going edge of the output pulse is called the turn-off delay. These definitions apply to an *inverting* logic circuit. In many circuits they are not equal, and the larger of the two is the "worst case" propagation delay. Figure 4–50 illustrates the propagation delays through a logic circuit. The delay times are usually defined between the 50 percent points on the respective pulse edges.

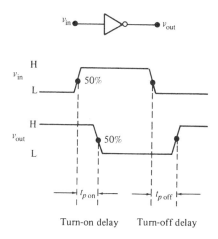

Figure 4–50. Propagation Delays Through a Logic Gate.

Power Dissipation

The *power dissipation* of a logic circuit is the DC supply voltage times the average current drawn from the supply. For example, if a gate operates from a +5 V DC supply, and draws 2 mA of current from that supply in the LOW state and 1 mA in the HIGH state, we can calculate the power dissipation as follows:

$$P_{LOW} = (5\text{ V})(2\text{ mA}) = 10\text{ mW}$$
$$P_{HIGH} = (5\text{ V})(1\text{ mA}) = 5\text{ mW}$$
$$P_{DISS} = P_{AVG} = \frac{P_{LOW} + P_{HIGH}}{2} = 7.5\text{ mW}$$

If the power dissipation of each logic circuit in a system is known, the capacity of the DC supply can be specified. For example, if we have a system composed of 100 gates and the power dissipation of each gate is 7.5 mW, the total power for the system is:

$$P_{TOT} = (100)(7.5\text{ mW}) = 750\text{ mW}$$

If a +5 V DC supply is required, then it must be capable of supplying *at least* the following current to the system:

$$I = \frac{P_{TOT}}{V_{CC}} = \frac{750\text{ mW}}{5\text{ V}} = 150\text{ mA}$$

Noise Margin

The *noise margin* of a logic circuit defines how well the circuit can withstand certain fluctuations of the voltage levels (noise) with which it must operate. This noise can come from several sources in a digital system. Some of the common sources are variations on the DC supply voltage; pickup of radiated energy; ground noise; thermal noise; and induced voltages from adjacent lines where currents are changing rapidly.

To learn about noise margins, we will define certain voltages associated with all logic circuits as follows:

$$V_{IL} = \text{LOW level input voltage}$$
$$V_{IH} = \text{HIGH level input voltage}$$
$$V_{OL} = \text{LOW level output voltage}$$
$$V_{OH} = \text{HIGH level output voltage}$$

Every logic circuit has certain limits on the values of these voltages within which it will operate properly. To illustrate this, let us discuss a logic gate that operates with 0 V as its ideal LOW and +5 V as its ideal HIGH. These voltages will vary because of circuit parameters, and the gate must be able to tolerate variations within certain specified limits. If, for example, a LOW level voltage range of 0 V to 0.5 V is acceptable to a gate, then the *maximum* LOW level input voltage $[V_{IL(\text{max})}]$ for that gate is +0.5 V. Any voltage above this value appears as a possible HIGH to the gate. The effect of a fluctuation due to noise of the LOW level input voltage on the output of the gate is illustrated in Figure 4–51.

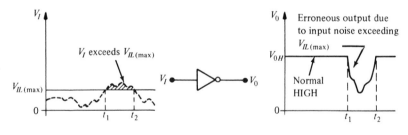

------ Fluctuating input voltage (noise)

Figure 4–51.   Potential Effect on a Gate Output of LOW Level Noise on the Input.

Now, if the type of logic gate we are using has a maximum LOW level *output* voltage $[V_{OL(\text{max})}]$ of +0.2 V, then there is a "safety" margin of 0.3 V between the maximum LOW level that a gate puts out and the maximum LOW level that a gate being driven can tolerate. This is called the LOW level *noise margin* and is expressed as

$$V_{NL} = V_{IL(\text{max})} - V_{OL(\text{max})} \tag{4-2}$$

Figure 4–52 illustrates the LOW level noise margin. In essence, there can be fluctuations due to noise on the line between the two gates without affecting the output of the second gate, as long as the peak value of these fluctuations does not exceed the LOW level noise margin (V_{NL}).

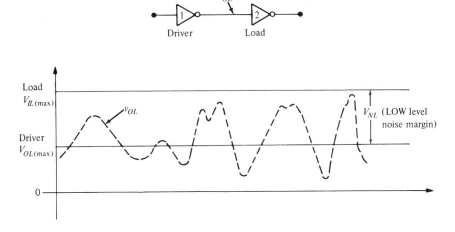

Figure 4-52.   LOW Level Noise Margin.

Now let us consider the HIGH level noise margin of a gate. If, for example, a HIGH level input voltage between +5 V and + 4 V is acceptable as a HIGH level to the gate, then the *minimum* HIGH level input voltage [$V_{IH(min)}$] is +4.0 V. Any voltage above this value is acceptable as a HIGH to the gate and any voltage below this value would appear as a possible LOW, as illustrated in Figure 4-53.

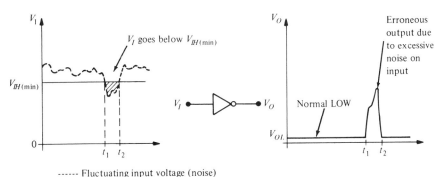

------ Fluctuating input voltage (noise)

Figure 4-53.   Potential Effect on a Gate Output of HIGH Level Noise on the Input.

Now, if the driving gate has a minimum HIGH level output voltage [$V_{OH(min)}$] of +4.5 V, then there is a HIGH level *noise margin* of 0.5 V, expressed as

$$V_{NH} = V_{OH(min)} - V_{IH(min)} \qquad (4\text{-}3)$$

Figure 4-54 illustrates the HIGH level noise margin, and shows that there can be fluctuations due to noise on the line between the two gates without affecting the output of the second gate, as long as the peak value of the noise does not exceed V_{NH}, the HIGH level noise margin.

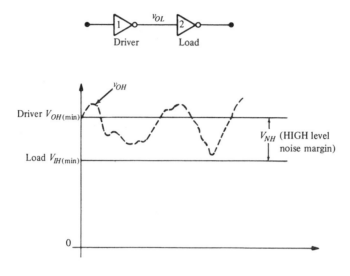

Figure 4-54. HIGH Level Noise Margin.

Fan-out

The *fan-out* of a gate is the *maximum* number of similar gates that it can drive and maintain its output within specified limits. In order to understand the meaning of fan-out as applied to logic circuits, let us first take a basic RTL inverter to illustrate the concept. Figure 4-55 shows an RTL inverter driving another RTL inverter. When the output of the driver (inverter 1) is HIGH, Q_1 is off and base current is supplied to the load (inverter 2) as indicated. The value of the HIGH voltage output of inverter 1 is dependent on its output resistance (approximately R_{C1}) and the input resistance of inverter 2 (approximately R_{B2}). For this condition, a voltage divider is created, as illustrated in Figure 4-56. The HIGH output

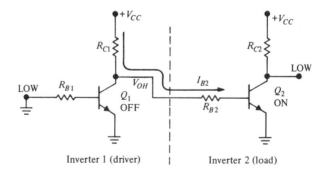

Figure 4-55. Illustration of RTL Loading.

voltage of inverter 1 is determined by using the voltage divider rule:

$$V_{OH} \cong \left(\frac{R_{B2}}{R_{B2} + R_{C1}}\right) V_{CC} \qquad (4-4)$$

Example 4–18 In Figure 4–56, if $V_{CC} = +5$ V, $R_{C1} = 1$ K, and $R_{B2} = 10$ K, determine V_{OH}.

Solution:

$$V_{OH} \cong \left(\frac{10 \text{ K}}{10 \text{ K} + 1 \text{ K}}\right) 5 \text{ V} = 4.55 \text{ V}$$

As you can see, the input resistance of inverter 2 has a loading effect on the output of inverter 1; that is, the HIGH output voltage level of inverter 1 is reduced from its no-load value of $+5$ V.

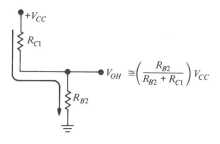

Figure 4–56. Approximate Loading Effect of Inverter 2 on Inverter 1.

Now, if inverter 1 is required to drive two inverters as shown in Figure 4–57, then V_{OH} is reduced further.

The input resistances of the two inverters being driven are effectively in parallel with each other. This reduces the total load resistance that inverter 1 must drive, thereby further loading it (as shown by substitution into the voltage divider equation using the values of Example 4–18):

$$V_{OH} \cong \left(\frac{\dfrac{R_B}{2}}{\dfrac{R_B}{2} + R_{C1}}\right) V_{CC} = \left(\frac{5 \text{ K}}{5 \text{ K} + 1 \text{ K}}\right) 5 \text{ V} = 4.17 \text{ V}$$

As additional loads are connected to the output of inverter 1, its HIGH level output voltage continues to be reduced.

Because of noise margin restrictions, as previously discussed, there is a minimum HIGH level output voltage that can be tolerated by the gates being driven. This establishes the maximum number of loads (other identical gates) that a given logic circuit can drive before its HIGH level output voltage is equal to or less than $V_{OH(\text{min})}$. The maximum number of loads that a logic circuit can drive, such that V_{OH} is equal to or greater than $V_{OH(\text{min})}$, is called the *fan-out* of the circuit.

Notice that we have considered only the loading effects when the output of the RTL circuit is at a HIGH level. We must also look at the loading effect, if any, at the LOW output level, and find out if this may be a limitation on the fan-out.

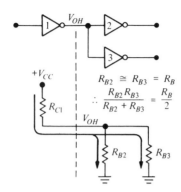

Figure 4–57.   Inverter Driving Two Parallel Loads.

Upon examination of Figure 4–55, it is evident that no current is required to be supplied to the load (inverter 2) and no current is accepted from the load when the output of inverter 1 is LOW. Therefore, there is no limitation on loading for the LOW output state of inverter 1. This is a current "sourcing" type of circuit because current is supplied from the driving circuit to the load circuit in the HIGH state; that is, the driving circuit is the source of current in this particular type of circuit, and the fan-out is limited by the HIGH output loading effects. The factors limiting fan-out may be different for different circuit implementations, as will be illustrated in the following discussion.

For other types of logic gates, such as DTL and TTL, the loading considerations are somewhat different, although fan-out is defined the same regardless of the particular implementation. Let us consider the loading effects for a DTL circuit (and the same can be applied to TTL because of the basic similarity). Consider the circuit of Figure 4–58(a). When the output of the driving circuit is at the HIGH level, the input diode of the load circuit is reverse biased; the equivalent is as shown in Figure 4–58(b). Because an open circuit (D_1 reverse biased) is presented to the driving circuit, no current is demanded; therefore, no loading on the driver occurs and $V_{OH} = V_{CC}$.

When the output of the driving circuit is LOW, as shown in Figure 4–58(c), Q_1 is on, D_1 is forward biased, and a current is drawn from the load to the driver. There is a limitation on the amount of current that the transistor Q_1 in the driver can accept and still remain saturated. This maximum current that the output transistor can "sink" sets the limit on how many circuits it can drive and, therefore, the fan-out.

For example, if the current from each load to the driver circuit in the LOW output state is 1 mA, and the maximum current that the driving transistor can accept is 10 mA, then the circuit cannot drive more than 10 load circuits and the fan-out is therefore equal to 10. This is illustrated in Figure 4–59. Most manufacturers of digital integrated circuits specify the fan-out of a particular circuit in terms of *unit load*. A unit load is basically defined as the amount of loading that a particular circuit presents to the *same* type of circuit that is driving it. For example, if the fan-out of a gate is specified as 10 unit loads (10 U.L.), this means that the gate can drive up to 10 gates of the same type.

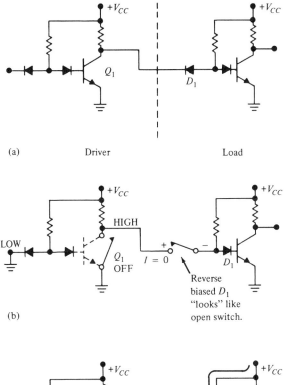

(a)   Driver   Load

(b)

(c)

Figure 4–58.   Illustration of DTL Loading.

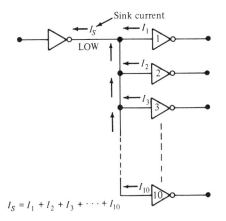

$$I_S = I_1 + I_2 + I_3 + \cdots + I_{10}$$

Figure 4–59.   LOW State Loading of Current Sinking Logic, Such as DTL or TTL.

Fan-in

The *fan-in* of a gate is simply the total number of inputs that it can accept. A two-input gate has a fan-in of two, a three-input gate has a fan-in of three, and so on.

PROBLEMS

4–1   The input waveform shown in Figure 4–60 is applied to an inverter. Sketch the output waveform in proper relationship to the input.

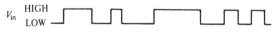

Figure 4–60.

4–2   A network of cascaded inverters is shown in Figure 4–61. If a HIGH is applied to point A, determine the logic levels at points B through F.

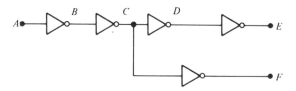

Figure 4–61.

4–3   Determine the output, X, for a two-input AND gate with the input waveforms shown in Figure 4–62.

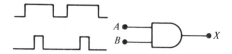

Figure 4–62.

4–4   Repeat Problem 4–3 for the waveforms in Figure 4–63.

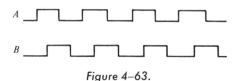

Figure 4–63.

4–5   The input waveforms are applied to a three-input AND gate as indicated in Figure 4–64. Determine the output waveform.

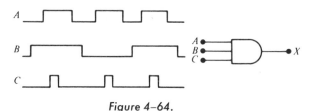

Figure 4–64.

4-6 The input waveforms are applied to a four-input gate as indicated in Figure 4–65. Determine the output waveform.

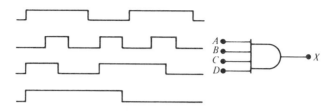

Figure 4–65.

4-7 Determine a second input waveform that can be used with input *A* in Figure 4–66 to produce the specified output from a two-input AND gate. (Choose an input unlike the output.)

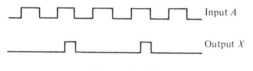

Figure 4–66.

4-8 Repeat Problem 4–3 for a two-input OR gate.

4-9 Determine the output for a two-input OR gate using the input waveforms in Problem 4–4.

4-10 Repeat Problem 4–5 for a three-input OR gate.

4-11 Repeat Problem 4–6 for a four-input OR gate.

4-12 For the five input waveforms in Figure 4–67, determine the corresponding output for an AND gate, and the corresponding output for an OR gate.

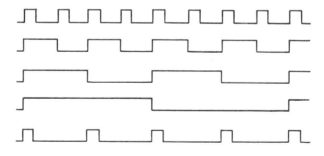

Figure 4–67.

4-13 Make a complete truth table for a three-input NAND gate.

4-14 Make a complete truth table for a four-input NAND gate.

4-15 Make a complete truth table for a three-input NOR gate.

4-16 Make a complete truth table for a four-input NOR gate.

4-17   For the set of input waveforms in Figure 4-68, determine the output for the gate shown.

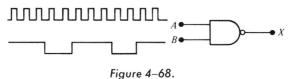

Figure 4-68.

4-18   Determine the gate output for the waveforms in Figure 4-69.

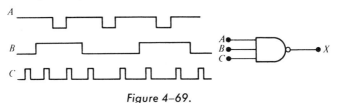

Figure 4-69.

4-19   Describe the circuit operation for all possible input conditions for each circuit in Figure 4-40.

4-20   Describe the circuit operation for all possible input conditions for each circuit in Figure 4-49.

4-21   Determine the output waveform in Figure 4-70.

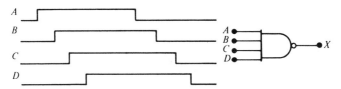

Figure 4-70.

4-22   Repeat Problem 4-17 for a two-input NOR gate.

4-23   Determine the output waveform in Figure 4-71.

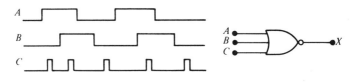

Figure 4-71.

4-24   Repeat Problem 4-21 for a four-input NOR gate.

4-25   As we learned, the two logic symbols shown in Figure 4-72 represent the same circuit. The difference between the two is strictly a matter of how we look at them from a functional viewpoint. For the NAND symbol we are looking for two HIGHs on the inputs to give us a LOW output. For the negative-OR, we are looking for at least one LOW on the inputs to give us a HIGH on the output. Using these two functional viewpoints, show that each gate will produce the same output for the given inputs.

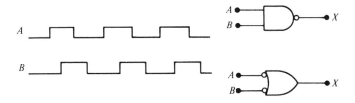

Figure 4-72.

4-26 For the NOR symbol, we are looking for at least one HIGH on the inputs to give us a LOW on the output. For the negative-AND we are looking for two LOWs on the inputs to give us a HIGH output. Using these two functional points of view, show that both gates in Figure 4-73 will produce the same output for the given inputs.

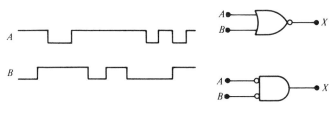

Figure 4-73.

4-27 If a logic gate operates on a DC supply voltage of +8 V and draws an average current of 4 mA, what is its power dissipation?

4-28 The minimum HIGH output of a gate is 4.4 V. The gate is driving a second gate that can tolerate a minimum HIGH input of 4.1 V. What is the HIGH level noise margin?

4-29 The maximum LOW output of a gate is specified as 0.3 V, and the maximum LOW input is specified as 0.6 V. What is the LOW level noise margin when these gates work together?

4-30 For the circuit in Figure 4-74, determine the fan-out if the minimum HIGH output is to be 9 V.

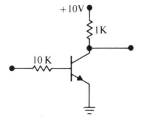

Figure 4-74.

4-31 In Problem 4–30, if the circuit drives 20 unit loads, what is V_{OH}?

4-32 For a TTL gate the maximum current that the output transistor can handle in the LOW state is 15 mA. When the input is LOW, the current out of the input terminal is 1.2 mA. Determine the fan-out for this type of gate.

4-33 What is the fan-in of the gate in Figure 4–75?

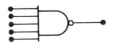

Figure 4–75.

CHAPTER 5

Boolean Algebra

Boolean algebra is essentially a set of rules, laws, and theorems by which logical operations can be expressed symbolically in equation form and manipulated mathematically. We are interested in its applications as a convenient and systematic way of expressing and analyzing the operation of digital circuits and systems.

5-1   Historical Background

It was in 1854 that George Boole published a classic book entitled *An Investigation of the Laws of Thought on Which are Founded the Mathematical Theories of Logic and Probabilities*. It was in this publication that a "logical algebra" that is commonly known today as Boolean algebra was developed.

The application of Boolean algebra to the analysis and design of digital logic circuits was first explored by Claude Shannon at MIT in a 1938 thesis entitled *A Symbolic Analysis of Relay and Switching Circuits*. Essentially, this paper described a method by which any circuit consisting of combinations of switches and relays could be represented by mathematical expressions. Today, of course, semiconductor circuits have for the most part replaced mechanical switches and relays. However, the same logical analysis is still valid, and a basic knowledge in this area is essential to the study of digital logic.

5-2   Symbology

In the applications of Boolean algebra in this book, we will use *capital letters* to represent variables and functions of variables. Any single variable or a function of several variables can have either a 1 or a 0 value. In Boolean algebra, the binary digits are utilized to represent the two levels that occur within digital logic circuits. A binary 1 will represent a HIGH level and a binary 0 will represent a LOW level when logical analysis using Boolean equations is performed. This is in keeping with our use in this text of the positive logic as explained in Chapter 1.

The complement of a variable is represented by a "bar" over the letter. For instance, for a variable represented by A, the complement of A is $\overline{A}$. So if $A = 1$, then $\overline{A} = 0$; or if $A = 0$, then $\overline{A} = 1$. The complement of a variable A is usually read "A bar" or "Not A". Sometimes a prime symbol rather than the bar symbol is used to denote the complement. For example the complement of A can be written as A'.

The logical AND function of two variables is represented either by a "dot" between the two variables, such as $A \cdot B$, or by simply writing the adjacent letters without the "dot", such as AB. We will normally use the latter notation because it is easier to write. The logical OR function of two variables is represented by a "$+$" between the two variables, such as $A + B$.

5–3   Addition and Multiplication

Addition in Boolean algebra involves variables having values of either a binary 1 or a binary 0. The basic rules for Boolean addition are listed below.

$$0 + 0 = 0$$
$$0 + 1 = 1$$
$$1 + 0 = 1$$
$$1 + 1 = 1$$

In the application of Boolean algebra to logic circuits, *Boolean addition is the same as the OR function.*

Multiplication in Boolean algebra utilizes the basic rules governing binary multiplication which were discussed in Chapter 2 and which are listed below:

$$0 \cdot 0 = 0$$
$$0 \cdot 1 = 0$$
$$1 \cdot 0 = 0$$
$$1 \cdot 1 = 1$$

Multiplication is the same as the AND function in Boolean algebra.

5–4   Expressions for the Basic Logic Functions

The operation of an inverter (NOT circuit) can be expressed with symbols as follows: if the input variable is called A and the output variable is called X, then $X = \overline{A}$. This expression "says" that the output is the complement of the input, so that if $A = 0$, then $X = 1$, and if $A = 1$, then $X = 0$. Figure 5–1 illustrates this further.

The operation of a two-input AND gate can be expressed in equation form as follows: if one input variable is A, the other input variable is B, and the output variable is X, then the Boolean expression for this basic gate function is $X = AB$. Figure 5–2(a) shows the gate with the input and output variables indicated.

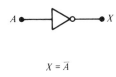

$$X = \overline{A}$$

Figure 5–1.

To extend the AND expression to more than two input variables, we simply use a new letter for each input variable. The function of a three-input AND gate, for example, can be expressed as $X = ABC$, where A, B, and C are the input variables. The expression for a four-input AND gate can be $X = ABCD$, and so on. Figures 5–2(b) and 5–2(c) show AND gates with three and four input variables, respectively.

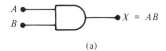

(a)

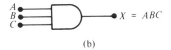

(b)

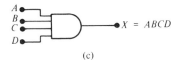

(c)

Figure 5–2.

An evaluation of AND gate operation can be made using the Boolean expressions for the output. For example, each variable on the inputs can be either a 1 or a 0; so for the two-input AND gate we can make the following substitutions in the equation for the output X:

$$A = 0, \quad B = 0: \quad X = AB = 0 \cdot 0 = 0$$
$$A = 0, \quad B = 1: \quad X = AB = 0 \cdot 1 = 0$$
$$A = 1, \quad B = 0: \quad X = AB = 1 \cdot 0 = 0$$
$$A = 1, \quad B = 1: \quad X = AB = 1 \cdot 1 = 1$$

The evaluation of this equation simply tells us that the output X of an AND gate is a 1 (HIGH) only when both inputs are 1s (HIGHs). A similar analysis can be made for any number of input variables.

The operation of a two-input OR gate can be expressed in equation form as follows: if one input is A, the other input B, and the output is X, then the Boolean expression is $X = A + B$. Figure 5–3(a) shows the gate logic symbol, with input and output variables labeled.

To extend the OR expression to more than two input variables, a new letter is used for each additional variable. For instance, the function of a three-input OR gate can be expressed as $X = A + B + C$. The expression for a four-input OR gate can be written as $X = A + B + C + D$, and so on. Figures 5–3(b) and 5–3(c) show OR gate logic symbols with three and four input variables, respectively.

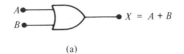

(a)

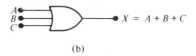

(b)

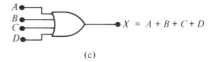

(c)

Figure 5–3.

OR gate operation can be evaluated using the Boolean expressions for the output X by substituting all possible combinations of 1 and 0 values for the input variables, as shown below for a two-input OR gate:

$$A = 0, \quad B = 0: \quad X = A + B = 0 + 0 = 0$$
$$A = 0, \quad B = 1: \quad X = A + B = 0 + 1 = 1$$
$$A = 1, \quad B = 0: \quad X = A + B = 1 + 0 = 1$$
$$A = 1, \quad B = 1: \quad X = A + B = 1 + 1 = 1$$

This evaluation shows that the output of an OR gate is a 1 (HIGH) when any of the inputs are 1 (HIGH). A similar analysis can be extended to OR gates with any number of input variables.

The Boolean expression for a two-input NAND gate is $X = \overline{AB}$. This expression says that the two input variables, A and B, are first ANDed and then complemented, as indicated by the "bar" over the AND expression. This is a logical description in equation form of the operation of a NAND gate with two inputs. If we evaluate this expression for all possible values of the input variables, the results are as follows:

$$A = 0, \quad B = 0: \quad X = \overline{A \cdot B} = \overline{0 \cdot 0} = \overline{0} = 1$$
$$A = 0, \quad B = 1: \quad X = \overline{A \cdot B} = \overline{0 \cdot 1} = \overline{0} = 1$$
$$A = 1, \quad B = 0: \quad X = \overline{A \cdot B} = \overline{1 \cdot 0} = \overline{0} = 1$$
$$A = 1, \quad B = 1: \quad X = \overline{A \cdot B} = \overline{1 \cdot 1} = \overline{1} = 0$$

So, you see that once a Boolean expression is determined for a given logic function, then that function can be evaluated for all possible values of the variables. The evaluation tells us exactly what the output of the logic circuit is for each of the input conditions, and therefore gives us a complete description of the circuit's logical operation. The NAND expression can be extended to more than two input variables by including additional letters to represent all of the variables.

Finally, the expression for a two-input NOR gate can be written as $X = \overline{A + B}$. This equation says that the two input variables are first ORed and then complemented, as indicated by the "bar" over the OR expression. Evaluating this expression, we get the following results:

$$A = 0, \quad B = 0: \quad X = \overline{A + B} = \overline{0 + 0} = \overline{0} = 1$$
$$A = 0, \quad B = 1: \quad X = \overline{A + B} = \overline{0 + 1} = \overline{1} = 0$$
$$A = 1, \quad B = 0: \quad X = \overline{A + B} = \overline{1 + 0} = \overline{1} = 0$$
$$A = 1, \quad B = 1: \quad X = \overline{A + B} = \overline{1 + 1} = \overline{1} = 0$$

5-5 Rules and Laws of Boolean Algebra

As in other areas of mathematics, there are certain well-developed rules and laws that must be followed in order to properly apply Boolean algebra. The most important of these are presented in this section.

Three of the basic laws of Boolean algebra are the same as in ordinary algebra: the *commutative laws,* the *associative laws,* and the *distributive laws.*

The commutative law of addition for two variables is written algebraically as

$$A + B = B + A \qquad \qquad (5\text{-}1)$$

This states that the order in which the variables are added (ORed) makes no difference. Remember, in Boolean algebra terminology as applied to logic circuits, addition and the OR function are the same. Figure 5-4 illustrates the commutative law as applied to the OR gate.

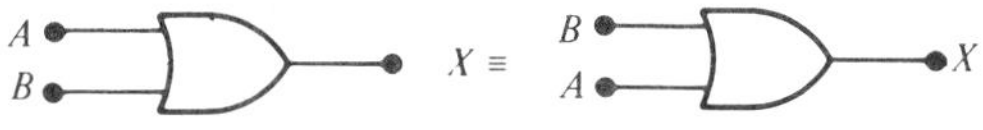

Figure 5-4. Application of Commutative Law of Addition.

The commutative law of multiplication of two variables is

$$AB = BA \qquad \qquad (5\text{-}2)$$

This states that the order in which the variables are multiplied (ANDed) makes no difference. Figure 5-5 illustrates this law as applied to the AND gate.

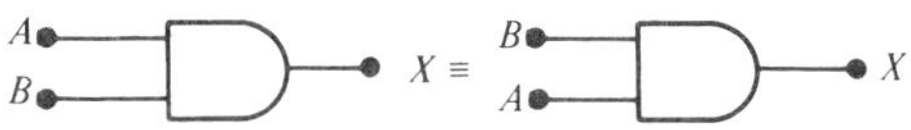

Figure 5-5. Application of Commutative Law of Multiplication.

The associative law of addition is stated as follows for three variables:

$$A + (B + C) = (A + B) + C \qquad (5\text{-}3)$$

This law states that in the addition (ORing) of several variables, the result is the same regardless of the grouping of the variables. Figure 5–6 illustrates this law as applied to OR gates.

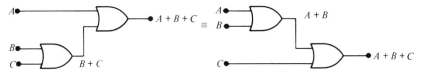

Figure 5–6. Application of Associative Law of Addition.

The associative law of multiplication is stated as follows for three variables:

$$A(BC) = (AB)C \qquad (5\text{-}4)$$

This law tell us that it makes no difference in what order the variables are grouped when multiplying (ANDing) several variables. Figure 5–7 illustrates this law as applied to AND gates.

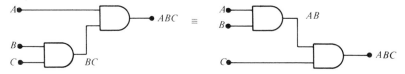

Figure 5–7. Application of Associative Law of Multiplication.

The distributive law is written for three variables as follows:

$$A(B + C) = AB + AC \qquad (5\text{-}5)$$

This law states that the sum (OR) of several variables when multiplied (ANDed) by a single variable is equivalent to the sum (OR) of the products (AND) of the single variable and each of the other variables. This law and the ones previously discussed should be familiar because they are the same in ordinary algebra. It should be noted that each of these laws can be extended to include any number of variables.

Table 5–1 lists several basic rules that are useful in manipulating and simplifying Boolean algebra expressions.

We will now look at rules 1 through 9 of Table 5–1 in terms of their application to logic gates. Rules 10 through 12 will be derived in terms of the simpler rules and the laws previously discussed.

Rule 1 can be understood by observing what happens when one input to an OR gate is always 0 and the other input, A, can take on either a 1 or a 0 value. If A is a 1, it is obvious that the output is a 1, which is equal to A. If A is a 0, the output is a 0, which is also equal to A. Therefore, it follows that a variable ORed with a 0 is equal to the value of the variable ($A + 0 = A$). This rule is further demonstrated in Figure 5–8.

TABLE 5-1.   Basic Rules of
Boolean Algebra.

1. $A + 0 = A$
2. $A + 1 = 1$
3. $A \cdot 0 = 0$
4. $A \cdot 1 = A$
5. $A + A = A$
6. $A + \overline{A} = 1$
7. $A \cdot A = A$
8. $A \cdot \overline{A} = 0$
9. $\overline{\overline{A}} = A$
10. $A + AB = A$
11. $A + \overline{A}B = A + B$
12. $(A + B)(A + C) = A + BC$

Note: A can represent a single variable
or combination of variables.

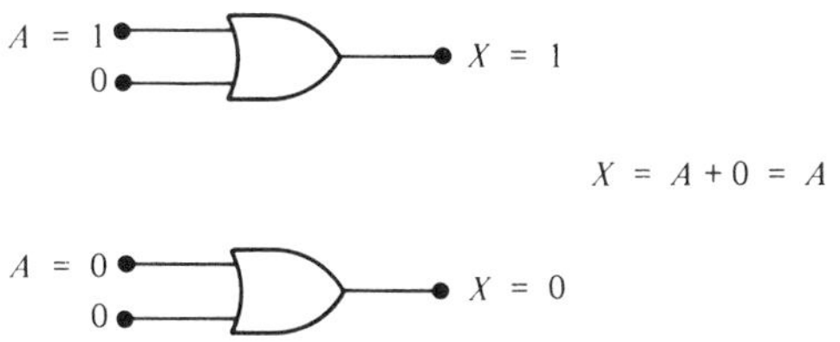

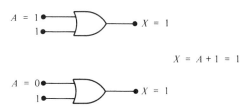

$$X = A + 0 = A$$

Figure 5-8.   Illustration of Rule 1.

Rule 2 is demonstrated when one input to an OR gate is always 1 and the other input, A, takes on either a 1 or a 0 value. A 1 on an input to an OR gate produces a 1 on the output, regardless of the value of the variable on the other input. Therefore, a variable ORed with a 1 is always equal to 1 ($A + 1 = 1$). This rule is illustrated in Figure 5-9.

$$X = A + 1 = 1$$

Figure 5-9.   Illustration of Rule 2.

Rule 3 is demonstrated when a 0 is ANDed with a variable. Of course, any time one input to an AND gate is 0, the output is 0, regardless of the value of the variable on the other input. A variable ANDed with a 0 always produces a 0 ($A \cdot 0 = 0$). This rule is illustrated in Figure 5-10.

Rule 4 can be verified by ANDing a variable with a 1. If the variable A is a 0, the output of the AND gate is a 0. If the variable A is a 1, the output of the AND gate is a 1 because both inputs are now 1s. Therefore, the AND function of a vari-

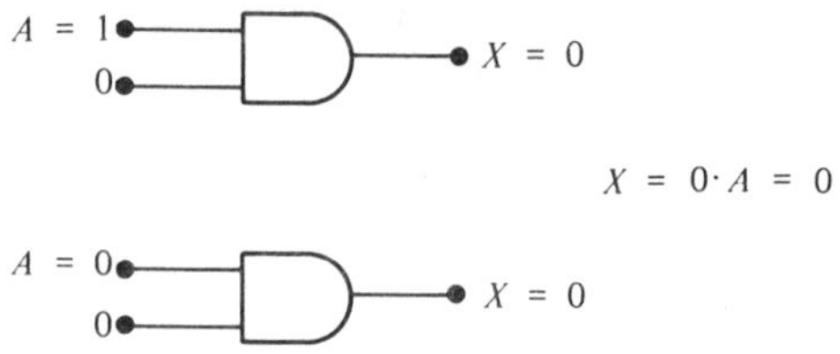

Boolean Algebra

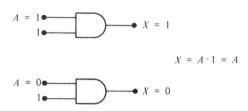

$$X = 0 \cdot A = 0$$

Figure 5–10. Illustration of Rule 3.

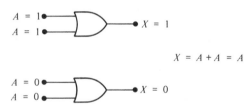

$$X = A \cdot 1 = A$$

Figure 5–11. Illustration of Rule 4.

able and a 1 is equal to the value of the variable ($A \cdot 1 = A$). This is shown in Figure 5–11.

Rule 5 states that if a variable is ORed with itself, the output is equal to the variable. For instance, if A is a 0, then $0 + 0 = 0$, and if A is a 1 then $1 + 1 = 1$. This is shown in relation to OR gate application in Figure 5–12.

$$X = A + A = A$$

Figure 5–12. Illustration of Rule 5.

Rule 6 can be explained as follows: if a variable and its complement are ORed, the result is always a 1. If A is a 0, then $0 + \bar{0} = 0 + 1 = 1$. If A is a 1, then $1 + \bar{1} = 1 + 0 = 1$. See Figure 5–13 for further illustration of this rule.

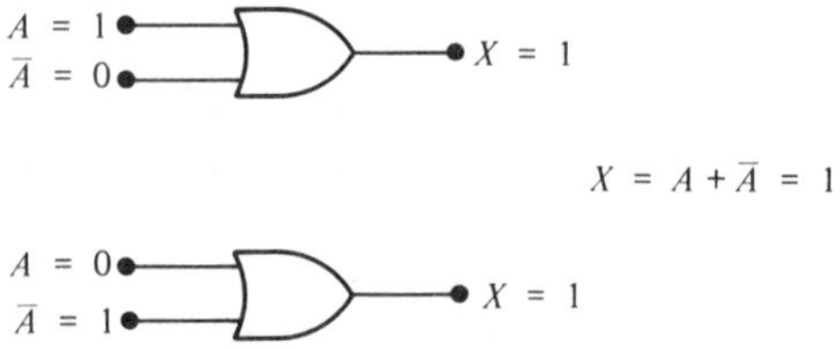

$$X = A + \bar{A} = 1$$

Figure 5–13. Illustration of Rule 6.

Rule 7 states that if a variable is ANDed with itself, the result is equal to the variable. For example, if $A = 0$, then $0 \cdot 0 = 0$, and if $A = 1$, then $1 \cdot 1 = 1$. For either case, the output of an AND gate is equal to the value of the input variable A. Figure 5–14 illustrates this rule.

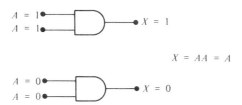

$$X = AA = A$$

Figure 5–14.   Illustration of Rule 7.

Rule 8 states that if a variable is ANDed with its complement, the result is 0. This is readily seen because either A or $\overline{A}$ will always be 0, and when a 0 is applied to the input of an AND gate, it insures that the output will be 0 also. Figure 5–15 will help illustrate this rule.

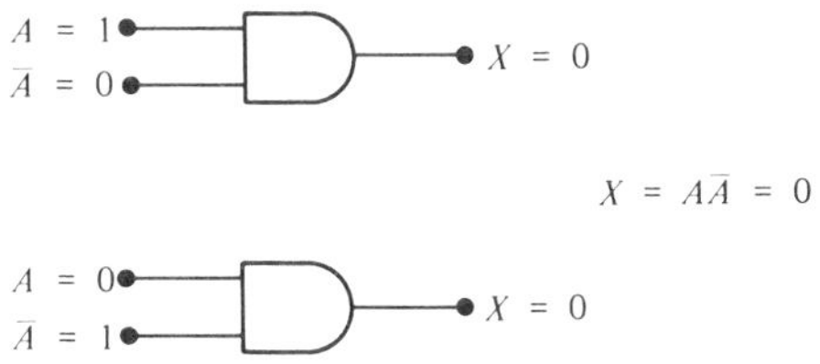

$$X = A\overline{A} = 0$$

Figure 5–15.   Illustration of Rule 8.

Rule 9 simply says that if a variable is complemented twice, the result is the variable itself. If we start with the variable A and complement (invert) it once, we get $\overline{A}$. If we then take $\overline{A}$ and complement (invert) it, we get A, which is the original variable. This is shown in Figure 5–16, using inverters.

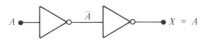

Figure 5–16.   Illustration of Rule 9.

The proof of rule 10 is shown by using the distributive law, rule 2, and rule 4 as follows:

$$A + AB = A(1 + B) \qquad \text{distributive law}$$
$$= A \cdot 1 \qquad \text{rule 2}$$
$$= A \qquad \text{rule 4}$$

The proof of Rule 11 is as follows:

$$A + \overline{A}B = (A + AB) + \overline{A}B \qquad \text{rule 10}$$
$$= (AA + AB) + \overline{A}B \qquad \text{rule 7}$$

$$= AA + AB + A\overline{A} + \overline{A}B \qquad \text{rule 8}$$
$$= (A + \overline{A})(A + B) \qquad \text{by factoring}$$
$$= 1 \cdot (A + B) \qquad \text{rule 6}$$
$$= A + B \qquad \text{rule 4}$$

Finally, the proof of rule 12 is as follows:

$$(A + B)(A + C) = AA + AC + AB + BC \qquad \text{distributive law}$$
$$= A + AC + AB + BC \qquad \text{rule 7}$$
$$= A(1 + AC) + AB + BC \qquad \text{distributive law and rule 4}$$
$$= A \cdot 1 + AB + BC \qquad \text{rule 2}$$
$$= A(1 + B) + BC \qquad \text{distributive law}$$
$$= A \cdot 1 + BC \qquad \text{rule 2}$$
$$= A + BC \qquad \text{rule 4}$$

5–6   DeMorgan's Theorems

DeMorgan, a logician and mathematician who was acquainted with Boole, proposed two theorems that are an important part of Boolean algebra. They are stated as follows in equation form:

$$\overline{AB} = \overline{A} + \overline{B} \qquad (5\text{–}6)$$
$$\overline{A + B} = \overline{A}\overline{B} \qquad (5\text{–}7)$$

The theorem expressed in Equation (5–6) can be stated as *"The complement of a product is equal to the sum of the complements."* This really says that the complement of two or more variables ANDed is the same as the OR of the complements of each individual variable.

The theorem expressed in Equation (5–7) can be stated as *"The complement of a sum is equal to the product of the complements."* This says that the complement of two or more variables ORed is the same as the AND of the complements of each individual variable.

We can see that the relation in Equation (5–6) is valid by substituting all possible values for the variables A and B, as follows:

$$A = 0, \quad B = 0: \quad \overline{AB} = \overline{0 \cdot 0} = \overline{0} \quad = 1$$
$$\overline{A} + \overline{B} = \overline{0} + \overline{0} = 1 + 1 = 1$$
$$A = 0, \quad B = 1: \quad \overline{AB} = \overline{0 \cdot 1} = \overline{0} \quad = 1$$
$$\overline{A} + \overline{B} = \overline{0} + \overline{1} = 1 + 0 = 1$$
$$A = 1, \quad B = 0: \quad \overline{AB} \quad = \overline{1 \cdot 0} = \overline{0} = 1$$
$$\overline{A} + \overline{B} = \overline{1} + \overline{0} = 0 + 1 = 1$$
$$A = 1, \quad B = 1: \quad \overline{AB} = \overline{1 \cdot 1} = \overline{1} \quad = 0$$
$$\overline{A} + \overline{B} = \overline{1} + \overline{1} = 0 + 0 = 0$$

This shows that the equality of Equation (5–6) holds for all possible combinations of variable values.

It can also be shown that the relation expressed by Equation (5–7) is valid by the same procedure.

$$A = 0, \quad B = 0: \quad \overline{A + B} = \overline{0 + 0} = \overline{0} = 1$$
$$\overline{A}\,\overline{B} = \overline{0}\cdot\overline{0} = 1\cdot 1 = 1$$
$$A = 0, \quad B = 1: \quad \overline{A + B} = \overline{0 + 1} = \overline{1} = 0$$
$$\overline{A}\,\overline{B} = \overline{0}\cdot\overline{1} = 1\cdot 0 = 0$$
$$A = 1, \quad B = 0: \quad \overline{A + B} = \overline{1 + 0} = \overline{1} = 0$$
$$\overline{A}\,\overline{B} = \overline{1}\cdot\overline{0} = 0\cdot 1 = 0$$
$$A = 1, \quad B = 1: \quad \overline{A + B} = \overline{1 + 1} = \overline{1} = 0$$
$$\overline{A}\,\overline{B} = \overline{1}\cdot\overline{1} = 0\cdot 0 = 0$$

This shows that the equality of Equation (5–7) is valid for all possible combinations of variable values. Although DeMorgan's theorems have been expressed here with only two variables, they can be extended to any number of variables, as will be illustrated in the following examples.

Example 5–1   Express the following complement-of-product terms as sum-of-complements using DeMorgan's theorem:

$$\overline{ABC}, \qquad \overline{ABCD}, \qquad \overline{ABCDEF}$$

Solutions:
Use DeMorgan's theorem as expressed in Equation (5–6).

$$\overline{ABC} = \overline{A} + \overline{B} + \overline{C}$$
$$\overline{ABCD} = \overline{A} + \overline{B} + \overline{C} + \overline{D}$$
$$\overline{ABCDEF} = \overline{A} + \overline{B} + \overline{C} + \overline{D} + \overline{E} + \overline{F}$$

Example 5–2   Express the following complement-of-sum terms as product-of-complements using DeMorgan's theorem:

$$\overline{A + B + C}$$
$$\overline{A + B + C + D}$$
$$\overline{A + B + C + D + E + F}$$

Solutions:
Use DeMorgan's theorem as expressed in Equation (5–7).

$$\overline{A + B + C} = \overline{A}\,\overline{B}\,\overline{C}$$
$$\overline{A + B + C + D} = \overline{A}\,\overline{B}\,\overline{C}\,\overline{D}$$
$$\overline{A + B + C + D + E + F} = \overline{A}\,\overline{B}\,\overline{C}\,\overline{D}\,\overline{E}\,\overline{F}$$

5–7   Forms of Boolean Expressions and Their Relation to Gate Networks

The form of a given Boolean (logic) expression indicates the type of gate network it describes. For example, let us take the expression $A(B + CD)$ and determine what sort of logic circuit it represents. First, there are four variables: A, B, C, and D. C is ANDed with D:CD; then this is ORed with B:$(B + CD)$. Then this is ANDed with A to produce the final function. Figure 5–17 illustrates the gate network represented by this particular Boolean expression and how it is formed.

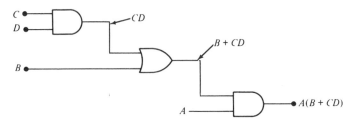

Figure 5–17.

As you have seen, the form of the Boolean expression does determine how many logic gates are used, what type of gates are needed, and how they are connected together; this will be explored further in Chapter 6. The more complex an expression, the more complex the gate network will be. It is therefore an advantage to simplify an expression as much as possible in order to have the simplest gate network. We will cover simplification methods in later sections of this chapter. There are also certain forms of Boolean expressions that are more desirable or more widely used than others; the two most important of these are the *sum-of-products* and the *product-of-sums* forms.

What does the sum-of-products form mean? First, let us review products in Boolean algebra. A product of two or more variables or their complements is simply the AND function of those variables. The product of two variables can be expressed as AB, the product of three variables as ABC, the product of four variables as $ABCD$, etc. Recall that a sum in Boolean algebra is the same as the OR function, so a sum-of-products expression is two or more AND functions ORed together. For instance, $AB + CD$ is a sum-of-products expression. Several other examples of expressions in sum-of-product form are as follows:

$$AB + BCD$$
$$ABC + DEF$$
$$A\overline{B}C + D\overline{E}FG + AEG$$
$$AB\overline{C} + \overline{D}EF + FGH + A\overline{F}G$$

A sum-of-products form can also contain a term with a single variable, such as $A + BCD + EFG$.

One reason the sum-of-products is a useful form of Boolean expression is the straightforward manner in which it can be implemented with logic gates. We have AND functions that are ORed, as Example 5–3 illustrates.

Example 5–3 Implement the expression $AB + BCD + EFGH$ with logic gates.

Solution:
See Figure 5–18.

An important characteristic of the sum-of-products form is that the corresponding implementation is always a *two-level* gate network; that is, the maximum number of gates through which a signal must pass in going from an input to the output is *two*.

The product-of-sums form can be thought of as the dual of the sum-of-products. It is, in terms of logic functions, the AND of two or more OR functions.

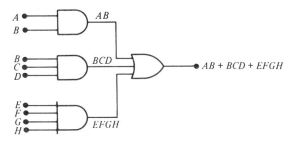

Figure 5-18.

For instance, $(A + B)(B + C)$ is a product-of-sums expression. Several other examples are:

$$(A + B)(B + C + D)$$
$$(A + B + C)(D + E + F)$$
$$(A + \overline{B} + C)(D + \overline{E} + F + G)(A + E + G)$$
$$(A + B + \overline{C})(\overline{D} + E + F)(F + G + H)(A + \overline{F} + G)$$

A product-of-sums expression can also contain a single variable term such as $A(B + C + D)(E + F + G)$.

This form also lends itself to straightforward implementation with logic gates because it simply involves ANDing two or more OR terms. A two-level gate network will always result, as the following example will show:

Example 5–4   Construct the following function with logic gates:

$$(A + B)(C + D + E)(F + G + H + I)$$

Solution:
See Figure 5–19.

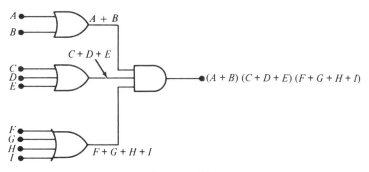

Figure 5-19.

5–8   Algebraic Simplification

Many times in the application of Boolean algebra, we have to reduce a particular expression to its simplest form or change its form to a more convenient one in order to implement the expression most efficiently. The approach taken in this section is to use the basic laws, rules, and theorems of Boolean algebra to manipulate and simplify an expression. This method depends on a thorough knowledge of Boolean algebra, and considerable practice in its application. A series of examples will serve to illustrate the technique.

Example 5–5   Simplify the expression $AB + A(B + C) + B(B + C)$, using Boolean algebra techniques.

Solution:

(This is not necessarily the only approach.)

Step 1.   Apply the distributive law to the second and third terms in the expression, as follows:

$$AB + AB + AC + BB + BC$$

Step 2.   Apply rule 7 ($BB = B$):

$$AB + AB + AC + B + BC$$

Step 3.   Apply rule 5 ($AB + AB = AB$):

$$AB + AC + B + BC$$

Step 4.   Factor B out of the last two terms, as follows:

$$AB + AC + B(1 + C)$$

Step 5.   Apply the commutative law and rule 2 ($1 + C = 1$):

$$AB + AC + B \cdot 1$$

Step 6.   Apply rule 4 ($B \cdot 1 = B$):

$$AB + AC + B$$

Step 7.   Factor B out of the first and third terms, as follows:

$$B(A + 1) + AC$$

Step 8.   Apply rule 2 ($A + 1 = 1$):

$$B \cdot 1 + AC$$

Step 9.   Apply rule 4 ($B \cdot 1 = B$):

$$B + AC$$

At this point we have simplified the expression as much as possible. It should be noted that once you gain experience in applying Boolean algebra, many individual steps can be combined. See if you can find an alternate approach.

Example 5–6   Simplify the expression $(A + B)(A + C) + AC$.

Solution:

(This is not necessarily the only approach.)

Step 1.   Expand the first two terms by multiplying:

$$AA + AC + BA + BC + AC$$

Step 2.   Apply rule 5 to the second and fifth terms ($AC + AC = AC$):

$$AA + AC + BA + BC$$

Step 3.   Apply rule 7 to the first term ($AA = A$):
$$A + AC + BA + BC$$

Step 4.   Apply rule 10 to the first two terms ($A + AC = A$):
$$A + BA + BC$$

Step 5.   Apply the commutative law to the second term:
$$A + AB + BC$$

Step 6.   Apply rule 10 to the first two terms ($A + AB = A$):
$$A + BC$$

We have reduced the original expression to its simplest form. Is there an easier way to do this? Simply apply rule 12 and rule 10.

Example 5–7   Simplify the expression $[A\overline{B}(C + BD) + \overline{A}\overline{B}]C$ as much as possible. Note that brackets or parentheses mean the same thing: simply that the terms are multiplied (ANDed) with each other.

Solution:

Step 1.   Apply the distributive law (multiply out) to the terms within the brackets:
$$(A\overline{B}C + A\overline{B}BD + \overline{A}\overline{B})C$$

Step 2.   Apply rule 8 to the second term in the parentheses:
$$(A\overline{B}C + A \cdot 0 \cdot D + \overline{A}\overline{B})C$$

Step 3.   Apply rule 3 to the second term:
$$(A\overline{B}C + 0 + \overline{A}\overline{B})C$$

Step 4.   Apply rule 1 within the parentheses:
$$(A\overline{B}C + \overline{A}\overline{B})C$$

Step 5.   Apply the distributive law:
$$A\overline{B}CC + \overline{A}\overline{B}C$$

Step 6.   Apply rule 7 to the first term:
$$A\overline{B}C + \overline{A}\overline{B}C$$

Step 7.   Factor out $\overline{B}C$:
$$\overline{B}C(A + \overline{A})$$

Step 8.   Apply rule 6:
$$\overline{B}C \cdot 1$$

Step 9.   Apply rule 4:
$$\overline{B}C$$

5–9 Forming a Boolean Expression

Let us assume, for example, that we wish to derive an expression for a specific logic function X, which is composed of three variables, A, B, and C. The following requirements are specified for the function:

X is a 1 only when

(a)   Both A and B are 1s.

(b) A is a 1, B is a 0, and C is a 1.

(c)   Both B and C are 0s.

X is a 0 for all other conditions.

This requirement can be stated as

X is a 1 if and only if
A is a 1 AND B is a 1,
OR *A is a 1 AND B is a 0 AND C is a 1,*
OR *B is a 0 AND C is a 0.*

Basically there are three AND functions ORed, which produces a *sum-of-products* expression. When a variable is required to be a 1 in order to make the function a 1, the true form of the variable is used in the expression. When a variable is required to be a 0 in order to make the function a 1, the complemented form of the variable is used in the expression.

We can write the previous statement in equation form as follows:

$$X = AB + A\bar{B}C + \bar{B}\bar{C}$$

To make sure that this expression meets our requirements, we will substitute the binary digit for each variable specified in requirement (b) above.

$$\begin{aligned} X &= 1 \cdot 0 + 1 \cdot \bar{0} \cdot 1 + \bar{0} \cdot \bar{1} \\ &= 1 \cdot 0 + 1 \cdot 1 \cdot 1 + 1 \cdot 0 \\ &= 0 + 1 + 0 \\ &= 1 \end{aligned}$$

This shows that X is a 1 for the specified conditions. You should verify that X is a 1 for the other requirements.

5–10 Karnaugh Maps

Another approach for reducing a Boolean expression to its simplest or minimum form is known as the *Karnaugh map* method, named after its originator. The method is systematic and easily applied. When properly used, it will *always* result in the minimum expression possible.

The effectiveness of algebraic simplification as discussed in Section 5–8 is dependent on your familiarity with the rules, laws, and theorems of Boolean algebra and on your ability and ingenuity in applying them. The Karnaugh map approach, therefore, has a distinct advantage, especially for more complex expressions where algebraic simplification is not immediately obvious or is extremely involved.

A Karnaugh map is composed of a number of adjacent "cells." Each cell represents one particular combination of variable values. Since the total number of possible combinations of n variables is 2^n, there must be 2^n cells in the Karnaugh map. For instance, for a function of two variables, the Karnaugh map has $2^2 = 4$ cells, and is constructed as shown in Figure 5–20.

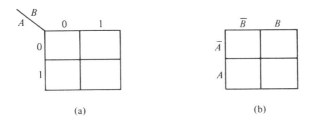

(a)   (b)

Figure 5–20.   Karnaugh Map for Two Variables.

The variables are A and B, and the labeling of the map is as follows for Figure 5–20(a): the upper left cell is for the case where $A = 0$ and $B = 0$; the upper right cell is for $A = 0$ and $B = 1$; the lower left cell is for $A = 1$ and $B = 0$; and the lower right cell is for $A = 1$ and $B = 1$. An alternate way of drawing the map is shown in Figure 5–20(b), and is essentially the same as that in (a) except that the cells are labeled with variable terms rather than 1 and 0 combinations. The upper left cell corresponds to the combination of variables $\bar{A}\bar{B}$, the upper right cells corresponds to $\bar{A}B$, and so on.

Now consider a Karnaugh map for a three-variable function. There must be $2^3 = 8$ cells, as shown in Figure 5–21. The variables are A, B, and C. In Figure 5–21(a), the upper left cell is for $A = 0$, $B = 0$, and $C = 0$; the lower left cell is for $A = 1$, $B = 0$, and $C = 0$; etc. In Figure 5–21(b), the upper left cell is for $\bar{A}\bar{B}\bar{C}$; the lower left cell is for $A\bar{B}\bar{C}$; etc. Both notations mean the same thing.

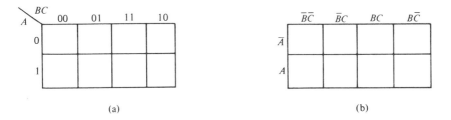

(a)   (b)

Figure 5–21.   Karnaugh Map for Three Variables.

A map for a four-variable function requires $2^4 = 16$ cells, and is constructed as shown in Figure 5–22.

At this point you should know basically how to draw a Karnaugh map for two-, three-, and four-variable functions. Functions of five or more variables can also be handled with Karnaugh maps, but the coverage of these is not included in this book. The difficulty of applying the Karnaugh map method increases considerably with an increasing number of variables.

Boolean Algebra

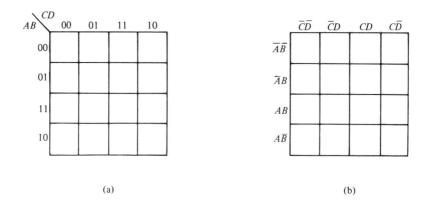

(a)   (b)

Figure 5–22. Karnaugh Map for Four Variables.

The next step is to learn how to use the Karnaugh map to minimize a Boolean expression. First, we will discuss how to "plot" a function on the map, and then cover the method of "factoring" the plotted function in order to get the minimum expression.

In plotting a function on the map, there are two approaches: 1s plotting, which results in a minimum sum-of-products expression; and 0s plotting, which results in a minimum product-of-sums expression.

The 1s plotting simply means that a 1 is inserted in *each* cell corresponding to a combination of variable values that cause the total function to be a 1. For example, let us take the two-variable function $X = A\overline{B} + AB(A + B)$, and determine when $X = 1$ by substitution of all possible combinations of the variables A and B:

$$A = 0, \quad B = 0: \quad X = 0 \cdot 1 + 0 \cdot 0(0 + 0) = 0$$
$$A = 0, \quad B = 1: \quad X = 0 \cdot 0 + 0 \cdot 1(0 + 1) = 0$$
$$A = 1, \quad B = 0: \quad X = 1 \cdot 1 + 1 \cdot 0(1 + 0) = 1$$
$$A = 1, \quad B = 1: \quad X = 1 \cdot 0 + 1 \cdot 1(1 + 1) = 1$$

The preceding substitutions tell us that $X = 1$ when $A = 1$ and $B = 0$ ($A\overline{B}$), or when $A = 1$ and $B = 1$ (AB). Now the 1s are inserted in the map in the cells corresponding to these conditions, as shown in Figure 5–23.

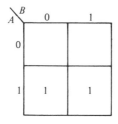

Figure 5–23.

Now that the 1s for this particular function are plotted on the Karnaugh map, we will turn our attention to "factoring" the 1s to get the minimum expression. But first the concept of *adjacent cells* needs to be defined. Adjacent cells on a Karnaugh map are those that differ by only the value of a single variable. For instance, in the map of Figure 5–23, two cells that are not adjacent are the upper left and the lower right because both variables differ between these two cells. The same is true for the upper right and lower left cells.

The rules for factoring 1s on a Karnaugh map are as follows:

1. Combine 1s appearing in adjacent cells into a *maximum* grouping of 1, 2, 4, 16, etc. In other words, form a group containing the largest number of cells possible, as long as it is a power of 2.

2. Form as many maximum-size groups as possible, until all 1s are included in at least one group. There can be overlapping groups if they include noncommon 1s.

3. Each group of 1s creates a minimized product term composed of all variables that have the *same* value (1 or 0) within the group. If a variable has a value of 1 in all cells in the group, the *true* form of the variable appears in the product term; if it has a value of 0 in all cells within the group, the *complement* form of the variable appears in the product term; and if it has both 1 and 0 values within the group, it does not appear in the product term.

4. The product terms represented by each group on the map are now summed (ORed), producing a minimum sum-of-products expression.

Going back to the example of Figure 5–23, the two 1s are adjacent and can be grouped together as shown in Figure 5–24. Notice that within this group of two 1s, only the variable A is single-valued in both cells. The variable B has a different value in each cell within the group, and is therefore eliminated. This group represents a term composed of only the variable A. There is only one group on the map, so the variable A is the only term in the expression. The original expression $X = A\bar{B} + AB(A + B)$ is thus reduced to $X = A$. See if you can arrive at the same result using the algebraic methods.

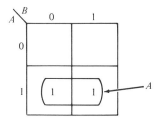

Figure 5–24.

You may think at this point that the algebraic method is easier to apply, so why bother with a Karnaugh map? The particular example just completed is a relatively simple one, and for it the algebraic method is probably easier. However, after you gain experience with simplification of more complex functions, the advantage of the Karnaugh map will be clear.

Now, let us take a look at 0s plotting. This means that a 0 is inserted in *each* cell corresponding to a combination of variable values that cause the total function to be 0. Considering the example we just went through for 1s plotting, $X = 0$ when $A = 0$ and $B = 0$, or when $A = 0$ and $B = 1$. This is shown on the Karnaugh map in Figure 5–25.

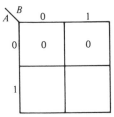

Figure 5–25.

The rules for factoring 0s in a Karnaugh map are as follows:

1. Combine 0s appearing in adjacent cells into a maximum grouping of 1, 2, 4, 16, etc. This is the same as with 1s factoring.

2. Form as many maximum-size groups as possible, until all 0s are included in at least one group. There can be overlapping groups if they include noncommon 0s. This is the same as for 1s factoring.

3. Each group of 0s creates a minimized product term composed of all variables that have the *same* value (1 or 0) within the group. If a variable has a value of 1 in all cells in the group, the *true* form of the variable appears in the product term; if it has a value of 0 in all cells within the group, the *complement* form of the variable appears in the product term; and if it has both 1 and 0 values within the group, it does not appear in the product term. This procedure is the same as with 1s factoring.

4. The product term represented by each group on the map is now summed, and the total function complemented to produce the minimum product-of-sums form. The complementation is an extra step in the 0s factoring that was not required in 1s factoring.

We will now work through several examples to illustrate the application of Karnaugh maps.

Example 5–8 Use the Karnaugh map method to simplify the expression $X = A(A\overline{B} + AB) + \overline{A}B$, and produce a minimum sum-of-products form.

Solution:

This is a two-variable function, so a map with four cells will be required. Since a sum-of-products expression is specified, we must use 1s factoring. The first step is to determine when $X = 1$ by subsituting values for the variables into the expression:

$$A = 0, \quad B = 0: \quad X = 0 + 0 = 0$$
$$A = 0, \quad B = 1: \quad X = 0 + 0 = 1$$
$$A = 1, \quad B = 0: \quad X = 0 + 1 = 1$$
$$A = 1, \quad B = 1: \quad X = 1 + 0 = 1$$

You should verify for yourself that these evaluations are correct.

The next step is to construct the Karnaugh map and insert the 1s in the proper cells, as shown in Figure 5–26. The 1s are combined in two groups as indicated, each representing the terms as shown in Figure 5–26. The resulting expression is $X = A + B$.

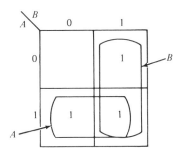

Figure 5–26.

Example 5–9   Minimize the expression $X = A\overline{B}C + \overline{A}BC + \overline{A}\,\overline{B}C + \overline{A}\,\overline{B}\,\overline{C} + A\overline{B}\,\overline{C}$ in sum-of-products form.

Solution:

Notice that this expression is already in a sum-of-products form from which the 1s can be plotted very easily. In this example we will use the alternate form of the Karnaugh map, as shown in Figure 5–27.

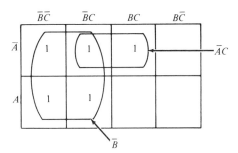

Figure 5–27.

Four of the 1s appearing in adjacent cells can be grouped. The remaining 1 is absorbed in an overlapping group. The group of four 1s produces a single variable term, $\overline{B}$. This is determined by observing that, within the group, $\overline{B}$ is the only variable that does not change from cell to cell. The group of two 1s produces a two-variable term, $\overline{A}C$. This is determined by observing that, within the group, the variables $\overline{A}$ and C do not change from one cell to the next. To get the minimized function, the two terms that are produced are summed (ORed) as $X = \overline{B} + \overline{A}C$.

Example 5–10 For the function in Example 5–9, use 0s factoring to arrive at a minimum product-of-sums expression.

Solution:
When the function X is not a 1, it is a 0. Therefore, 0s are placed in the cells not occupied by 1s, as shown on the map in Figure 5–28.

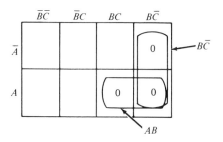

Figure 5–28.

The 0s can be grouped in pairs, and the resulting product terms are as indicated. The complement of the function X is then equal to the sum of the product terms: $\overline{X} = AB + B\overline{C}$. To get the function X, we take the complement as shown in the following steps:

$$X = \overline{AB + B\overline{C}}$$
$$X = (\overline{AB})(\overline{B\overline{C}})$$
$$X = (\overline{A} + \overline{B})(\overline{B} + C)$$

This final expression is the minimum product-of-sums form of the original.

Before proceeding to the next example, some additional discussion of adjacent cells is necessary. The Karnaugh map can be visualized as being "rolled" to form a cylinder in first the horizontal and then the vertical direction. If this is done, you can see that each cell along the outer perimeter of the map is adjacent to the corresponding cell on the opposite side of the map. Note that there is a change of only one variable between such cells. For instance, in Figure 5–22 the $\overline{A}\overline{B}C\overline{D}$

cell is adjacent to the $\overline{A}\,\overline{B}C\overline{D}$, the $\overline{A}\,\overline{B}CD$ cell is adjacent to the $A\overline{B}CD$ cell, and so on. The diagrams in Figure 5–29 illustrate the cell adjacency concept for a three- and a four-variable Karnaugh map.

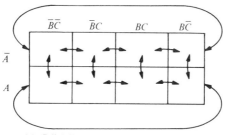

(a) Cell-Adjacency for a Three-Variable Map

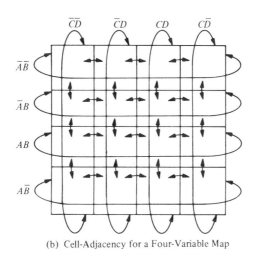

(b) Cell-Adjacency for a Four-Variable Map

Figure 5–29.

Example 5–11 Reduce the following four-variable function to its minimum sum-of-products form:

$$X = \overline{A}\,\overline{B}\,\overline{C}\,\overline{D} + \overline{A}B\overline{C}\,\overline{D} + AB\overline{C}\,\overline{D} + A\overline{B}\,\overline{C}\,\overline{D} + \overline{A}\,\overline{B}CD + A\overline{B}CD + \overline{A}BC\overline{D}$$
$$+ \overline{A}BC\overline{D} + ABC\overline{D} + A\overline{B}C\overline{D}$$

Solution:

A group of eight can be factored as shown in Figure 5–30 because the 1s in the outer columns are adjacent. A group of four is formed by the "wrap-around" adjacency of the cells to pick up the remaining two 1s. The minimum form of the original equation is $X = \overline{D} + \overline{B}C$.

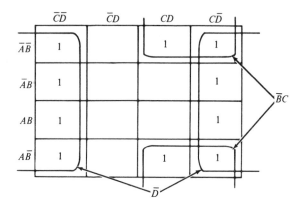

Figure 5–30.

PROBLEMS

5–1   Evaluate the following expressions for all possible values of the variables:

(a) $X = AB$ (b) $X = ABC$ (c) $X = A + B$
(d) $X = A + B + C$ (e) $X = AB + C$ (f) $X = \overline{A} + B$
(g) $X = A\overline{B}\overline{C}$ (h) $X = AB + \overline{A}C$ (i) $X = A(B + C)$
(j) $X = \overline{A}(\overline{B} + \overline{C})$

5–2   Evaluate the following expressions for all possible values of the variables:

(a) $X = (A + B)C + B$ (b) $X = (\overline{A + B})C$
(c) $X = A\overline{B}C + AB$ (d) $X = (A + B)(\overline{A} + B)$
(e) $X = (A + BC)(\overline{B} + \overline{C})$

5–3   Write the Boolean expression for each of the logic circuits in Figure 5–31.

Figure 5–31.

5-4   Repeat Problem 5-3 for the following circuits in Figure 5-32.

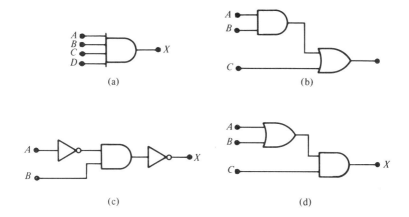

(a)   (b)

(c)   (d)

Figure 5-32.

5-5   Construct a truth table for each of the following Boolean expressions:
    (a) $A + B$ (b) AB (c) $AB + BC$
    (d) $(A + B)C$ (e) $(A + B)(\bar{B} + C)$

5-6   Apply DeMorgan's theorems to each expression.
    (a) $\overline{A + \bar{B}}$ (b) $\overline{\bar{A}B}$ (c) $\overline{A + B + C}$
    (d) $\overline{ABC}$ (e) $\overline{A(B + C)}$ (f) $\overline{AB + CD}$
    (g) $\overline{AB + CD}$ (h) $\overline{(A + \bar{B})(\bar{C} + D)}$

5-7   Apply DeMorgan's theorems to each expression.

    (a) $\overline{A\bar{B}(C + \bar{D})}$ (b) $\overline{AB(CD + EF)}$
    (c) $\overline{(A + \bar{B} + C + \bar{D}) + \overline{ABC\bar{D}}}$ (d) $\overline{(\bar{A} + B + C + D)(\overline{A\bar{B}\bar{C}D})}$
    (e) $\overline{A\bar{B}(CD + \bar{E}F)(\overline{AB} + \overline{CD})}$

5-8   Apply DeMorgan's theorems to the following:
    (a) $\overline{(\overline{ABC})(\overline{EFG})} + \overline{(\overline{HIJ})(\overline{KLM})}$ (b) $\overline{(A + \overline{B\bar{C}} + CD) + \overline{\overline{BC}}}$

    (c) $\overline{\overline{(A + B)(C + D)}(\overline{E + F})(\overline{G + H})}$

5-9   Convert the following expressions to sum-of-product forms:
    (a) $(A + B)(C + D)$ (b) $(A + \bar{B}C)D$
    (c) $(A + C)(ABC + ACD)$

5-10   Convert the following expressions to sum-of-product forms:
    (a) $AB + CD(A\bar{B} + CD)$ (b) $AB(\bar{B}\bar{C} + BC)$
    (c) $A + B[AC + (B + \bar{C})D]$

5-11   Identify each of the following expressions as a product-of-sums or a sum-of-products:

(a) $ABC + \bar{A}BC + ABC$ (b) $A + B\bar{C}$

(c) $(A + B)(\bar{B} + \bar{C} + D)$ (d) $\bar{A}(B + C)(A + C)$

(e) $ABCDE + AD + \bar{A}E$ (f) $B + C$

(g) $\bar{A} + B + C + D + E + BC$

(h) $(A + B + C)(\bar{A} + B + \bar{C})(A + B + D)$

5-12   Write an expression for each of the gate networks in Figure 5–33, and identify the form.

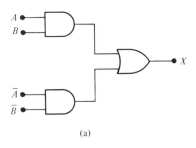

(a)

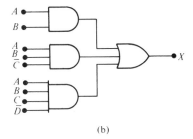

(b)

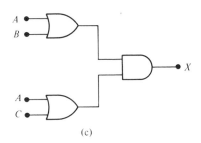

(c)

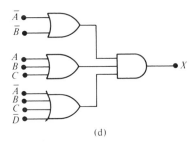

(d)

Figure 5–33.

5-13   Simplify as much as possible the following expressions, using Boolean algebra techniques:

(a) $A(A + B)$ (b) $A(\bar{A} + AB)$ (c) $BC + \cdot \bar{B}C$

(d) $A(A + \bar{A}B)$ (e) $\bar{A}B\bar{C} + \bar{A}BC + \bar{A}B\bar{C}$

5-14   Simplify the following expressions, using Boolean algebra:

(a) $(A + \bar{B})(A + C)$ (b) $\bar{A}B + \bar{A}B\bar{C} + \bar{A}BCD + \bar{A}B\bar{C}\bar{D}E$

(c) $AB + \bar{A}BC + A$ (d) $(A + \bar{A})(AB + AB\bar{C})$

(e) $AB + (\bar{A} + \bar{B})C + AB$

5-15   Simplify each expression, using Boolean algebra:

(a) $BD + B(D + E) + \bar{D}(D + F)$ (b) $\bar{A}\bar{B}C + \overline{(A + B + \bar{C})} + \bar{A}\bar{B}\bar{C}D$

(c) $(B + BC)(B + \bar{B}C)(B + D)$ (d) $ABCD + AB(\overline{CD}) + (\overline{AB})CD$

(e) $ABC[AB + \bar{C}(BC + AC)]$

5-16   Write a Boolean expression for the following statement:

X is a 1 only if A is a 1 and B is a 1

or if A is a 0 and B is a 0.

5-17 Write a Boolean expression for the following statement:
X is a 1 only if A, B, and C are all 1s
or if only one of the variables is a 0.

5-18 Write a Boolean expression for the following conditions:
X is a 0 if any two of the three variables A, B, and C are 1s. X is a 1 for all other conditions.

5-19 Draw the logic circuit represented by each of the following expressions:
(a) $A + B + C$ (b) ABC (c) $AB + C$
(d) $AB + CD$ (e) $\overline{A}B(C + \overline{D})$

5-20 Draw the logic circuit represented by each expression.
(a) $A\overline{B} + \overline{A}B$ (b) $AB + \overline{A}B + \overline{A}BC$ (c) $A + B[C + D(B + \overline{C})]$

5-21 What size Karnaugh map (number of cells) is required if the number of variables in an expression is
(a) 2 (b) 3 (c) 4 (d) 5

5-22 Using the Karnaugh map method, simplify the following expressions to their minimum sum-of-products form:
(a) $X = \overline{A}\overline{B} + A\overline{B}$ (b) $X = \overline{A}\overline{B} + \overline{A}B$
(c) $X = A\overline{B} + AB$ (d) $X = \overline{A}\overline{B} + A\overline{B} + AB$
(e) $X = A(\overline{B} + AB)$ (f) $X = \overline{A}\overline{B} + AB$

5-23 Use a Karnaugh map to find the minimum sum-of-products form for each expression.
(a) $X = \overline{A}\overline{B}\overline{C} + \overline{A}\overline{B}C + A\overline{B}C$
(b) $X = AC(\overline{B} + C)$
(c) $X = \overline{A}(BC + B\overline{C}) + A(BC + B\overline{C})$
(d) $X = \overline{A}\overline{B}C + A\overline{B}C + \overline{A}B\overline{C} + AB\overline{C}$
(e) $X = A + B\overline{C}$

5-24 Use a Karnaugh map to simplify each function to a minimum sum-of-products form.
(a) $X = \overline{A}\overline{B}\overline{C} + A\overline{B}C + \overline{A}BC + AB\overline{C}$
(b) $X = AC[\overline{B} + A(B + \overline{C})]$
(c) $X = DE\overline{F} + \overline{D}E\overline{F} + \overline{D}E\overline{F}$

5-25 For the expressions in Problem 5-23, reduce each to its minimum product-of-sums form, using a Karnaugh map.

5-26 For the expressions in Problem 5-24, reduce each to its minimum product-sum form, using a Karnaugh map.

5-27 Use a Karnaugh map to reduce each expression to a minimum sum-of-products.
(a) $X = A + B\overline{C} + CD$
(b) $X = \overline{A}\overline{B}\overline{C}D + \overline{A}\overline{B}CD + ABCD + ABC\overline{D}$
(c) $X = \overline{A}B(\overline{C}D + \overline{C}D) + AB(\overline{C}D + \overline{C}D) + A\overline{B}\overline{C}D$
(d) $X = (\overline{A}\overline{B} + A\overline{B})(CD + C\overline{D})$
(e) $X = \overline{A}\overline{B} + A\overline{B} + \overline{C}D + C\overline{D}$

5-28 Repeat Problem 5-27 for the following expressions:
(a) $X = \overline{A}(\overline{B}\overline{C}D + \overline{B}C\overline{D}) + A(\overline{B}\overline{C}D + \overline{B}C\overline{D})$
(b) $X = B[(\overline{A}\overline{C}D + \overline{A}C\overline{D}) + (A\overline{C}D + AC\overline{D})]$
(c) $X = \overline{A}B + A\overline{B} + \overline{A}B\overline{C}D + \overline{A}BCD$

CHAPTER 6

Combinational Logic

In Chapter 4, logic gates were studied on an individual basis. In this chapter we will combine various types of gates in order to produce specified logic functions. When logic gates are connected to generate a specified output for certain combinations of input variables with no storage involved, the resulting network is normally called *combinational* or *combinatorial* logic because it combines the input variables in such a way as to *immediately produce the desired output.*

6–1 Analysis of Combinational Functions

In analyzing a combinational logic function, we are concerned with determining in what way the output is dependent on the inputs; i.e., for what combinations of inputs is the output a HIGH or a LOW? The analysis is approached in two ways: first, the output of the combinational logic is determined for a given set of input values, based on the logical operation of the individual gates in the network; second, a logic equation for the output function is derived and, using this equation, the output value is determined for various combinations of input values, using Boolean algebra.

We will begin by using a series of examples of combinational logic with AND and OR gates as the basic elements. Figure 6–1 shows a combinational logic network composed of two AND gates and one OR gate with three input variables, A, B, and C.

Each of the input variables can be a HIGH (1) or a LOW (0). Because there are three input variables, there are eight possible combinations of the variables ($2^3 = 8$). To illustrate an analysis procedure, we will assign one of the eight possible input combinations and see what the corresponding output value is.

First, we make each input variable a LOW, and examine the output of each gate in the network in order to arrive at the final output, X. If the inputs to gate G_1 are both LOW, the output of gate G_1 is LOW. Also, the output to gate G_2 is LOW because its inputs are LOW. As a result of the LOWs on the outputs of

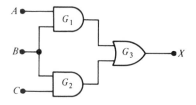

Figure 6-1.

gates G_1 and G_2, both inputs to gate G_3 are LOW and, therefore, its output is
LOW. We have determined that the output function of the logic circuit of Figure
6-1 is LOW when all of its inputs are LOW. This condition is illustrated in Figure
6-2(a), and the remaining seven possible input combinations are illustrated in
Figures 6-2(b) through 6-2(h). You should verify for yourself each of these
conditions.

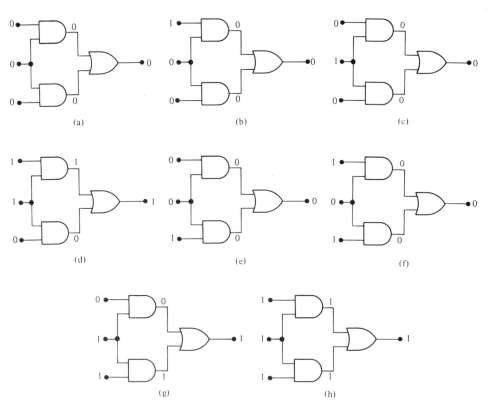

Figure 6-2.   All of the Possible Logic State Conditions for the Circuit of Figure 6-1.

Now, as a second method of analyzing the logical operation of the circuit of
Figure 6-1, we develop a logic equation for the output function and, using
Boolean algebra, evaluate the equation for each of the eight combinations of input
variables.

Since gate G_1 is an AND gate and its two inputs are A and B, its output can be expressed as AB. Gate G_2 is an AND gate and its two inputs are B and C, so its output can be expressed as BC. Gate G_3 is an OR gate and its two inputs are the outputs of gates G_1 and G_2, so its output can be expressed as $AB + BC$. The output of gate G_3 is the output function of the logic network, so $X = AB + BC$. Figure 6–3 shows the logic functions at each point in the circuit.

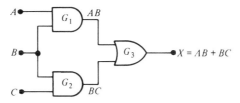

Figure 6–3.

We now evaluate the output equation by substituting into it the various combinations of input variable values. HIGH is a 1 and LOW is a 0.

$$A = 0, \quad B = 0, \quad C = 0: \quad X = 0 \cdot 0 + 0 \cdot 0 = 0 + 0 = 0$$
$$A = 0, \quad B = 0, \quad C = 1: \quad X = 0 \cdot 0 + 0 \cdot 1 = 0 + 0 = 0$$
$$A = 0, \quad B = 1, \quad C = 0: \quad X = 0 \cdot 1 + 1 \cdot 0 = 0 + 0 = 0$$
$$A = 0, \quad B = 1, \quad C = 1: \quad X = 0 \cdot 1 + 1 \cdot 1 = 0 + 1 = 1$$
$$A = 1, \quad B = 0, \quad C = 0: \quad X = 1 \cdot 0 + 0 \cdot 0 = 0 + 0 = 0$$
$$A = 1, \quad B = 0, \quad C = 1: \quad X = 1 \cdot 0 + 0 \cdot 1 = 0 + 0 = 0$$
$$A = 1, \quad B = 1, \quad C = 0: \quad X = 1 \cdot 1 + 1 \cdot 0 = 1 + 0 = 1$$
$$A = 1, \quad B = 1, \quad C = 1: \quad X = 1 \cdot 1 + 1 \cdot 1 = 1 + 1 = 1$$

In the above analyses, we have determined that the output is HIGH (1) for only three input conditions. It is LOW (0) for all the rest. We have essentially created a truth table for the circuit of Figure 6–1 which is shown in a more conventional form in Table 6–1.

TABLE 6–1.   Truth Table
for the Circuit of Figure 6–1.

Inputs			Output
A	B	C	X
0	0	0	0
0	0	1	0
0	1	0	0
0	1	1	1
1	0	0	0
1	0	1	0
1	1	0	1
1	1	1	1

Next, we will analyze the logic circuit of Figure 6–4, which is similar to the previous circuit except that the input B is complemented (inverted). Let us start

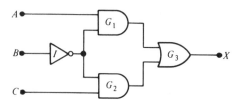

Figure 6–4.

with all input variables LOW. Since input B is inverted, one input to gate G_1 is HIGH, but because the A input is LOW, the output of gate G_1 is LOW. For gate G_2, one of its inputs is HIGH but the other is LOW, so its output is LOW. The LOWs on the outputs of gates G_1 and G_2 are also on the inputs of gate G_3, making its output LOW. This condition is illustrated in Figure 6–5(a). The remaining seven possible input conditions are illustrated in Figures 6–5(b) through 6–5(h). You should verify each of these conditions before proceeding further.

Now we develop a logic equation for the output function and again, using Boolean algebra, evaluate the equation for each of the eight combinations of input variable values for the circuit of Figure 6–4.

The output of the inverter, I, can be expressed as $\bar{B}$. Since gate G_1 is an AND gate and its two inputs are A and $\bar{B}$, its output can be expressed as $A\bar{B}$. Gate G_2 is an AND gate and its two inputs are $\bar{B}$ and C, so its output can be expressed as $\bar{B}C$. Gate G_3 is an OR gate and its two inputs are the outputs of gates G_1 and G_2, so its output can be expressed as $A\bar{B} + \bar{B}C$. The output of gate G_3 is the final output of the entire logic network, so $X = A\bar{B} + \bar{B}C$. Figure 6–6 shows the logic functions at each point in the circuit.

We now evaluate the output equation by substituting into it the various combinations of input variable values.

$$A = 0, \quad B = 0, \quad C = 0: \quad X = 0 \cdot 1 + 1 \cdot 0 = 0 + 0 = 0$$
$$A = 0, \quad B = 0, \quad C = 1: \quad X = 0 \cdot 1 + 1 \cdot 1 = 0 + 1 = 1$$
$$A = 0, \quad B = 1, \quad C = 0: \quad X = 0 \cdot 0 + 0 \cdot 0 = 0 + 0 = 0$$
$$A = 0, \quad B = 1, \quad C = 1: \quad X = 0 \cdot 0 + 0 \cdot 1 = 0 + 0 = 0$$
$$A = 1, \quad B = 0, \quad C = 0: \quad X = 1 \cdot 1 + 1 \cdot 0 = 1 + 0 = 1$$
$$A = 1, \quad B = 0, \quad C = 1: \quad X = 1 \cdot 1 + 1 \cdot 1 = 1 + 1 = 1$$
$$A = 1, \quad B = 1, \quad C = 0: \quad X = 1 \cdot 0 + 0 \cdot 0 = 0 + 0 = 0$$
$$A = 1, \quad B = 1, \quad C = 1: \quad X = 1 \cdot 0 + 0 \cdot 1 = 0 + 0 = 0$$

In the above analyses we have determined the logical operation for the circuit of Figure 6–4, from which the truth table shown in Table 6–2 can be generated. Notice that both approaches to the analysis agree in the final results.

As a third example, we will analyze a circuit that is somewhat more complex than the previous two. Keep in mind that we will have considered only three examples of combinational logic analysis in this section. We could continue almost indefinitely with innumerable combinations of gates and input variables, but the three discussed in this section serve as a basis to illustrate the approaches that can

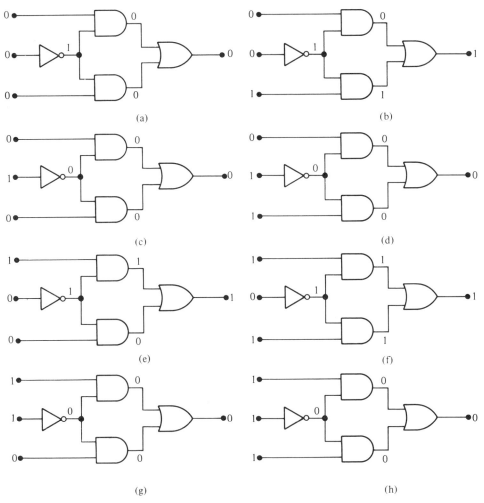

Figure 6-5.   All Possible Logic State Conditions for the Circuit of Figure 6-4.

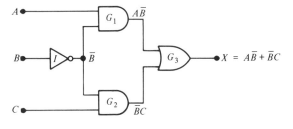

Figure 6-6.

be taken to analyze the operation of a given logic circuit. Further exercises in logic circuit analysis are presented as problems at the end of the chapter.

Figure 6-7 is the third combinational logic circuit that we will analyze.

TABLE 6–2.   Truth Table
for the Circuit of Figure 6–4.

Inputs			Outputs
A	B	C	X
0	0	0	0
0	0	1	1
0	1	0	0
0	1	1	0
1	0	0	1
1	0	1	1
1	1	0	0
1	1	1	0

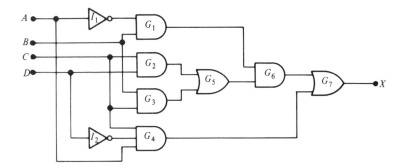

Figure 6–7.

This circuit has four input variables, which result in a total of 16 possible input combinations (2^4 = 16). In order to determine the complete logical function of this circuit, we would have to go through each of the 16 combinations and determine the resulting output for each. As you might think, this is a quite lengthy process, so we will assign only two out of the 16 possible combinations and analyze the circuit operation for each. The remaining combinations will be left for the reader as a problem at the end of the chapter.

For the first input combination let us use the following: A = 0, B = 1, C = 0, D = 1. Since A is a LOW, the output of inverter I_1 is a HIGH. This makes both inputs to gate G_1 HIGH and causes its output to be HIGH. The C input to gate G_2 is LOW, making the output of G_2 LOW. The C input to gate G_3 is LOW, causing its output to be LOW. Because the output of gates G_2 and G_3 are both LOW, the output of gate G_5 is also LOW. The output of gate G_1 and the output of gate G_5 become the inputs to gate G_6, making its output LOW. The outputs of gates G_6 and G_4 are inputs to gate G_7 and, since both are LOW, the output of G_7 is LOW. These logic levels are indicated on the logic diagram of Figure 6–8.

Next, we set the inputs to the following levels: A = 1, B = 0, C = 1, D = 0. If A is HIGH, the output of inverter I_1 is LOW, making the output of gate G_1 LOW. Since D is LOW, the output of gate G_2 is LOW, and since B is LOW, the output of gate G_3 is LOW. The output of inverter I_2 is HIGH because D is LOW; this means that all of the inputs to gate G_4 are HIGH, making its output HIGH.

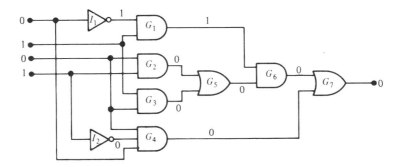

Figure 6–8.

The output of gate G_4 is fed into gate G_7, causing the output to be HIGH. These logic levels are indicated in Figure 6–9. If we continued this analysis for the 14 remaining input combinations, the results would be a complete definition of the logical operation of this circuit.

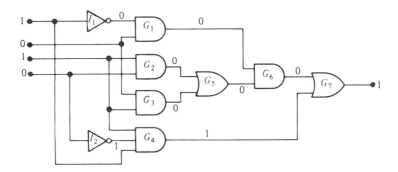

Figure 6–9.

Having gone through the logic level analysis for two possible input combinations, we now develop the logic equation for the circuit of Figure 6–7 and evaluate the operation using Boolean algebra. The development of the logic equation is as follows:

Output of inverter I_1 = $\overline{A}$

Output of gate G_1 = $\overline{A}B$

Output of gate G_2 = CD

Output of gate G_3 = BC

Output of inverter I_2 = $\overline{D}$

Output of gate G_4 = $AC\overline{D}$

Output of gate G_5 = (Output of G_2) + (Output of G_3)

$\qquad\qquad$ = $CD + BC$

Output of gate G_6 = (Output of G_1) · (Output of G_5)

$\qquad\qquad$ = $\overline{A}B(CD + BC)$

Final output X = Output of gate G_7
$$= (\text{Output of } G_4) + (\text{Output of } G_5)$$
$$= AC\overline{D} + \overline{A}B(CD + BC)$$

If each of the sixteen possible input combinations is now substituted into the equation for this circuit (equation for the final output X), a complete logical description or truth table operation is derived. Figure 6-10 illustrates the logic functions at each point in the circuit.

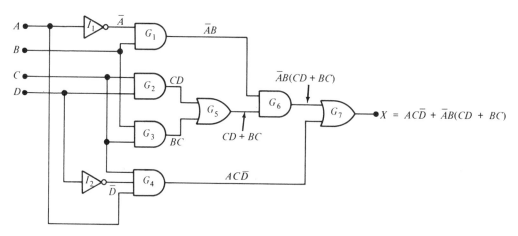

Figure 6-10.

Listed below are the 16 input states and the corresponding equation substitution and evaluation for the output equation $X = AC\overline{D} + \overline{A}B(CD + BC)$.

$A = 0, \quad B = 0, \quad C = 0, \quad D = 0: \quad X = 0 \cdot 0 \cdot 1 + 1 \cdot 0 \cdot (0 \cdot 0 + 0 \cdot 0)$
$$= 0 + 0(0 + 0) = 0 + 0 = 0$$

$A = 0, \quad B = 0, \quad C = 0, \quad D = 1: \quad X = 0 \cdot 0 \cdot 0 + 1 \cdot 0 \cdot (0 \cdot 1 + 0 \cdot 0)$
$$= 0 + 0(0 + 0) = 0 + 0 = 0$$

$A = 0, \quad B = 0, \quad C = 1, \quad D = 0: \quad X = 0 \cdot 1 \cdot 1 + 1 \cdot 0 \cdot (1 \cdot 0 + 0 \cdot 1)$
$$= 0 + 0(0 + 0) = 0 + 0 = 0$$

$A = 0, \quad B = 0, \quad C = 1, \quad D = 1: \quad X = 0 \cdot 1 \cdot 0 + 1 \cdot 0 \cdot (1 \cdot 1 + 0 \cdot 1)$
$$= 0 + 0(1 + 0) = 0 + 0 = 0$$

$A = 0, \quad B = 1, \quad C = 0, \quad D = 0: \quad X = 0 \cdot 0 \cdot 1 + 1 \cdot 1 \cdot (0 \cdot 0 + 1 \cdot 0)$
$$= 0 + 1(0 + 0) = 0 + 0 = 0$$

$A = 0, \quad B = 1, \quad C = 0, \quad D = 1: \quad X = 0 \cdot 0 \cdot 0 + 1 \cdot 1 \cdot (0 \cdot 1 + 1 \cdot 0)$
$$= 0 + 1(0 + 0) = 0 + 0 = 0$$

$A = 0, \quad B = 1, \quad C = 1, \quad D = 0: \quad X = 0 \cdot 1 \cdot 1 + 1 \cdot 1 \cdot (1 \cdot 0 + 1 \cdot 1)$
$$= 0 + 1(0 + 1) = 0 + 1 = 1$$

$A = 0, \quad B = 1, \quad C = 1, \quad D = 1: \quad X = 0 \cdot 1 \cdot 0 + 1 \cdot 1 \cdot (1 \cdot 1 + 1 \cdot 1)$
$$= 0 + 1(1 + 1) = 0 + 1 = 1$$

$A = 1, \quad B = 0, \quad C = 0, \quad D = 0: \quad X = 1 \cdot 0 \cdot 1 + 0 \cdot 0 \cdot (0 \cdot 0 + 0 \cdot 0)$
$$= 0 + 0(0 + 0) = 0 + 0 = 0$$

$A = 1, \quad B = 0, \quad C = 0, \quad D = 1: \quad X = 1 \cdot 0 \cdot 0 + 0 \cdot 0 \cdot (0 \cdot 1 + 0 \cdot 0)$
$$= 0 + 0(0 + 0) = 0 + 0 = 0$$

$$A = 1, \quad B = 0, \quad C = 1, \quad D = 0: \quad X = 1\cdot1\cdot1 + 0\cdot0\cdot(1\cdot0 + 0\cdot1)$$
$$= 1 + 0(0 + 0) = 1 + 0 = 1$$
$$A = 1, \quad B = 0, \quad C = 1, \quad D = 1: \quad X = 1\cdot1\cdot0 + 0\cdot0\cdot(1\cdot1 + 0\cdot1)$$
$$= 0 + 0(1 + 0) = 0 + 0 = 0$$
$$A = 1, \quad B = 1, \quad C = 0, \quad D = 0: \quad X = 1\cdot0\cdot1 + 0\cdot1\cdot(0\cdot0 + 1\cdot0)$$
$$= 0 + 0(0 + 0) = 0 + 0 = 0$$
$$A = 1, \quad B = 1, \quad C = 0, \quad D = 1: \quad X = 1\cdot0\cdot0 + 0\cdot1\cdot(0\cdot1 + 1\cdot0)$$
$$= 0 + 0(0 + 0) = 0 + 0 = 0$$
$$A = 1, \quad B = 1, \quad C = 1, \quad D = 0: \quad X = 1\cdot1\cdot1 + 0\cdot1\cdot(1\cdot0 + 1\cdot1)$$
$$= 1 + 0(0 + 1) = 1 + 0 = 1$$
$$A = 1, \quad B = 1, \quad C = 1, \quad D = 1: \quad X = 1\cdot1\cdot0 + 0\cdot1\cdot(1\cdot1 + 1\cdot1)$$
$$= 0 + 0(1 + 1) = 0 + 0 = 0$$

The analysis tells us the output level for each combination of input levels, and gives us a complete description of the logical operation of the circuit of Figure 6–7. This operation can be expressed more concisely in a truth table, as shown in Table 6–3.

TABLE 6–3. Truth Table for the Circuit of Figure 6–7.

Inputs				Output
A	B	C	D	X
0	0	0	0	0
0	0	0	1	0
0	0	1	0	0
0	0	1	1	0
0	1	0	0	0
0	1	0	1	0
0	1	1	0	1
0	1	1	1	1
1	0	0	0	0
1	0	0	1	0
1	0	1	0	1
1	0	1	1	0
1	1	0	0	0
1	1	0	1	0
1	1	1	0	1
1	1	1	1	0

In this section we have examined several examples of approaches to the analysis of combinational logic circuits. We have found that for a given set of input values, you can proceed from the inputs toward the output to determine the logic level at each gate output, and thereby arrive at the logic level for the final output. The second approach was to determine the logic expression for each gate output in order to arrive at the equation for the final output. Once this equation is determined, you can evaluate the output for any combination of inputs by substituting the appropriate values into the equation and using Boolean algebra rules. Both ways produce the same results.

6–2 Implementing Logic Equations with Gates

In this section we will start with an equation that describes a logic function and from it determine the circuit required to implement or produce the function. As before, several examples will be used to illustrate a general procedure.

First, let us review two basic conventions regarding logic equations:

1. Whenever a + appears between two or more terms, such as $X + Y + Z$, we know that this means these terms are ORed.
2. Whenever two or more terms appear as $X \cdot Y \cdot Z$ or simply XYZ, this means that they are ANDed.

Equation (6–1) is the first in the series of examples that will be considered.

$$X = AB + CDE \qquad\qquad (6\text{–}1)$$

A brief inspection reveals that this function is composed of two terms, AB and CDE, with a total of five variables. The first term is formed by ANDing A with B, and the second term is formed by ANDing C, D, and E. These two terms are then ORed to form the function X. These operations are indicated in the structure of the equation following:

$$X = A\overset{\displaystyle\diagup}{B} + C\overset{\displaystyle\diagup}{D}E \quad\text{AND (second level operation)}$$

OR (first level operation)

It should be noted that, in this particular equation, the AND operations forming the two individual terms, AB and CDE, must be performed *before* the terms can be ORed. The OR operation is the last to be performed before the final output function is produced; therefore, it is called a *first-level* operation, meaning that it is performed by the first-level gate(s), starting at the output and working back toward the inputs. The AND operations are performed at the second-level gates from the output, and are therefore *second-level* operations.

To implement the logic function, a two-input AND gate is required to form the term AB and a three-input AND gate is needed to form the term CDE. A two-input OR gate is then required to combine the two AND terms. The resulting logic circuit is shown in Figure 6–11.

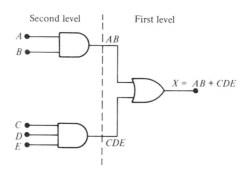

Figure 6–11. Logic Circuit for Equation (6–1).

Now, let us look at a second example expressed in Equation (6–2):

$$X = \overline{A}BC + A\overline{BC} \qquad\qquad (6\text{–}2)$$

This equation requires three levels of logic gates. How do we know this? First, observe that the equation is composed of two terms in sum-of-products form (the two AND terms, $\overline{A}BC$ and $A\overline{BC}$, are ORed). The OR operation requires one level of logic (one OR gate). The first term in the equation is formed by ANDing the three variables $\overline{A}$, B, and C. The second term is formed by ANDing the three variables A, $\overline{B}$, and $\overline{C}$. These two operations are performed at the second level of logic (the two AND gates whose outputs are connected to the first level OR gate inputs). The formation of the complement of each variable requires a third level of logic consisting of inverters; these inverters are at the third level because the complements must be formed before each term is formed. The logic gates required to implement this function are as follows: three inverters to form the $\overline{A}$, $\overline{B}$, and $\overline{C}$ variables, two three-input AND gates to form the terms $\overline{A}BC$ and $A\overline{BC}$, and one 2-input OR gate to form the final output function, $\overline{A}BC + A\overline{BC}$. Figure 6–12 illustrates the implementation of this logic function. Do not confuse levels of logic we have discussed here with input and output levels of a gate (HIGH or LOW).

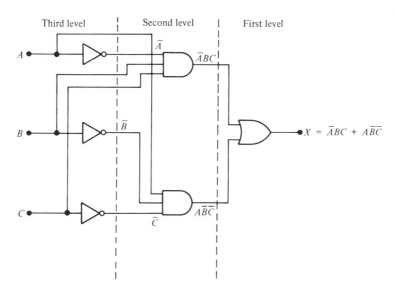

Figure 6–12.   Logic Circuit for Equation (6–2).

In this case where there is a three-level logic function, notice that between input A and the output X there are three logic gates; the same is true for input B and input C. In general, the number of logic levels in a given network is the greatest number of gates through which a signal has to pass in going from an input to the output.

As a third and final example, we will implement Equation (6–3):

$$X = AB(C\overline{D} + EF) \qquad\qquad (6\text{–}3)$$

A breakdown of this equation shows that the term AB and the term $C\overline{D} + EF$ are

ANDed. The term AB is formed by ANDing the variables A and B. The term $C\overline{D} + EF$ is formed by first ANDing C and $\overline{D}$, ANDing E and F, and then ORing these two terms. This structure is indicated in relation to the equation as follows:

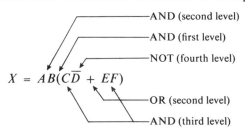

Before the output function X can be formed, we must have the term AB, which is formed by ANDing input A with input B. Also, we must have the term $C\overline{D} + EF$; but before we can get this term, we must have the terms $C\overline{D}$ and EF; but before we can get the term $C\overline{D}$, we must have $\overline{D}$. So, as you can see, there is a "chain" of logic operations that must be done in the proper order before the output function itself is realized. In other words, the output function is composed of several sub-functions.

The logic gates required to implement Equation (6–3) are as follows:

1.   One inverter to form $\overline{D}$.
2.   Two two-input AND gates to form $C\overline{D}$ and EF.
3.   One two-input OR gate to form $C\overline{D} + EF$.
4.   One two-input AND gate to form AB.
5.   One two-input AND gate to form X.

The logic circuit which produces this function is shown in Figure 6–13(a).

This is a good time to illustrate the fact that there is, in many cases, more than one way to implement a given function. For instance, the AND gate in Figure 6–13(a) that forms the function AB can be eliminated and the inputs A and B brought into the first-level AND gate, as shown in Figure 6–13(b). The resulting output is exactly the same, and the circuit used is simpler. Note that both circuits in Figure 6–13 have a maximum of four logic levels.

6–3   Gate Minimization

In this section we will be concerned with reducing the number of logic gates required to produce a given function. In many applications it is desirable to use the minimum number of gates in the simplest configuration possible to implement a given logic function. This simplification may be desirable for several reasons, such as economy or cost, limitations of available power, or minimization of delay times by reduction of logic levels. Here we will examine *two* basic methods of gate minimization. First, the rules and laws of Boolean algebra will be applied in order to simplify the logic equation for the circuit in question, and thereby reduce the number of gates required to implement the function. Second, the Karnaugh map method will be used to minimize the logic function.

To begin, Boolean algebra will be applied to the equation for the circuit of Figure 6–7, which is redrawn for convenience in Figure 6–14.

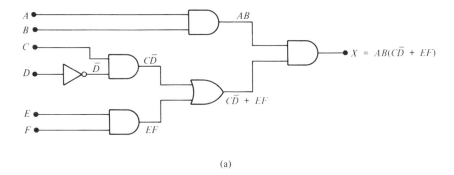

(a)

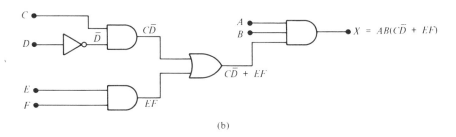

(b)

Figure 6–13.   Logic Circuit for Equation (6–3).

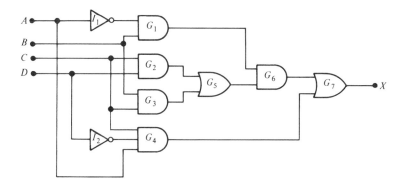

Figure 6–14.

The logic equation for this circuit was developed in Section 6–1 and is restated in Equation (6–4):

$$X = AC\overline{D} + \overline{A}B(CD + BC) \qquad (6\text{-}4)$$

We know that five AND gates, two OR gates, and two inverters are needed to implement this function *as it appears* in Equation (6–4). We do not yet know whether this is the simplest form of the equation, but we are going to find out. When we apply the rules of Boolean algebra in an attempt to simplify the equation, our cleverness and imagination are major factors in determining the outcome because an equation can normally be written in many ways and still express the same logic function. In general, we must use all the rules, laws, and theorems at our command in such a way as to arrive at what appears to be the simplest form of the equation. Inspection of Equation (6–4) shows that a possible first step in simplification is to apply the distributive law to the second term by multiplying the term $CD + BC$ by $\overline{A}B$. The result is

$$X = AC\overline{D} + \overline{A}BCD + \overline{A}BBC \qquad (6\text{–}5)$$

The rule that $BB = B$ can be applied to the third term of Equation (6–5), yielding

$$X = AC\overline{D} + \overline{A}BCD + \overline{A}BC \qquad (6\text{–}6)$$

In Equation (6–6), notice that C is common to each term, so it can be factored out using the distributive law:

$$X = C(A\overline{D} + \overline{A}BD + \overline{A}B) \qquad (6\text{–}7)$$

Now notice that $\overline{A}B$ appears in the last two terms within the parentheses of Equation (6–7) and can be factored out of those two terms:

$$X = C[A\overline{D} + \overline{A}B(D + 1)] \qquad (6\text{–}8)$$

Since $D + 1 = 1$, Equation (6–8) becomes

$$X = C(A\overline{D} + \overline{A}B) \qquad (6\text{–}9)$$

It appears that this equation cannot be reduced any further, although it can be written in a slightly different way (sum-of-products form) by application of the distributive law:

$$X = AC\overline{D} + \overline{A}BC \qquad (6\text{–}10)$$

This equation can be implemented with two three-input AND gates, two inverters, and one two-input OR gate, as shown in Figure 6–15.

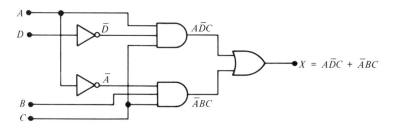

Figure 6–15. Logic Circuit for Equation (6–10).

Compare this circuit with the circuit in Figure 6–14 that has nine logic gates including inverters. The minimized circuit is equivalent, but has only five gates including inverters. At this point you should verify that the logic circuit of Figure 6–15 is indeed equivalent to the logic circuit of Figure 6–14 in terms of the logical operation. Also, the number of gate levels has been reduced from a maximum of four in Figure 6–14 to three in Figure 6–15. Fewer levels mean a shorter propagation delay through the circuit.

The second approach to minimization of Equation (6–4) is the Karnaugh map method. This method is a more systematic means of arriving at a minimum expression and, if factoring is done properly, we can be assured that the result is a minimum expression. The function of Equation (6–4) is rewritten here for convenience:

$$X = AC\overline{D} + \overline{A}B(CD + BC) \tag{6–11}$$

Recall from Chapter 5 that if a 1 is entered into each cell of the map corresponding to a combination of variables that result in the function being true ($X = 1$), and if the cells are factored properly, the resulting expression is in a minimum sum-of-products form. We first have to determine when the function X is a 1; this is done by generating a truth table for the function. A truth table for this particular function has already been developed and is shown in Table 6–3. There we find that $X = 1$ for the following input conditions:

1. $A = 1$, $B = 0$, $C = 1$, $D = 0$
2. $A = 0$, $B = 1$, $C = 1$, $D = 0$
3. $A = 1$, $B = 1$, $C = 1$, $D = 0$
4. $A = 0$, $B = 1$, $C = 1$, $D = 1$

Since there are four variables, a sixteen-cell Karnaugh map is required, and appears in Figure 6–16. A 1 is entered into the cell on the map corresponding to each of the four conditions stated previously.

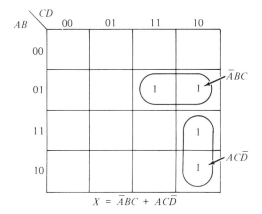

Figure 6–16.   Karnaugh Map for the Function of Equation (6–11).

By factoring the map as indicated, the two terms shown are the result, and the function can be expressed in the minimum sum-of-products form as

$$X = \overline{A}BC + AC\overline{D} \tag{6–12}$$

This equation is the same as Equation (6–10), which was derived by using laws and rules of Boolean algebra. The map method of logic simplification is especially useful when the logic function is long and cumbersome. The application of Boolean rules to extremely complex functions tends to become tedious and more of a trial-and-error process, whereas the map method is very systematic and always yields a minimum result.

6–4   The NAND Gate as a Universal Logic Element

Up to this point only combinational circuits using AND and OR gates have been considered. To implement many functions, both AND and OR gates in combination are required. In this section we will see that NAND gates can be used to produce *any* logic function. For this reason, they are referred to as *universal* gates.

The NAND gate can be used to generate the NOT function, the AND function, and the OR function. An inverter can be made from a NAND gate by connecting all of the inputs and creating, in effect, a single common input, as shown in Figure 6–17(a) for a two-input gate. An AND function can be generated using only NAND gates, as shown in Figure 6–17(b). Also, an OR function can be produced with NAND gates, as illustrated in Figure 6–17(c).

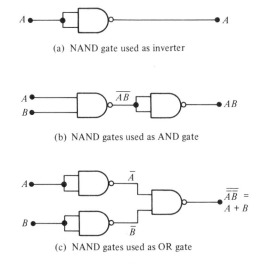

(a) NAND gate used as inverter

(b) NAND gates used as AND gate

(c) NAND gates used as OR gate

Figure 6–17.   Universal Application of NAND Gates.

In Figure 6–17(b), a NAND gate is used to invert (complement) a NAND output to form the AND function, as indicated in the following equation:

$$X = \overline{\overline{AB}} = AB \qquad (6\text{--}13)$$

In Figure 6–17(c), NAND Gates G_1 and G_2 are used to invert the two input variables before they are applied to NAND gate G_3. The final output is derived as follows by application of DeMorgan's theorem:

$$X = \overline{\overline{A}\,\overline{B}} = A + B \qquad (6\text{--}14)$$

As you already know, the function of the NAND gate can be expressed in two ways by applying DeMorgan's theorem. If inputs A and B are applied to a NAND gate as shown in Figure 6–18, the output can be expressed as indicated.

By applying DeMorgan's theorem to the output function, we can also express it as follows:

$$X = \overline{AB} = \overline{A} + \overline{B} \qquad (6\text{--}15)$$

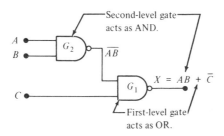

$$\overline{AB} = \overline{A} + \overline{B}$$

Figure 6–18.   Dual Expression of NAND Output.

Therefore, the function of the NAND gate can be stated in two ways:

1. The output of a NAND gate is equal to the complement of the AND of the input variables.
2. The output of a NAND gate is equal to the OR of the input variable complements.

Rules for multilevel NAND logic can be developed, using these two statements of NAND operation as the basis. To develop a set of rules, we will begin with a two-level NAND circuit, as shown in Figure 6–19.

Figure 6–19.

For this circuit, the logic equation is developed as follows by repeated application of DeMorgan's theorem:

$$X = (\overline{\overline{AB}})C = \overline{(\overline{A} + \overline{B})C} = AB + \overline{C} \qquad (6\text{–}16)$$

Notice that the input variables, A and B, appear as an AND term in the final output equation, and the input variable C appears complemented and ORed with the term AB. Gate G_2 is a second-level gate (second gate back from the output) and gate G_1 is a first-level gate (first gate from output). Notice also that the individual variables, A and B, are inputs to the second-level gate G_2, and that the term AB and the individual variable C are the inputs to the first-level gate G_1. From this observation, two general rules for multilevel NAND logic can be stated:

1. All odd-numbered logic levels (1, 3, 5, etc.) "act" as OR gates, with single-input variables complemented.
2. All even-numbered levels (2, 4, 6, etc.) "act" as AND gates, with single-input variables left as is.

To demonstrate these rules further, we will consider several cases of multilevel NAND logic. Figure 6–20 shows a three-level circuit with single-input variables at each level.

The logic equation for this circuit is developed as follows:

$$X = (\overline{\overline{\overline{ABC}}})D = \overline{\overline{ABC} + \overline{D}} = (\overline{A} + \overline{B})C + \overline{D} \qquad (6\text{–}17)$$

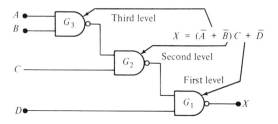

Figure 6–20.

In this case, A and B are the inputs at the third level, C is an input at the second level, and D is an input at the first level. Notice in the final form of Equation (6–17) that the complements of the inputs to the third-level gate G_3 appear to be ORed, as rule 1 states. The single-input C to the second-level gate G_2 appears uncomplemented and ANDed with the third-level term, as rule 2 states. The single-input D to the first-level gate G_1 appears complemented and ORed with the second- and third-level terms, in accordance with rule 1. The relationship to Equation (6–17) is shown as follows:

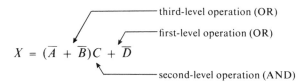

$$X = (\overline{A} + \overline{B})C + \overline{D}$$

Figure 6–21 shows a four-level NAND circuit, again with single-input variables at each level, as a means of demonstrating the basic multilevel NAND logic rules.

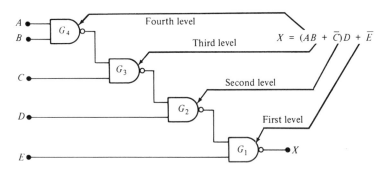

Figure 6–21.

The logic equation for this circuit is developed by direct application of the two rules. We will write the output equation one step at a time:

1. Gate G_4 (fourth level) acts as an AND gate with the input variables left as is, resulting in

$$AB$$

2. Gate G_3 (third level) acts as an OR gate with single-input variable C com-

plemented, resulting in

$$AB + \overline{C}$$

3. Gate G_2 (second level) acts as an AND gate with the single-input variable D left as is, resulting in

$$(AB + \overline{C})D$$

4. Gate G_1 (first level) acts as an OR gate with the single-input variable E complemented, resulting in

$$(AB + \overline{C})D + \overline{E}$$

The output of gate G_1 is the final output function X. A series of examples will serve to further illustrate the analysis of multilevel NAND logic.

Example 6–1   For the circuit shown in Figure 6–22, determine the output equation using the two rules for multilevel NAND logic.

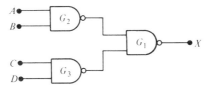

Figure 6–22.

Solution:
This is a two-level circuit, with gates G_2 and G_3 at the second level and gate G_1 at the first level. Gates G_2 and G_3 both act as AND gates, with their input variables appearing *as is* in the output equation. Gate G_1 has no single-input variables, so it simply acts as an OR gate to the second-level terms. The resulting equation is

$$X = AB + CD$$

Example 6–2   Repeat the problem of Example 6–1 for the circuit in Figure 6–23.

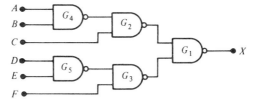

Figure 6–23.

Solution:
Gates G_4 and G_5 are third-level gates, and therefore appear as OR gates with their input variables complemented. The terms in the output function due to these gates will be

$$(\overline{A} + \overline{B}) \quad \text{and} \quad (\overline{D} + \overline{E})$$

Gates G_2 and G_3 are second-level gates, and therefore appear as AND gates. They AND the third-level terms with the single-input variables, C and F. The terms in the output function due to these gates will be

$$(\overline{A} + \overline{B})C \quad \text{and} \quad (\overline{D} + \overline{E})F$$

Gate G_1 is the first-level gate and acts as an OR gate for the second-level terms (its inputs), thus generating the final output function as follows:

$$X = (\overline{A} + \overline{B})C + (\overline{D} + \overline{E})F$$

Let us now take a somewhat different approach to the analysis of multilevel NAND logic. This approach is based on the two rules previously discussed, and might make the job easier for you.

We can relate the dual function of the NAND gate (discussed in Chapter 4, Section 4–4) to each of these two rules as follows: since all odd levels "act" as OR gates with single-input variables complemented, the *negative-OR* function of the NAND gate can be used at these levels; since all even level gates "act" as AND gates with single-input variables uncomplemented, the *NAND* function can be used at these levels.

To illustrate how this method can be applied to multilevel NAND logic, let us take the network shown in Figure 6–24(a) and proceed as follows. Replace each *odd*-level NAND gate with its equivalent *negative-OR* symbol, as shown in Figure 6-24(b). Remember, nothing has been changed operationally in the circuit; we are simply replacing the NAND gate symbol with an *equivalent* one. Now, rather than having to remember the two rules, we can simply write the output expression based on the logical function at each level, as indicated by the gate symbol at that level.

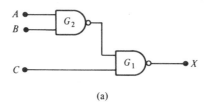

(a)

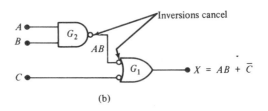

(b)

Figure 6–24.

For Figure 6–24, the output expression is

$$X = AB + \overline{C}$$

Notice that the "bubble" output of G_2 is connected to the "bubble" input of G_1. This, of course, indicates a double inversion that effectively cancels the inversion at each of the levels. In this approach, all connections between levels are "bubble-to-bubble" going from even levels to odd levels, and straight connections going from odd levels to even levels. Thus, all inversions can be discarded when writing the output expression, except for single-input variables at odd levels.

Next, let us take the three-level network of Figure 6-25(a) to further illustrate this method. By replacing gates G_1 and G_3 (odd-level gates) with their negative-OR equivalents, we have the equivalent network of Figure 6–25(b). From this we can easily write the output expression as follows:

$$X = (\bar{A} + \bar{B})C + \bar{D}$$

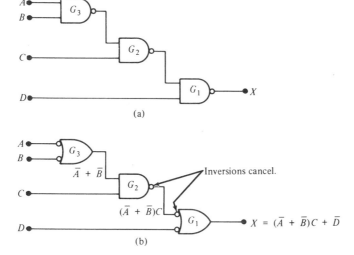

Figure 6–25.

Example 6–3   Write the output expression for the circuit in Figure 6–26, using the procedure just discussed.

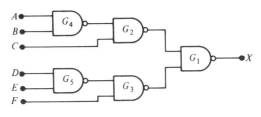

Figure 6–26.

Solution:

Redrawing the network in Figure 6–26 with equivalent negative-OR symbols at the odd levels, we get Figure 6–27. Writing the expression for X directly from the indicated logic operation at each level gives us

$$X = (\overline{A} + \overline{B})C + (\overline{D} + \overline{E})F$$

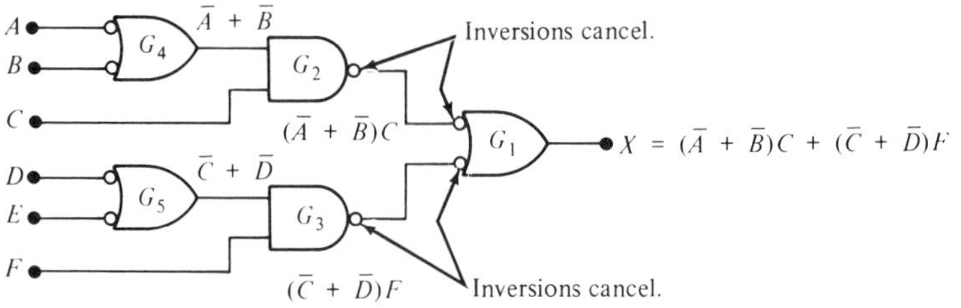

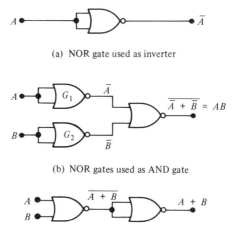

Figure 6–27.

6–5   The NOR Gate as a Universal Logic Element

As with the NAND gate, the NOR gate can be used to generate the AND, OR, and NOR functions. A NOT circuit or inverter can be made from a NOR gate by connecting all of its inputs to effectively create a single input, as shown in Figure 6–28(a) with a two-input example. An AND gate can be constructed using NOR gates, as shown in Figure 6–28(b). In this case the NOR gates G_1 and G_2 are used as inverters, and the final output is derived using DeMorgan's theorem as follows:

$$X = \overline{\overline{A} + \overline{B}} = AB \qquad\qquad (6\text{–}18)$$

Also, an OR gate can be produced from NOR gates, as illustrated in Figure 6–28(c).

(a) NOR gate used as inverter

(b) NOR gates used as AND gate

(c) NOR gates used as OR gate

Figure 6–28.   Universal Application of NOR Gates.

If the inputs A and B are applied to the NOR gate in Figure 6–29, the output will be as shown. By applying DeMorgan's theorem to the output function, we get:

$$X = \overline{A + B} = \overline{A}\,\overline{B} \tag{6-19}$$

Notice that the output can be expressed in two ways:

1. The output of a NOR gate is equal to the complement of the OR of the input variables.
2. The output of the NOR gate is equal to the AND of the input variable complements.

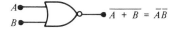

Figure 6-29. Dual Expression of NOR Output.

Rules for multilevel NOR logic can be developed, using these two statements of NOR operation as the basis. To develop a set of rules, we will begin with a two-level NOR circuit, as shown in Figure 6-30.

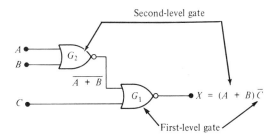

Figure 6-30.

For this circuit, the logic equation is developed as follows by application of DeMorgan's theorem:

$$X = \overline{\overline{(A + B)} + C} = \overline{\overline{(A + B)}}\,\overline{C} = (A + B)\overline{C} \tag{6-20}$$

Notice that the input variables A and B appear as an OR term in the final output equation, and the input variable C appears complemented and ANDed with the term $A + B$. Gate G_2 is a second-level gate (second back from output), and gate G_1 is a first-level gate (first gate from output). Notice also that the individual variables A and B are inputs to the second-level gate, and that the term $A + B$ and the individual variable C are the inputs to the first-level gate. From this observation, two rules for multilevel NOR logic can be stated:

1. All odd-numbered logic levels (1, 3, 5, etc.) act as AND gates, with single-input variables complemented.
2. All even-numbered levels (2, 4, 6, etc.) act as OR gates, with single-input variables left as is.

To demonstrate these rules further, we will consider several cases of multilevel NOR Logic. Figure 6-31 shows a three-level circuit with single-input variables at each level.

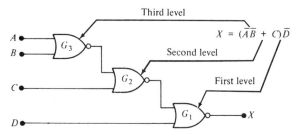

Figure 6–31.

For this circuit, the logic equation is developed as follows, using DeMorgan's theorem:

$$X = \overline{[\overline{(\overline{A + B}) + C}] + D} = \overline{(\overline{A + B})\overline{C} + D}$$
$$= [(A + B)\overline{C}]\overline{D} = (\overline{A}\overline{B} + C)\overline{D} \qquad (6\text{--}21)$$

In this case, A and B are inputs at the third level, C is an input at the second level, and D is an input at the first level. Notice in the final form of Equation (6–21) that the complements of the inputs to the third-level gate G_3 appear ANDed, as rule 1 states. The single-input variable C to the second-level gate G_2 appears uncomplemented and ORed with the third-level term $\overline{A}\overline{B}$, in accordance with rule 2. The single-input variable D to the first-level gate G_1 appears complemented and ANDed with the term $\overline{A}\overline{B} + C$, in accordance with rule 1. The operations we have just discussed are presented in relation to Equation (6–21) as follows:

$$X = (\overline{A}\overline{B} + C)\overline{D}$$

third level (AND)
first level (AND)
second level (OR)

Figure 6–32 shows a four-level NOR circuit, again with single-input variables at each level for illustration of the rules.

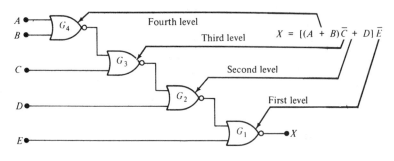

Figure 6–32.

The logic equation for this circuit is developed by direct application of the two rules for multilevel NOR logic. We will write the output equation one step at a time.

1. Gate G_4 (fourth level) acts as an OR gate, with the single-input variables A and B left as is, resulting in

$$A + B$$

2. Gate G_3 (third level) acts as an AND gate, with the single-input variable C complemented, resulting in

$$(A + B)\overline{C}$$

3. Gate G_2 (second level) acts as an OR gate, with the single-input variable D left as is, resulting in

$$(A + B)\overline{C} + D$$

4. Gate G_1 (first level) acts as an AND gate, with the single-input variable E complemented, resulting in

$$[(A + B)\overline{C} + D]\overline{E}$$

The output of gate G_1 is the final output X. A couple of examples will serve to further illustrate the analysis procedure for multilevel NOR logic.

Example 6–4   For the circuit shown in Figure 6–33, determine the output equation using the two rules for multilevel NOR logic.

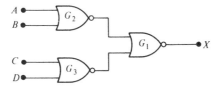

Figure 6–33.

Solution:
This is a two-level circuit, with gates G_2 and G_3 at the second level and gate G_1 at the first level. Gates G_2 and G_3 both act as OR gates, with their input variables appearing *as is* in the output equation. Gate G_1 has no single-input variable, so it simply acts as an AND gate to the second-level terms. The resulting equation is

$$X = (A + B)(C + D)$$

Example 6–5   Repeat the problem in Example 6–4 for the circuit shown in Figure 6–34.

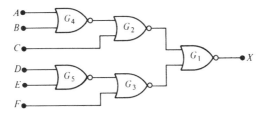

Figure 6–34.

Solution:

Gates G_4 and G_5 are third-level gates, and therefore appear as AND gates with their input variables complemented. The terms in the output function due to these gates will be

$$\overline{A}\overline{B} \quad \text{and} \quad \overline{D}\overline{E}$$

Gates G_2 and G_3 are second-level gates, and therefore appear as OR gates. They OR the third-level terms with the single-input variables C and F, as shown in the following expressions:

$$\text{for gate } G_2: \quad \overline{A}\overline{B} + C$$
$$\text{for gate } G_3: \quad \overline{D}\overline{E} + F$$

Gate G_1 is the first-level gate, and acts as an AND gate for the second- and third-level terms, thus generating the final output function as follows:

$$X = (\overline{A}\overline{B} + C)(\overline{D}\overline{E} + F)$$

An approach to multilevel NOR analysis similar to that used for multilevel NAND logic is available using the *negative-AND* equivalent of the NOR function, which was discussed in Section 4–5. This method is based on the two rules for multilevel NOR logic previously covered.

We can relate the dual function of the NOR gate to each of the two rules as follows. Since all odd levels act as AND gates with single-input variables complemented, the *negative-AND* function of the NOR gate can be used as these levels; and since all even-level gates act as AND gates with single-input variables uncomplemented, the NOR function can be used at these levels.

To illustrate how this method can be applied to multilevel NOR logic, let us take the network shown in Figure 6–35(a) and proceed as follows. Replace each *odd*-level NOR gate with its equivalent *negative-AND* symbol, as shown in Figure 6–35(b). Keep in mind that nothing has been changed operationally in the circuit; we are simply replacing the NOR gate symbol with an *equivalent* one. Now, we can write the expression for the output based on the logical function at each level, as indicated by the gate symbol at that level.

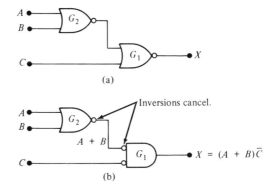

Figure 6–35.

The output expression for the network of Figure 6–35 is

$$X = (A + B)\overline{C}$$

Notice that the "bubble" output of gate G_2 is connected to the "bubble" input of gate G_1. The "bubble-to-bubble" connection going from an even to an odd level effectively cancels the inversions. So, in this case, gate G_2 effectively looks like an OR gate. The only inversions occur for single-input variables at odd levels.

Next, let us take the three-level network of Figure 6–36(a) in order to further illustrate this method of analysis. By replacing gates G_1 and G_3 (odd-level gates) with their equivalent negative-AND symbols, we have the equivalent network of Figure 6–36(b). From this we can write the output expression as follows:

$$X = (\overline{A}\overline{B} + C)\overline{D}$$

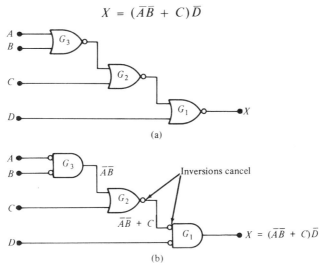

(a)

(b)

Figure 6–36.

Example 6–6   Write the output expression for the circuit in Figure 6–37, using the method just discussed.

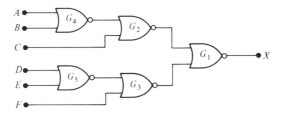

Figure 6–37.

Solution:
Redraw the network with the equivalent negative-AND symbols at the odd levels, as shown in Figure 6–38. Writing the expression for X directly from the indicated

logic operation at each level, we get

$$X = (\overline{AB} + C)(\overline{DE} + F)$$

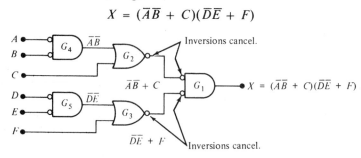

Figure 6–38.

6–6 Pulsed Operation

The operation of combinational logic circuits with pulsed inputs follows the same logical operation we have discussed in the previous sections. The output of the logic network is dependent on the states of the inputs at any given time; if the inputs are changing levels according to a time-varying pattern (waveform), the output depends on what states the inputs happen to be at any instant. The following is a summary of the operation of individual gates for use in analyzing combinational circuits with time-varying inputs.

1. The only time that the output of an AND gate is HIGH is when *all* inputs are HIGH *at the same time.*
2. The output of an OR gate is HIGH any time at least *one* of its inputs is HIGH. The output is LOW only when *all* inputs are LOW *at the same time.*
3. The only time the output of a NAND gate is LOW is when *all* inputs are HIGH *at the same time.*
4. The output of a NOR gate is LOW any time at least *one* of its inputs is HIGH. The output is HIGH only when *all* inputs are LOW *at the same time.*

The logical operation of any gate is the same regardless of whether its inputs are pulsed or not. The nature of the inputs (pulsed or constant levels) does not alter the truth table operation of a gate. A few examples will serve to illustrate the analysis of combinational logic circuits with pulsed inputs.

Example 6–7 Determine the output waveform for the circuit in Figure 6–39(a), with the inputs as shown.

Solution:
X is shown in the proper time relationship to the inputs in Figure 6–39(b).

Example 6–8 Determine the output waveform for the circuit in Figure 6–40(a), if the input waveforms are as indicated.

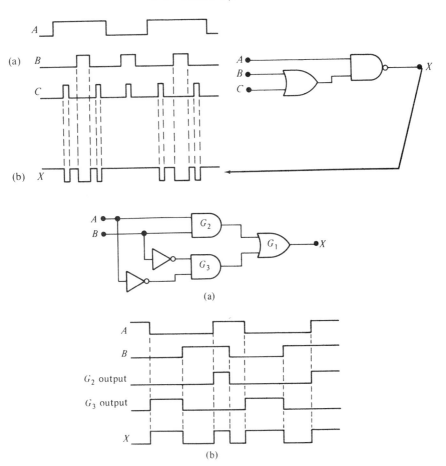

Figure 6–40.

Solution:
When both inputs are HIGH or when both inputs are LOW, the output is HIGH.
This is sometimes called a coincidence circuit. The intermediate outputs of gates
G_2 and G_3 are used to develop the final output. See Figure 6–40(b).

Example 6–9   If the waveform shown in Figure 6–41(a) is applied to the input A,
what waveform must be applied to B to generate the indicated output?

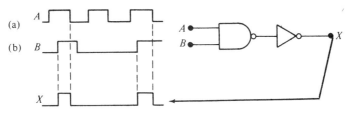

Figure 6–41.

Solution:

Input *B* could be as shown in Figure 6–41(b). This would produce the required output.

PROBLEMS

6–1   Determine the truth table for each of the circuits in Figure 6–42.

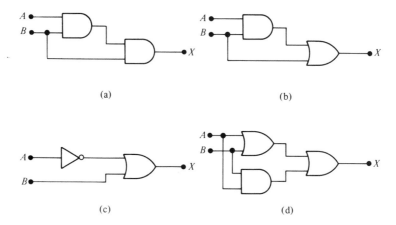

(a)   (b)

(c)   (d)

Figure 6–42.

6–2   Determine the truth table for each circuit in Figure 6–43.

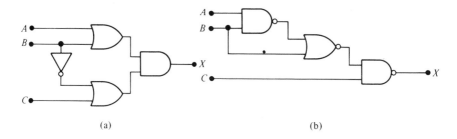

(a)   (b)

Figure 6–43.

6–3   Determine the truth table for the circuit shown in Figure 6–44.

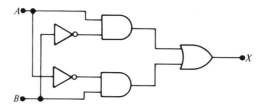

Figure 6–44.

6–4 Write the logic expression for the output of each of the circuits in Figure 6–45.

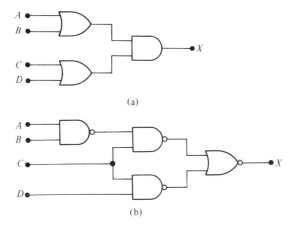

(a)

(b)

Figure 6–45.

6–5 Write the logic expression for the output of each of the circuits in Figure 6–46, below and at the top of the next page.

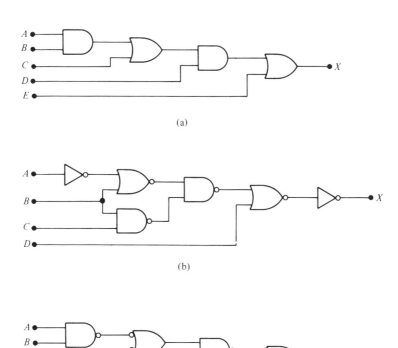

(a)

(b)

(c)

Figure 6–46.

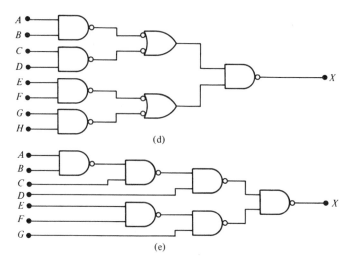

(d)

(e)

Figure 6–46, cont.

6–6 Develop the truth table for the circuit in Figure 6–47 by writing and evaluating the output expression for each input variable combination.

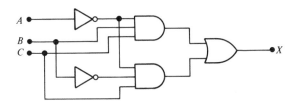

Figure 6–47.

6–7 Complete the truth table for the circuit of Figure 6–7.

6–8 Repeat Problem 6–6 if each AND gate is replaced by a NAND gate and each OR gate is replaced by a NOR gate.

6–9 Simplify the circuit in Figure 6–48 as much as possible, and verify that the simplified circuit is equivalent to the original by showing that their truth tables are identical.

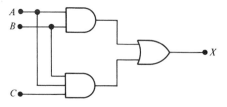

Figure 6–48.

6–10 Repeat Problem 6–9 for the circuit in Figure 6–49.

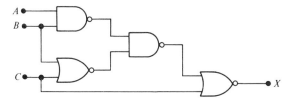

Figure 6-49.

6-11 Use AND and OR gates to implement the following logic expressions:
  (a) $X = AB$ (b) $X = A + B$
  (c) $X = AB + C$ (d) $X = ABC + D$
  (e) $X = A + B + C$ (f) $X = ABCD$
  (g) $X = A(CD + B)$ (h) $X = AB(C + DEF) + CE(A + B + F)$

6-12 Use AND gates, OR gates, and inverters to implement the following logic expressions:
  (a) $X = AB + \overline{B}C$
  (b) $X = A(B + \overline{C})$
  (c) $X = A\overline{B} + AB$
  (d) $X = \overline{ABC} + B(EF + \overline{G})$
  (e) $X = A[BC(A + B + C + D)]$
  (f) $X = B(C\overline{D}E + \overline{E}FG)(\overline{AB} + C)$

6-13 Use NAND and NOR gates to implement the following logic expressions:
  (a) $X = \overline{A}B + CD + \overline{(A + B)}(ACD + \overline{B}E)$
  (b) $X = AB\overline{C}\overline{D} + D\overline{E}F + \overline{A}F$
  (c) $X = \overline{A}[B + \overline{C}(D + E)]$

6-14 Show how the following expressions can be implemented using only NAND gates.
  (a) $X = ABC$ (b) $X = \overline{ABC}$
  (c) $X = A + B$ (d) $X = A + B + \overline{C}$
  (e) $X = \overline{AB} + \overline{CD}$ (f) $X = (A + B) \cdot (C + D)$
  (g) $X = AB[C(\overline{DE} + \overline{AB}) + \overline{BCE}]$

6-15 Repeat Problem 6-14 using only NOR gates.

6-16 Minimize the gates required to implement the functions in each part of Problem 6-12.

6-17 Minimize the gates required to implement the functions in each part of Problem 6-13.

6-18 Minimize the gates required to implement the functions in each part of Problem 6-14.

6-19 Use NAND gates to implement the function plotted on the Karnaugh map in Figure 6-50.

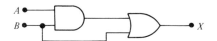

Figure 6–50.

6–20 Given the logic circuit and the input waveforms in Figure 6–51, determine and sketch the output waveform.

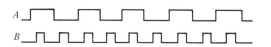

Figure 6–51.

6–21 For the logic circuit in Figure 6–52, determine and sketch the output waveform in proper relation to the inputs.

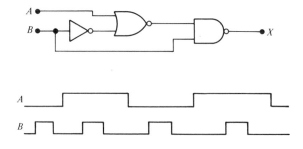

Figure 6–52.

6–22 Assume that you require the output waveform in Figure 6–53, and one input is given as indicated. Determine the other input needed to generate the desired output for the logic circuit shown.

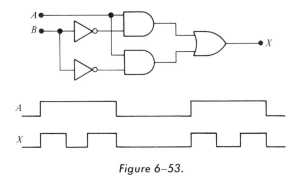

Figure 6–53.

6-23 For the input waveforms in Figure 6–54, what logic circuit would generate the output waveform shown?

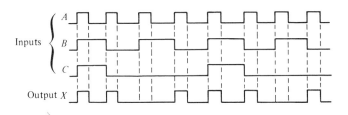

Figure 6–54.

6-24 Repeat Problem 6–23 for the waveforms in Figure 6–55.

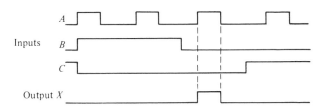

Figure 6–55.

6-25 Develop the logic circuit required to meet the following requirements:
A light in a room is to be operated from two switches, one at the back door and one at the front door. The light is to be on if the front switch is on and the back switch off, or if the front switch is off and the back switch is on. The light is to be off if both switches are off or if both switches are on. Let a HIGH output represent the on condition and a LOW output represent the off condition.

6-26 For the circuit in Figure 6–56, sketch the waveforms at each point indicated in the proper relationship to each other.

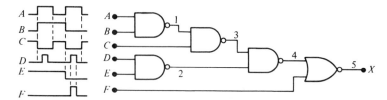

Figure 6-56.

6-27 For the logic circuit and the input waveforms in Figure 6–57, the indicated output waveform is observed. Determine if this is the correct output waveform.

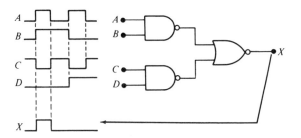

Figure 6-57.

6-28 The output waveform in Figure 6–58 is incorrect for the inputs that are applied to the circuit. Assuming that one gate in the circuit has failed with its output either a constant HIGH or a constant LOW, determine the faulty gate(s), and the failure mode that would cause the erroneous output(s).

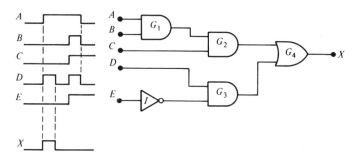

Figure 6-58.

6-29 Repeat Problem 6–28 for the circuit in Figure 6–59, with input and output waveforms as shown.

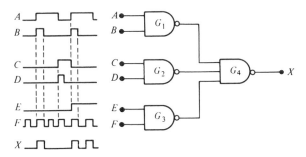

Figure 6–59.

6-30 Assuming a propagation delay through each gate of 10 nanoseconds (ns), determine if the *desired* output waveform in Figure 6–60 will be generated properly with the given inputs.

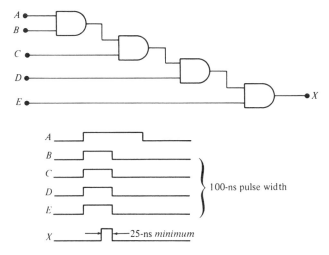

Figure 6–60.

6-31 Given the integrated circuit package configuration in Figure 6–61 (see next page), show the IC packages you would use and how you would interconnect them to achieve the logic network shown.

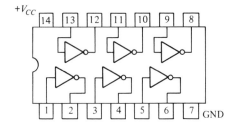

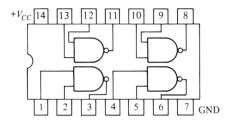

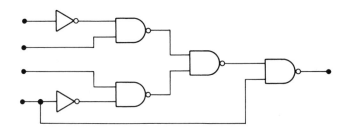

Figure 6–61.

CHAPTER 7

Functions of Combinational Logic

In this chapter, several basic and widely used functions of combinational logic are introduced. These include exclusive-OR, adders, subtractors, multipliers, comparators, decoders and display devices, encoders, code converters, multiplexers, and parity checkers. These particular functions are stressed because they find wide application in a variety of digital systems, and most are normally available in integrated circuit form.

7–1 The Exclusive-OR

The expression for the exclusive-OR function is given in Equation (7–1):

$$X = A\overline{B} + \overline{A}B \qquad (7\text{--}1)$$

An evaluation of this equation will illustrate the meaning of the term "exclusive-OR."

$$A = 0, \quad B = 0: \quad X = 0 \cdot 1 + 1 \cdot 0 = 0 + 0 = 0$$
$$A = 0, \quad B = 1: \quad X = 0 \cdot 0 + 1 \cdot 1 = 0 + 1 = 1$$
$$A = 1, \quad B = 0: \quad X = 1 \cdot 1 + 0 \cdot 0 = 1 + 0 = 1$$
$$A = 1, \quad B = 1: \quad X = 1 \cdot 0 + 0 \cdot 1 = 0 + 0 = 0$$

Notice that X is a 1 if A is a 1 OR if B is a 1, *but not if both are 1*. Stated another way, X is a 1 only when A and B are not equal.

Now let us see how this function can be implemented with logic gates. An examination of Equation (7–1) tells us that A is ANDed with $\overline{B}$, $\overline{A}$ is ANDed with B, and then the two terms are ORed as shown in Figure 7–1 (which illustrates one way to implement an exclusive-OR function).

Because the exclusive-OR is an important and widely used logic function, standard symbols have been adopted for both the gate and the operation itself. Rather than using the logic diagram of Figure 7–1, a single-gate symbol (as shown in Figure 7–2) is used to designate the exclusive-OR function. Figure 7–2 also

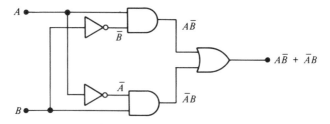

Figure 7-1.   A Method of Implementing the Exclusive-OR Function.

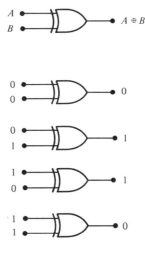

Figure 7-2.   Standard Symbol for the Exclusive-OR Gate with Logical Operation
              for All Possible Input Conditions.

shows the outputs for all possible input combinations. Also, the operation symbol
⊕ is used to indicate an exclusive-OR operation, so that Equation (7–1) can
be written as

$$X = A\overline{B} + \overline{A}B = A \oplus B \qquad (7\text{--}2)$$

In the sections ahead, we will see several applications of the exclusive-OR
gate.

7-2   Basic Adders

Recall the basic rules for binary addition as stated in Chapter 2:

$$0 + 0 = 0$$
$$0 + 1 = 1$$
$$1 + 0 = 1$$
$$1 + 1 = 10$$

These operations are performed by a logic circuit called a *half-adder*. The half-adder accepts two binary digits on its inputs and produces two binary digits on its outputs, a *sum* bit and a *carry* bit, as shown by the block diagram in Figure 7-3.

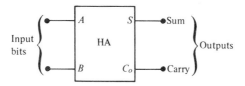

Figure 7-3.   Typical Block Diagram Symbol for a Half-Adder.

TABLE 7-1.   Half-
Adder Truth Table.

A	B	C_o	S
0	0	0	0
0	1	0	1
1	0	0	1
1	1	1	0

S = sum
C_o = carry
A and B = input variables

From the logical operation of the half-adder as expressed in Table 7-1, expressions can be derived for the sum and carry outputs as functions of the inputs. Notice that the carry output C_o is a 1 only when both A and B are 1s; therefore, C_o can be expressed as the AND of the input variables.

$$C_o = AB \qquad (7-3)$$

Now observe that the sum output S is a 1 only if the input variables are not equal. The sum can therefore be expressed as the exclusive-OR of the input variables.

$$S = A \oplus B \qquad (7-4)$$

From these two expressions, the implementation required for the half-adder function is apparent. The carry output is produced with an AND gate with A and B on the inputs, and the sum output is generated with an exclusive-OR gate as shown in Figure 7-4.

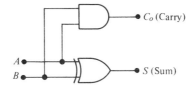

Figure 7-4.   Half-Adder Logic Diagram.

The second basic category of adder is the *full-adder*. The full-adder accepts *three* inputs, and generates a sum output and a carry output. So, the basic difference in a full-adder and a half-adder is that the full-adder accepts an additional input that allows for handling input carries. A block diagram illustrates the full-adder in Figure 7–5, and the truth table in Table 7–2 shows its operation.

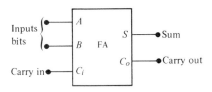

Figure 7–5. Typical Block Diagram Symbol for a Full-Adder.

TABLE 7–2.   Truth
Table for a Full Adder.

A	B	C_i	C_o	S
0	0	0	0	0
0	0	1	0	1
0	1	0	0	1
0	1	1	1	0
1	0	0	0	1
1	0	1	1	0
1	1	0	1	0
1	1	1	1	1

C_i = input carry
C_o = output carry
S = sum
A and B = input variables

The full-adder must sum the two input bits and the input carry bit. From the half-adder we know that the sum of the input bits A and B is the exclusive-OR of those two variables, $A \oplus B$. In order to add the input carry C_i to the input bits, it must be exclusive-ORed with $A \oplus B$, yielding the equation for the *sum* output of the full-adder.

$$S = (A \oplus B) \oplus C_i \qquad (7-5)$$

This means that to implement the full-adder sum function, two exclusive-OR gates can be used. The first must generate the term $A \oplus B$, and the second has as its inputs the output of the first gate and the input carry as illustrated in Figure 7–6(a).

The output *carry* is a 1 for the full-adder if both inputs to the first exclusive-OR gate are 1s or if both inputs to the second exclusive-OR gate are 1s. You can verify this by studying Table 7–2. The output carry of the full-adder is therefore A ANDed with B, $A \oplus B$ ANDed with C_i, and these two terms ORed as expressed in Equation (7–6). This function is implemented and combined with the sum logic to form a complete full-adder circuit, as shown in Figure 7–6(b).

$$C_o = AB + (A \oplus B) \cdot C_i \qquad (7-6)$$

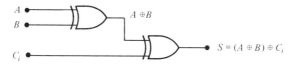

(a) Logic Required to Form the Sum of Three Bits

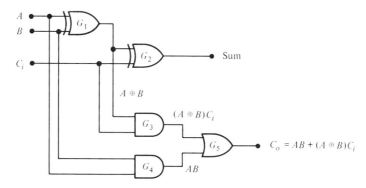

(b) Complete Logic Circuitry for a Full-Adder

Figure 7–6.   Full-Adder Logic.

Figure 7–7 shows the logic states within the full-adder for two input conditions. The rest are left as problems at the end of the chapter.

Note that in Figure 7–6(b) there are two half-adders, connected as shown in the partial block diagram of Figure 7–8(a), with their output carries ORed. A simplified block diagram as shown in Figure 7–8(b) will normally be used to represent the full-adder (FA).

Example 7–1   Determine an alternate method for implementing the full-adder.

Solution:
Going back to Table 7–2, we can write sum-of-product expressions for both S and C_o by observing the input conditions that make them 1s. The expressions are as follows:

$$S = \overline{A}\,\overline{B}C_i + \overline{A}B\overline{C}_i + A\overline{B}\,\overline{C}_i + ABC_i$$
$$C_o = \overline{A}BC_i + A\overline{B}C_i + AB\overline{C}_i + ABC_i$$

These two functions can be implemented with AND/OR logic as shown in Figure 7–9. This is simply another way to build a full-adder.

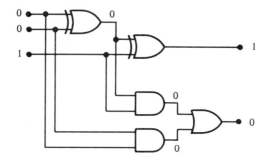

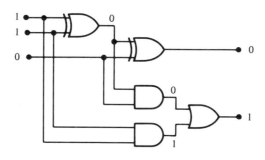

Figure 7–7.   Full-Adder Operation for Two Different Input Conditions.

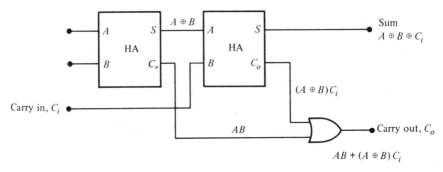

(a) Arrangement of Two Half-Adders to Form a Full-Adder

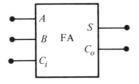

(b) Full-Adder Block Diagram

Figure 7–8.

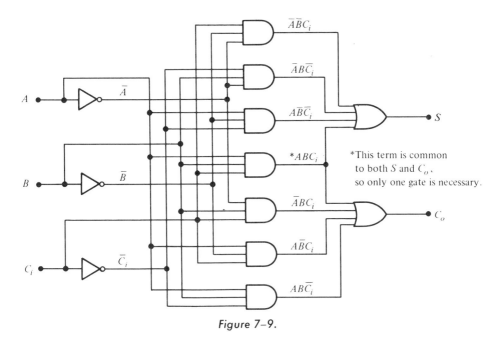

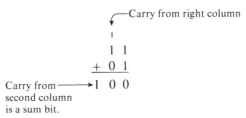

Figure 7–9.

7–3   Parallel Binary Adder

As we have seen, a single full-adder is capable of adding two one-bit numbers and an input carry. In order to add binary numbers with more than one bit, additional full-adders must be utilized. Refer to Chapter 2; when one binary number is added to another, each column generates a sum and a 1 or 0 carry to the next higher order column, as illustrated with two-bit numbers:

$$
\begin{array}{r}
\text{Carry from right column} \\
1 \\
1\ 1 \\
+\ 0\ 1 \\
\hline
1\ 0\ 0
\end{array}
$$

Carry from → second column is a sum bit.

To implement the addition of binary numbers with logic circuits, a full-adder is required for each column. So, for two-bit numbers, two adders are needed; for three-bit numbers, three adders are used; and so on. The carry output of each adder is connected to the carry input of the next higher order adder, as shown in Figure 7–10 for a two-bit adder. It should be pointed out that a half-adder can be used for the least significant position, or the carry input of a full-adder is made 0 because there is no carry into the least significant bit position.

In Figure 7–10, the least significant bits of the two numbers are represented by A_0 and B_0. The next higher order bits are represented by A_1 and B_1. The three complete sum bits are S_0, S_1, and S_2. Notice that the output carry from the full-adder is the most significant sum bit.

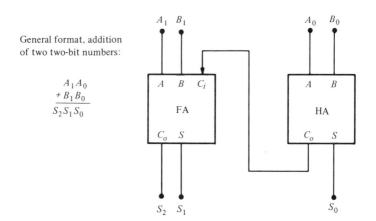

General format, addition
of two two-bit numbers:

$$\begin{array}{r} A_1 A_0 \\ + B_1 B_0 \\ \hline S_2 S_1 S_0 \end{array}$$

Figure 7–10.   Block Diagram of a Two-Bit Parallel Adder.

A four-bit parallel adder is shown in Figure 7–11 and an eight-bit parallel adder in Figure 7–12. Again, the least significant bits in each number being added go into the right-most adder; the higher order bits are applied as shown to the successively higher order adders, with the most significant bits in each number being applied to the left-most full-adder. The carry output of each adder is connected to the carry input of the next higher order adder.

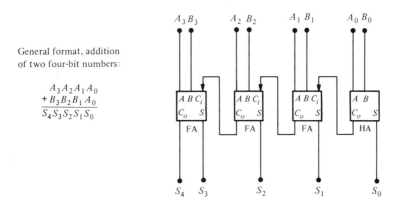

General format, addition
of two four-bit numbers:

$$\begin{array}{r} A_3 A_2 A_1 A_0 \\ + B_3 B_2 B_1 A_0 \\ \hline S_4 S_3 S_2 S_1 S_0 \end{array}$$

Figure 7–11.   Block Diagram of a Four-Bit Parallel Adder.

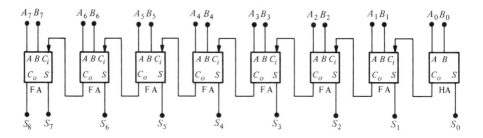

Figure 7–12.   Block Diagram of an Eight-Bit Parallel Adder.

Example 7-2   Verify that the two-bit parallel adder in Figure 7-13 properly performs the following addition:

$$\begin{array}{r} 11 \\ +10 \\ \hline 101 \end{array}$$

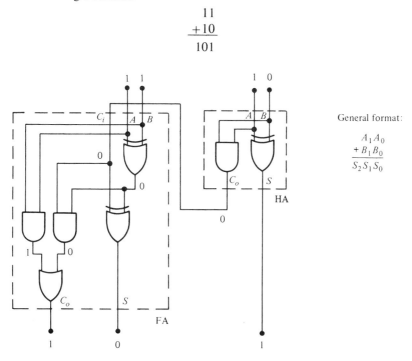

General format:

$$\begin{array}{r} A_1 A_0 \\ + B_1 B_0 \\ \hline S_2 S_1 S_0 \end{array}$$

Figure 7-13.

Solution:

The logic levels at each point in the circuit for the given input numbers are determined from the truth table of each gate. By following these levels through the circuit as indicated on the logic diagram, we find that the proper levels appear on the sum outputs.

Example 7-3   For the four-bit parallel adder shown in Figure 7-14(a) in block diagram form, determine the sum and carry outputs of each adder for the inputs indicated.

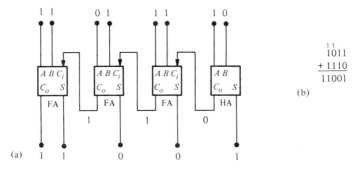

$$\begin{array}{r} {\scriptstyle 1\,1} \\ 1011 \\ + 1110 \\ \hline 11001 \end{array}$$

(b)

Figure 7-14.

Solution:
According to the truth table operation of the adders, the sum and carry outputs are as shown. This total sum output agrees with the sum arrived at by long hand addition in Figure 7-14(b).

Adders will be discussed in further detail in Chapter 12 where additional techniques and more sophisticated methods will be explored. The purpose of the discussion in this chapter is to introduce you to the basic way in which combinational logic can perform addition.

7-4   Basic Subtractors

Recall the basic rules for direct subtraction that were presented in Chapter 2:

$$0 - 0 = 0 \quad \text{with a borrow of } 0$$
$$1 - 0 = 1 \quad \text{with a borrow of } 0$$
$$0 - 1 = 1 \quad \text{with a borrow of } 1$$
$$1 - 1 = 0 \quad \text{with a borrow of } 0$$

These operations are performed by a logic circuit called a *half-subtractor*. The half-subtractor accepts two binary digits on its inputs and produces two binary digits (a *difference* bit and a *borrow* bit) on its outputs, as shown in the block diagram of Figure 7-15.

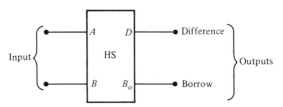

Figure 7-15.   **Block Diagram for a Half-Subtractor.**

Expressions for the logical operation of the half-subtractor can be developed from Table 7-3. First observe that the borrow output B_o is a 1 only when A is a 0 and B is a 1; therefore (B_o) can be expressed as the complement of A ANDed with B, as shown in Equation (7-7):

$$B_o = \overline{A} B \tag{7-7}$$

Now notice that the difference output (D) is a 1 only when the input variables are not equal. The difference can therefore be expressed as the exclusive-OR of the input variables A and B.

$$D = A \oplus B \tag{7-8}$$

Notice that the expression for the difference of two variables and the sum of two variables is the same. From the expression for the output borrow and the expression for the difference, the implementation should be apparent. The output borrow is generated by an AND gate with $\overline{A}$ and B on its inputs. The $\overline{A}$ is achieved

TABLE 7-3.
Truth Table for
Half-Subtractor.

Inputs		Outputs	
A	B	B_o	D
0	0	0	0
0	1	1	1
1	0	0	1
1	1	0	0

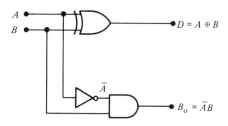

Figure 7-16.   Logic Diagram for a Half-Subtractor.

by inverting the variable A. The difference is generated by an exclusive-OR gate. The resulting logic implementation is shown in Figure 7-16.

The second basic category of subtractor is the *full-subtractor*. The full-subtractor accepts three inputs, and generates a difference output and a borrow output. The difference between a full-subtractor and a half-subtractor, then, is that the full-subtractor accepts an additional input that allows for handling *borrow inputs*. A block diagram of a full-subtractor appears in Figure 7-17(a), and the truth table is given in Table 7-4.

TABLE 7-4.   Truth Table
for a Full-Subtractor.

Inputs			Outputs	
A	B	B_i	D	B_o
0	0	0	0	0
0	0	1	1	1
0	1	0	1	1
0	1	1	0	1
1	0	0	1	0
1	0	1	0	0
1	1	0	0	0
1	1	1	1	1

$A, B \equiv$ input variables
$B_i \equiv$ input borrow
$B_o \equiv$ output borrow
$D \equiv$ difference

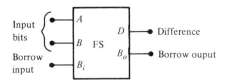

(a) Full-Subtractor Block Diagram

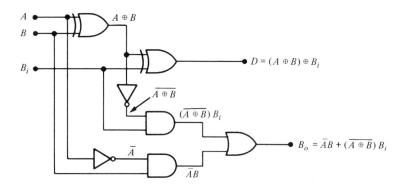

(b) Logic Diagram of a Full-Subtractor

Figure 7–17.

The expression for the difference is given in Equation (7–9), and the output borrow expression is stated in Equation (7–10):

$$D = (A \oplus B) \oplus B_i \qquad\qquad (7\text{–}9)$$

$$B_o = \overline{A}B + (\overline{A \oplus B})B_i \qquad\qquad (7\text{–}10)$$

The implementation is shown in Figure 7–17(b).

Example 7–4 Determine an alternate method for implementing a full-subtractor.

Solution:

Referring to Table 7–4, we can write the sum-of-product expressions for the difference and the borrow outputs by observing the input conditions that make them 1s. The expressions are as follows:

$$D = \overline{A}\,\overline{B}B_i + \overline{A}B\overline{B_i} + A\overline{B}\,\overline{B_i} + ABB_i$$

$$B_o = \overline{A}\,\overline{B}B_i + \overline{A}B\overline{B_i} + \overline{A}BB_i + ABB_i$$

These two functions can be implemented with AND/OR logic as shown in Figure 7–18, resulting in a full-subtractor. Notice that the two expressions are the same except for the third term.

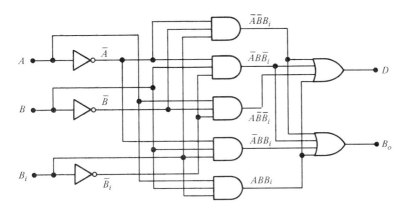

Figure 7–18.

7–5 Parallel Binary Subtractor

A full-subtractor can handle only two one-bit numbers and an input borrow. In order to subtract binary numbers with more than one bit, additional full-subtractors are required. When a 0 is subtracted from a 1 in a given column, no borrow is required from the next higher order column. However, if in a given column a 1 is subtracted from a 0, then we must borrow a 1 from the next higher order column. The output borrow of the column being subtracted becomes the input borrow of the next higher order column.

A block diagram of a four-bit subtractor is shown in Figure 7–19. It is composed of one half-subtractor and three full-subtractors. The half-subtractor is used in the least significant bit position because there is no input borrow.

General format, subtraction of two four-bit numbers:

$$A_3 A_2 A_1 A_0$$
$$\underline{- B_3 B_2 B_1 B_0}$$
$$D_3 D_2 D_1 D_0$$

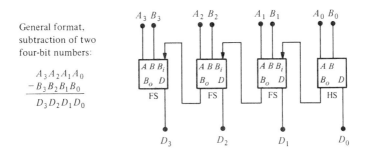

Figure 7–19. Block Diagram of a Four-Bit Parallel Subtractor.

Example 7-5   Verify that the two-bit parallel subtractor in Figure 7–20 properly performs the following operation:

$$\begin{array}{r} 10 \\ -01 \\ \hline 01 \end{array}$$

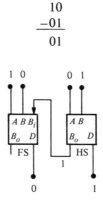

Figure 7–20.

Solution:

The logic levels at each point in the diagram for the given input numbers are determined from the truth table of the subtractor used. The subtractor does produce the proper result.

Other methods of subtraction will be introduced in Chapter 12. The purpose of this discussion is to introduce you to the basic way in which combinational logic can be used to perform the subtraction operation.

7–6   The Comparator

The basic function of a *comparator* is to *compare* the magnitudes of two quantities in order to determine the relationship of those quantities. In its simplest form, a comparator circuit determines if two numbers are equal.

The exclusive-OR gate is a basic comparator because its output is a 1 if its two input bits are not equal and a 0 if the inputs are equal. Figure 7–21 shows the exclusive-OR as a two-bit comparator.

In order to compare numbers containing two bits each, an additional exclusive-OR gate is necessary. The two least significant bits (LSBs) of the two numbers are compared by gate G_1, and the two most significant bits (MSBs) are compared by gate G_2, as shown in Figure 7–22. If the two numbers are equal, their corresponding bits are the same and the output of each exclusive-OR gate is a 0. If either of the corresponding sets of bits are not equal, a 1 occurs on that exclusive-OR gate output. In order to produce a *single* output indicating an equality or inequality of two numbers, two inverters and an AND gate can be utilized as shown in Figure 7–22. The output of each exclusive-OR gate is inverted and applied to the AND gate input. When the two numbers are equal, the corresponding bits of each number are equal, producing a 1 on both inputs to the AND gate and

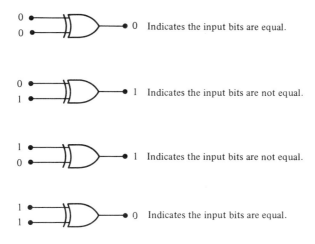

Figure 7–21.   Basic Comparator Operation.

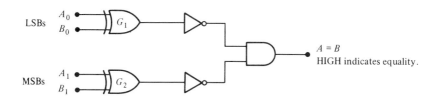

General format: Number $A \rightarrow A_1 A_0$
Number $B \rightarrow B_1 B_0$

Figure 7–22.   Logic Diagram for Comparison of Two Two-Bit Numbers.

thus a 1 on the output. When the two numbers are not equal, one or both sets of corresponding bits are unequal, and a 0 appears on at least one input to the AND gate to produce a 0 on its output. Thus the output of the AND gate indicates equality (1) or nonequality (0) of the two numbers. Example 7–6 illustrates this operation for two specific cases.

Example 7–6   Apply each of the following sets of numbers to comparator inputs, and determine the output by following the logic levels through the circuit: 10 and 10; 11 and 10.

Solutions:
The output is 1 for inputs 10 and 10, as shown in Figure 7–23(a). The output is 0 for inputs 11 and 10, as shown in Figure 7–23(b).

The basic comparator circuit can be expanded to any number of bits, as illustrated in Figure 7–24(a) for two four-bit numbers and Figure 7–24(b) for two five-bit numbers. The AND gate sets the condition that all corresponding bits of the two numbers must be equal if the two numbers are equal.

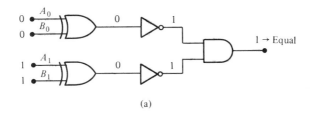

(a)

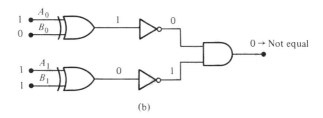

(b)

Figure 7-23.

Many comparators, including those commonly available in integrated circuit form, provide additional outputs that indicate which of the two numbers being compared is the larger. That is, there is an output that indicates when number A is greater than number B ($A > B$) and an output that indicates when number A is less than number B ($A < B$), as shown in the block diagram of Figure 7–25 for a four-bit comparator.

A method of implementing these two additional output functions is shown in Figure 7–26. In order to understand the logic circuitry required for the $A > B$ and $A < B$ functions, let us examine two binary numbers and determine what characterizes an inequality of the numbers.

For our purposes we will use two four-bit binary numbers with the general format $A_3 A_2 A_1 A_0$ for one number, which we will call number A, and $B_3 B_2 B_1 B_0$ for the other number, which we will call number B. To determine an inequality of numbers A and B, we first examine the highest order bit in each number. The following conditions are possible:

1. $A_3 = 1$ and $B_3 = 0$ indicates number A is greater than number B.
2. $A_3 = 0$ and $B_3 = 1$ indicates number A is less than number B.
3. If $A_3 = B_3$, then we must examine the next lower bit position for an inequality.

The three observations are valid for each bit position in the numbers. The general procedure is to check for an inequality in a bit position, starting with the highest order. When such an inequality is found, the relationship of the two numbers is established and any other inequalities in lower order bit positions *must be ignored* because it is possible for an opposite indication to occur—the highest

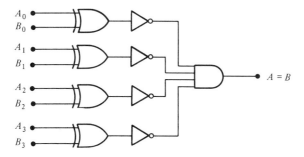

(a) Logic Diagram for the Comparison of Two Four-Bit Numbers

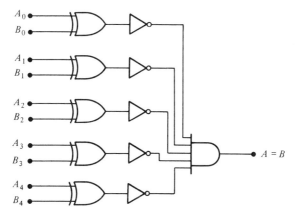

(b) Logic Diagram for the Comparison of Two Five-Bit Numbers

Figure 7-24.

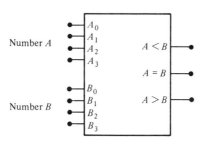

Figure 7-25.   Typical Block Diagram for a Four-Bit Comparator with Inequality Indication Capability.

order indication *must take precedence*. To illustrate, let us assume that number A is 0111 and number B is 1000. Comparison of bits A_3 and B_3 indicates that $A < B$ because $A_3 = 0$ and $B_3 = 1$. However, comparison of bits A_2 and B_2 indicates $A > B$ because $A_2 = 1$ and $B_2 = 0$. The same is true for the remaining lower order bits. In this case priority must be given to A_3 and B_3 because they determine the proper inequality condition.

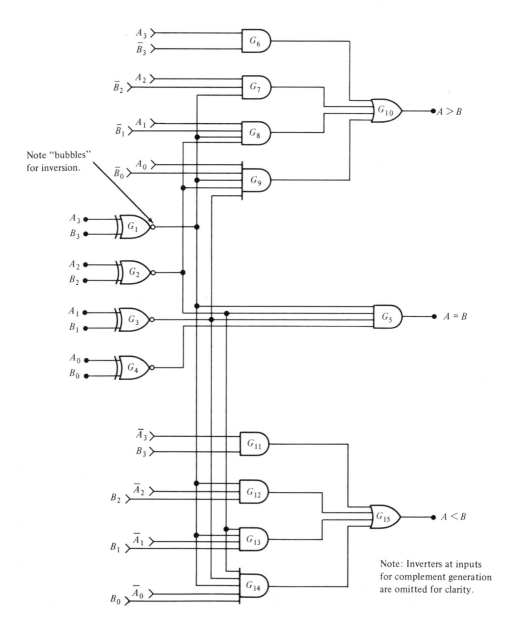

Figure 7–26.   Logic Diagram for a Four-Bit Comparator.

Figure 7-26 shows a method of comparing two four-bit numbers and generating an $A > B$ and an $A < B$ output in addition to the $A = B$ function. The $A > B$ condition is determined by gates G_6 through G_{10}. Gate G_6 checks for $A_3 = 1$ and $B_3 = 0$, and its function is expressed as $A_3\bar{B}_3$; gate G_7 checks for $A_2 = 1$ and $B_2 = 0$ $(A_2\bar{B}_2)$; gate G_8 checks for $A_1 = 1$ and $B_1 = 0$ $(A_1\bar{B}_1)$; gate G_9 checks for $A_0 = 1$ and $B_0 = 0$ $(A_0\bar{B}_0)$. These conditions all indicate that number A is greater than number B. The output of each of these gates is ORed by gate G_{10} to produce the $A > B$ output.

Notice that the output of gate G_1 is connected to inputs of gates G_7, G_8, and G_9. This provides a priority inhibit so that if the proper inequality occurs in bits A_3 and B_3, the lower order bit checks will be inhibited. A priority inhibit is also provided by gate G_2 to gates G_8 and G_9, and by gate G_3 to gate G_9.

Gates G_{11} through G_{15} check for an $A < B$ condition. Each AND gate checks a given bit position for the occurrence of a 0 in the number A and a 1 in number B. Each AND gate output is ORed by gate G_{15} to provide the $A < B$ output. Priority inhibiting is provided as previously discussed.

Examples 7-7 and 7-8 show the comparison of specific numbers and indicate the logic levels throughout the circuitry for each case. You should also go through the analysis with numbers of your own choosing to verify the operation.

Example 7-7 Analyze the comparator operation for the numbers $A = 1010$ and $B = 1001$.

Solutions:

Figure 7-27 shows all logic levels within the comparator for the specified inputs.

Example 7-8 Analyze the comparator operation for the numbers $A = 1011$ and $B = 1100$.

Solutions:

Figure 7-28 shows all the logic levels within the comparator for the specified inputs.

7-7 Decoders

The process of taking some type of complex code and determining what it represents in terms of a recognizable number or character is called *decoding*. A decoder is a combinational logic circuit that performs the decoding function, and produces an output that indicates the "meaning" of the input code.

A common example of the decoding function occurs when we are given a binary number and determine the decimal number it represents. We have, in essence, "decoded" the binary number because we have converted it into a more familiar form (decimal). In this section, several types of important logic decoders are examined.

The Basic Binary Decoder

Let us imagine that a certain digital circuit has four output terminals, and is capable of producing any four-bit binary number on these outputs with a LOW

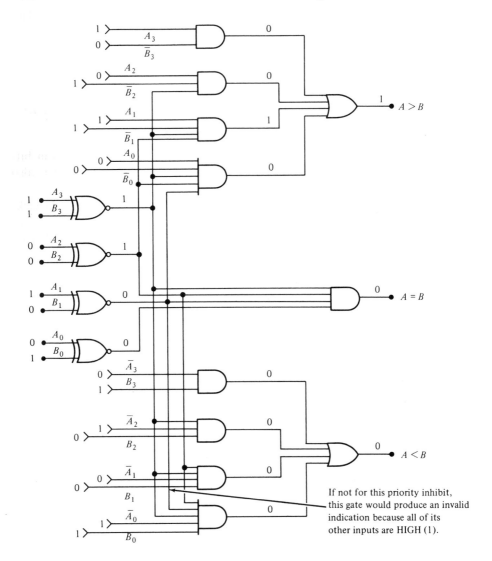

Figure 7–27.

representing a 0 and a HIGH representing a 1. Suppose we wish to determine when a binary 1001 occurs on the outputs of this digital circuit. An AND gate can be used as the basic decoding element because it produces a HIGH output only when all of its inputs are HIGH. Therefore, we must make sure that all of the inputs to the AND gate are HIGH when the binary number 1001 occurs; this can be done by complementing the two middle bits (the 0s), as shown in Figure 7–29.

When the AND gate output goes HIGH, we know that the binary number 1001 corresponding to decimal digit 5 is on the decoder inputs.

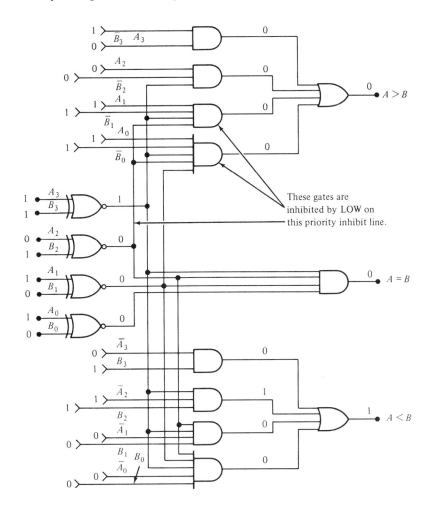

Figure 7–28.

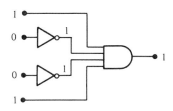

Figure 7–29.   Basic Decoding Logic for 1001 with an Active HIGH Output.

If a NAND gate is used in place of the AND gate, as shown in Figure 7–30, a LOW output will indicate the presence of the proper binary code.

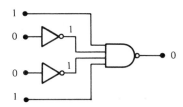

Figure 7–30.   Basic Decoding Logic for 1001 with an Active LOW Output.

The logic equation for the decoder of Figure 7–29 is developed as illustrated in Figure 7–31. You should verify that the output function is 0 except when $A = 1$, $B = 0$, $C = 0$, and $D = 1$ are applied to the inputs.

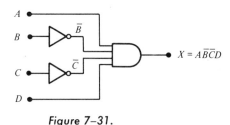

Figure 7–31.

Example 7–9   Determine the logic required to decode the binary number 1011 (decimal 11) by producing a HIGH indication.

Solution:

The decoding function can be formed by complementing only the variables that appear as 0 in the binary number as follows:

$$X = A\overline{B}CD$$

This function can be implemented by connecting the true (uncomplemented) variables A, C, and D directly to the inputs of an AND gate, and inverting the variable B before applying it to the AND gate input. The decoding logic is shown in Figure 7–32.

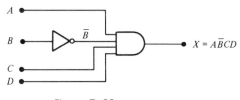

Figure 7–32.

Four-Bit Binary Decoder

In order to decode all possible combinations of four binary digits, 16 decoding gates are required ($2^4 = 16$). This type of decoder is commonly called a *1-of-16*

TABLE 7–5.   Four-Bit Binary
Decoding Functions.

Decimal Digit	Binary Code A B C D	Logic Function X
0	0 0 0 0	$\overline{A}\,\overline{B}\,\overline{C}\,\overline{D}$
1	0 0 0 1	$\overline{A}\,\overline{B}\,\overline{C}\,D$
2	0 0 1 0	$\overline{A}\,\overline{B}\,C\,\overline{D}$
3	0 0 1 1	$\overline{A}\,\overline{B}\,C\,D$
4	0 1 0 0	$\overline{A}\,B\,\overline{C}\,\overline{D}$
5	0 1 0 1	$\overline{A}\,B\,\overline{C}\,D$
6	0 1 1 0	$\overline{A}\,B\,C\,\overline{D}$
7	0 1 1 1	$\overline{A}\,B\,C\,D$
8	1 0 0 0	$A\,\overline{B}\,\overline{C}\,\overline{D}$
9	1 0 0 1	$A\,\overline{B}\,\overline{C}\,D$
10	1 0 1 0	$A\,\overline{B}\,C\,\overline{D}$
11	1 0 1 1	$A\,\overline{B}\,C\,D$
12	1 1 0 0	$A\,B\,\overline{C}\,\overline{D}$
13	1 1 0 1	$A\,B\,\overline{C}\,D$
14	1 1 1 0	$A\,B\,C\,\overline{D}$
15	1 1 1 1	$A\,B\,C\,D$

because each of the four-bit code words will activate only one of the 16 total outputs. It is also sometimes referred to as a *4-line-to-16-line* decoder because there are four inputs and 16 outputs. A list of the 16 binary code words and their corresponding decoding functions is given in Table 7–5.

If an active LOW output is desired for each decoded number, the entire decoder can be implemented with NAND gates and inverters as follows:

First, since each variable and its complement are required in the decoder as

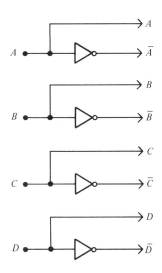

Figure 7–33.   Generation of Complements for Decoder.

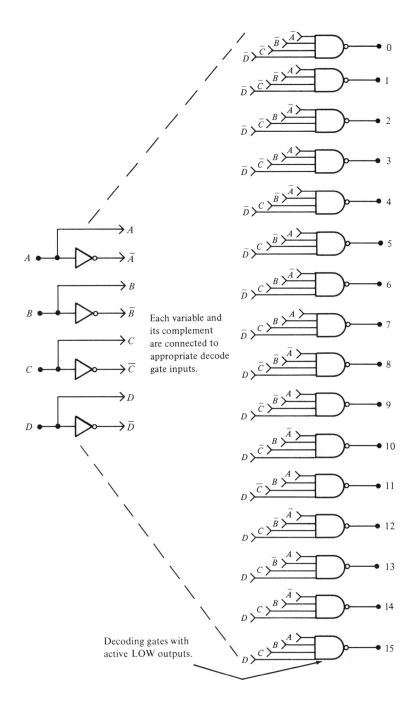

Each variable and its complement are connected to appropriate decode gate inputs.

Decoding gates with active LOW outputs.

Figure 7–34. Logic for a 1-of-16 Decoder.

196

seen from Table 7-5, they can be generated once and then used for all decoding gates as required, rather than duplicating an inverter each place a complement is used. This arrangement, shown in Figure 7-33, indicates that each variable and its complement are available to be connected to the inputs of the proper decoding gates.

In order to decode each of the 16 binary code words, 16 NAND gates are required (AND gates can be used to produce active HIGH outputs). The decoding gate arrangement is illustrated in Figure 7-34.

Rather than reproduce the complex logic diagram for the decoder each time it is required in a schematic, a simpler representation is normally used. A typical logic symbol for a four-bit binary decoder (1-of-16) is shown in Figure 7-35, and the corresponding truth table is given in Table 7-6.

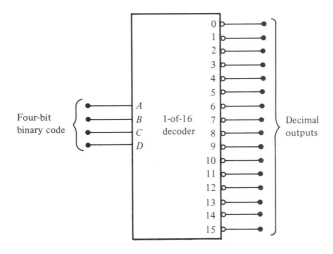

Figure 7-35.   Typical Block Diagram Symbol for a 1-of-16 Decoder.

Some refinement in the basic decoder appears in Figure 7-36(a), which is a complete logic diagram showing all of the internal logic connections. The additional inverters on the inputs are required to prevent excessive loading of the driving source(s). Each input is connected only to the input of one inverter, rather than to the inputs of several NAND gates as in Figure 7-34. There is also an enable function provided on this particular device. This is implemented with a NOR gate used as a negative AND. A LOW level on each input, $\overline{E}_0$ and $\overline{E}_1$, is required in order to make the enable gate output HIGH. The enable gate output is connected to an input of *each* NAND gate, so it must be HIGH for the gates to be enabled. If the enable gate is not activated (LOW on both inputs), then all 16 decoder outputs will be HIGH *regardless* of the states of the four input variables, A_0, A_1, A_2, and A_3. The logic symbol is shown in Figure 7-36(b). The pin numbers on both figures indicate the pins on the integrated circuit package to which the corresponding logic points are internally connected.

TABLE 7–6.   One-of-16 Decoder Truth Table.

Inputs	Outputs															
ABCD	0	1	2	3	4	5	6	7	8	9	10	11	12	13	14	15
0000	0	1	1	1	1	1	1	1	1	1	1	1	1	1	1	1
0001	1	0	1	1	1	1	1	1	1	1	1	1	1	1	1	1
0010	1	1	0	1	1	1	1	1	1	1	1	1	1	1	1	1
0011	1	1	1	0	1	1	1	1	1	1	1	1	1	1	1	1
0100	1	1	1	1	0	1	1	1	1	1	1	1	1	1	1	1
0101	1	1	1	1	1	0	1	1	1	1	1	1	1	1	1	1
0110	1	1	1	1	1	1	0	1	1	1	1	1	1	1	1	1
0111	1	1	1	1	1	1	1	0	1	1	1	1	1	1	1	1
1000	1	1	1	1	1	1	1	1	0	1	1	1	1	1	1	1
1001	1	1	1	1	1	1	1	1	1	0	1	1	1	1	1	1
1010	1	1	1	1	1	1	1	1	1	1	0	1	1	1	1	1
1011	1	1	1	1	1	1	1	1	1	1	1	0	1	1	1	1
1100	1	1	1	1	1	1	1	1	1	1	1	1	0	1	1	1
1101	1	1	1	1	1	1	1	1	1	1	1	1	1	0	1	1
1110	1	1	1	1	1	1	1	1	1	1	1	1	1	1	0	1
1111	1	1	1	1	1	1	1	1	1	1	1	1	1	1	1	0

1 = HIGH, 0 = LOW

The BCD Decoder

The BCD decoder converts each BCD code word (8421 code) into one of ten possible decimal digit indications. It is typically referred to as a *1-of-10* or a *4-line-to-10-line* decoder, although other types of decoders also fall into this category (such as an Excess-3 decoder).

The method of implementation is essentially the same as for the 1-of-16 decoder previously discussed, except that only *ten* decoding gates are required because the BCD code represents only the ten decimal digits 0 through 9. A list of

TABLE 7–7.   BCD Decoding Functions.

	BCD Code				Logic Function
Decimal Digit	A	B	C	D	X
0	0	0	0	0	$\overline{A}\,\overline{B}\,\overline{C}\,\overline{D}$
1	0	0	0	1	$\overline{A}\,\overline{B}\,\overline{C}\,D$
2	0	0	1	0	$\overline{A}\,\overline{B}\,C\,\overline{D}$
3	0	0	1	1	$\overline{A}\,\overline{B}\,C\,D$
4	0	1	0	0	$\overline{A}\,B\,\overline{C}\,\overline{D}$
5	0	1	0	1	$\overline{A}\,B\,\overline{C}\,D$
6	0	1	1	0	$\overline{A}\,B\,C\,\overline{D}$
7	0	1	1	1	$\overline{A}\,B\,C\,D$
8	1	0	0	0	$A\,\overline{B}\,\overline{C}\,\overline{D}$
9	1	0	0	1	$A\,\overline{B}\,\overline{C}\,D$

LOGIC DIAGRAM

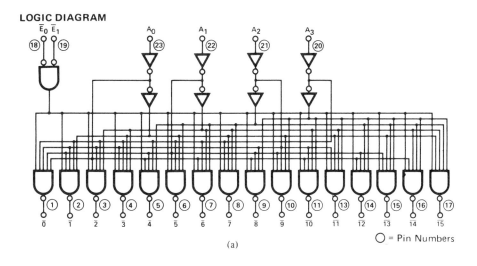

(a)

◯ = Pin Numbers

LOGIC SYMBOL

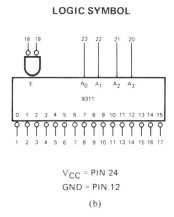

V_{CC} = PIN 24
GND = PIN 12

(b)

Figure 7–36.   One-of-16 Decoder. Courtesy of Fairchild Semiconductor.

the ten BCD code words and their corresponding decoding functions is given in Table 7–7. Each of these decoding functions is implemented with NAND gates to provide active LOW outputs, as shown in Figure 7–37. If an active HIGH output is required, AND gates are used for decoding. Notice that the logic is identical to the first ten decoding gates in the 1-of-16 decoder (Figure 7–34). A complete logic diagram is shown in Figure 7–38(a), and a typical logic symbol appears in Figure 7–38(b).

BCD-to-Seven-Segment Decoder

This type of decoder accepts the BCD code on its inputs and provides outputs to energize seven-segment display devices in order to produce a digital readout.

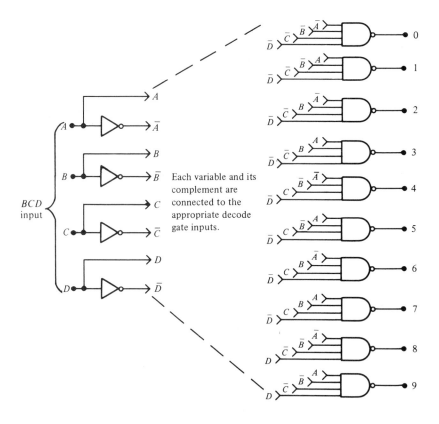

Figure 7–37. Logic for a BCD Decoder.

Before proceeding with a discussion of this decoder, let us examine the basics of a seven-segment display device.

Figure 7–39 shows a common display format composed of seven light-emitting elements or segments. By lighting certain combinations of these segments, each of the ten decimal digits can be produced. Figure 7–40 illustrates this method of digital display for each of the ten digits by using a darker segment to represent one that is illuminated. To produce a 1, segments *f* and *e* are illuminated; to produce a 2, segments *a*, *b*, *g*, *e*, and *d* are used; and so on.

There are several ways in which seven-segment type displays are currently implemented; these include the light-emitting diode (LED), incandescent, and liquid crystal. The LED type is perhaps the most widely used. The basic operation, regardless of type, is as follows. Each segment is activated by either a HIGH or a LOW voltage level, depending on the particular device characteristics. For instance, Figure 7–41(a) shows an LED display that requires an active LOW to illuminate a segment, and Figure 7–41(b) shows an arrangement that requires an active HIGH to illuminate a segment. In either case, the particular light-emitting diode is forward biased and conducts current, causing light to be emitted. The physics of the diode which bring about light emission are beyond the scope of this book.

LOGIC DIAGRAM

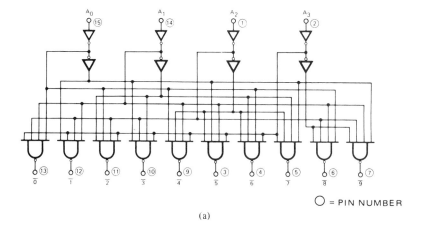

O = PIN NUMBER

(a)

LOGIC SYMBOL

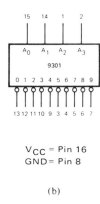

V_{CC} = Pin 16
GND = Pin 8

(b)

Figure 7–38.   BCD Decoder. Courtesy Fairchild Semiconductor.

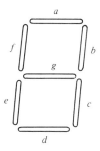

Figure 7–39.   Seven-Segment Display Format Showing Arrangement of Segments.

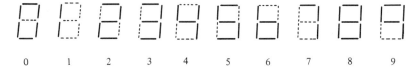

0   1   2   3   4   5   6   7   8   9

Figure 7–40.   Display of Decimal Digits with a Seven-Segment Device.

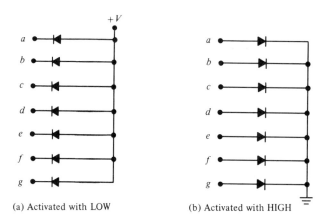

(a) Activated with LOW   (b) Activated with HIGH

Figure 7–41.   Schematic of Seven-Segment LED Devices.

The activated segments for each of the ten decimal digits are listed in Table 7–8.

We will now examine the decoding logic required to produce the format for a seven-segment display with a BCD input. Notice that segment a is activated for digits 0, 2, 3, 5, 7, 8, and 9. Segment b is activated for digits 0, 2, 3, 4, 7, 8, and 9; and so on. If we let the BCD inputs to the decoder be represented by the general form $ABCD$, a Boolean expression can be found for each segment in the display, and this will tell us the logic circuitry required to drive or activate each seg-

TABLE 7–8.
Seven-Segment
Display Format.

Digit	Segments
0	a, b, c, d, e, f
1	e, f
2	a, b, g, e, d
3	a, b, c, d, g
4	b, c, f, g
5	a, c, d, f, g
6	c, d, e, f, g
7	a, b, c
8	a, b, c, d, e, f, g
9	a, b, c, f, g

ment. For example, the equation for segment a is as follows:

$$a = \overline{A}\,\overline{B}\,\overline{C}\,\overline{D} + \overline{A}\,B\,\overline{C}\,\overline{D} + \overline{A}\,B\,C\,\overline{D} + \overline{A}\,B\,C\,\overline{D} + \overline{A}\,B\,C\,D + A\,\overline{B}\,\overline{C}\,\overline{D} + A\,\overline{B}\,\overline{C}\,D$$

This equation says that segment a is activated or "true" if the BCD code is 0 "OR 2 OR 3 OR 5 OR 7 OR 8 OR 9." Table 7–9 lists the logic function for each of the seven segments.

TABLE 7–9. Seven-Segment Decoding Functions.

Segment	Used in Digits	Function
a	0, 2, 3, 5, 7, 8, 9	$\overline{A}\,\overline{B}\,\overline{C}\,\overline{D} + \overline{A}\,B\,\overline{C}\,\overline{D} + \overline{A}\,B\,C\,\overline{D} + \overline{A}\,B\,C\,\overline{D} + \overline{A}\,B\,C\,D + A\,\overline{B}\,\overline{C}\,\overline{D} + A\,\overline{B}\,\overline{C}\,D$
b	0, 2, 3, 4, 7, 8, 9	$\overline{A}\,\overline{B}\,\overline{C}\,\overline{D} + \overline{A}\,B\,\overline{C}\,\overline{D} + \overline{A}\,B\,\overline{C}\,D + \overline{A}\,B\,C\,\overline{D} + \overline{A}\,B\,C\,D + A\,\overline{B}\,\overline{C}\,\overline{D} + A\,\overline{B}\,\overline{C}\,D$
c	0, 3, 4, 5, 6, 7, 8, 9	$\overline{A}\,\overline{B}\,\overline{C}\,\overline{D} + \overline{A}\,B\,\overline{C}\,D + \overline{A}\,B\,C\,\overline{D} + \overline{A}\,B\,C\,\overline{D} + \overline{A}\,B\,C\,D + A\,\overline{B}\,\overline{C}\,\overline{D} + A\,\overline{B}\,\overline{C}\,D + A\,\overline{B}\,C\,\overline{D}$
d	0, 2, 3, 5, 6, 8	$\overline{A}\,\overline{B}\,\overline{C}\,\overline{D} + \overline{A}\,B\,\overline{C}\,\overline{D} + \overline{A}\,B\,C\,\overline{D} + \overline{A}\,B\,C\,\overline{D} + \overline{A}\,B\,C\,D + A\,\overline{B}\,\overline{C}\,\overline{D}$
e	0, 1, 2, 6, 8	$\overline{A}\,\overline{B}\,\overline{C}\,\overline{D} + \overline{A}\,\overline{B}\,\overline{C}\,D + \overline{A}\,B\,\overline{C}\,\overline{D} + \overline{A}\,B\,C\,D + A\,\overline{B}\,\overline{C}\,\overline{D}$
f	0, 1, 4, 5, 6, 8, 9	$\overline{A}\,\overline{B}\,\overline{C}\,\overline{D} + \overline{A}\,\overline{B}\,\overline{C}\,D + \overline{A}\,B\,\overline{C}\,\overline{D} + \overline{A}\,B\,C\,\overline{D} + \overline{A}\,B\,C\,D + A\,\overline{B}\,\overline{C}\,\overline{D} + A\,\overline{B}\,\overline{C}\,D$
g	2, 3, 4, 5, 6, 8, 9	$\overline{A}\,\overline{B}\,C\,\overline{D} + \overline{A}\,B\,C\,D + \overline{A}\,B\,\overline{C}\,\overline{D} + \overline{A}\,B\,\overline{C}\,D + \overline{A}\,B\,C\,\overline{D} + A\,\overline{B}\,\overline{C}\,\overline{D} + A\,\overline{B}\,\overline{C}\,D$

From the expressions in Table 7–9, the logic for the BCD-to-Seven-Segment decoder can be implemented. Each of the ten BCD code words is decoded, and then the decoding gates are ORed as dictated by the logic expression for each segment. For instance, segment a requires that the decoded BCD digits 0, 2, 3, 5, 7, 8, and 9 be ORed, as shown in Figure 7–42.

Segment b requires that the decoded BCD digits 0, 2, 3, 4, 7, 8, and 9 be ORed. Segment c is activated by ORing the outputs of the 0, 3, 4, 5, 6, 7, 8, and 9 decode gates, etc. The total logic for the BCD-to-Seven-Segment decoder is shown in Figure 7-43, and a typical block diagram representation in Figure 7-44.

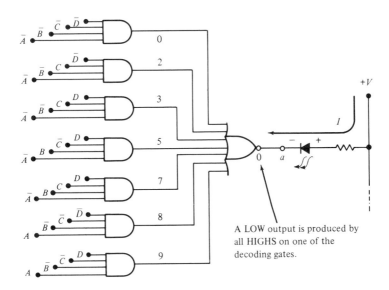

A LOW output is produced by all HIGHS on one of the decoding gates.

Figure 7–42. Decoding Logic for the a Segment.

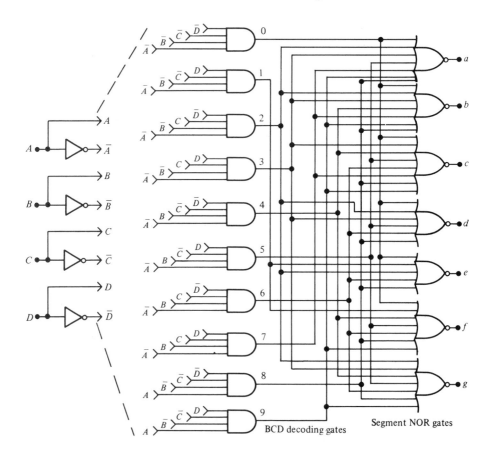

Figure 7–43. Logic for Complete Seven-Segment Decoder.

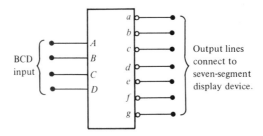

Figure 7–44. Typical Block Diagram Symbol for a Seven-Segment Decoder.

An additional feature found on many seven-segment decoders is the *zero suppression logic*. This extra function is useful in multidigit displays because it is used to blank out unnecessary zeros in the display. For instance, the number 0006.400 would be displayed as 6.4, which is read more easily. Blanking of the zeros on the front of the number is called *leading edge zero suppression* and the blanking of the zeros after the number is called *trailing edge zero suppression*.

A block diagram of a four-digit display is shown in Figure 7–45, and will be used to illustrate the requirements for leading edge zero suppression. Notice that two additional functions have been added to each BCD-to-Seven-Segment decoder, a *blanking input* (BI) and a *blanking output* (BO). The highest order digit position is always blanked if a 0 code appears on its BCD inputs *and* the blanking input is HIGH. Each lower order digit position is blanked if a 0 code appears on its BCD inputs *and* the next higher order digit is a 0 as indicated by a HIGH on its blanking output. The blanking output (BO) of any decoder indicates that it has a BCD 0 on its inputs and *all* higher order digits are also 0. The blanking output of each stage is connected to the blanking input of the next lower order stage, as shown in the diagram. As an example, in Figure 7–45 the highest order digit is 0, which is therefore blanked. Also, the next digit is a 0, and because the highest order digit is 0, it is also blanked. The remaining two digits are displayed.

For the fractional portion of a display (the digits to the right of the decimal point), trailing edge zero suppression is used; that is, the lowest order digit is blanked if it is 0 and each digit that is 0 *and* is followed by 0s in all the lower order positions is blanked. To illustrate, Figure 7–46 shows a block diagram in which there are three digits to the right of the decimal point. In this example, the lowest order digit is blanked because it is a 0. The next digit is also 0 *and* its blanking input is HIGH, so it is blanked. The highest order digit is a 5, and it is displayed. Notice that the blanking output of each decoder stage is connected to the blanking input of the next higher order stage.

One basic way of implementing the 0 suppression function is shown in Figure 7–47. Here, a NAND gate G_1 has been added to the basic BCD-to-Seven-Segment decoder, and is shown within the dashed lines. Gate G_1 detects the presence of a BCD 0 on the inputs *and* a HIGH on the blanking input. The complement of the output of this gate is the blanking output, and it is used to inhibit the decoding logic in the BCD-to-Seven-Segment decoder so that the seven-segment display device will not be activated. In this particular implementation, the output of gate G_1 is connected to an input of each BCD decoding gate so that when a blanking condition occurs (a 0 code on the inputs and BI = 1), each decoding gate is inhibited by the LOW on the output of gate G_1. As a result, the output lines from the decoder that drive or activate the display segments are held HIGH, preventing the segments from lighting (we are using active LOW outputs in this case).

Decoders can be designed for any type of special code using an approach similar to that discussed in this section. Excess-3 and Gray decoders are examples.

7–8   Encoders

An *encoder* is a combinational logic circuit that essentially performs a "reverse" decoder function. An encoder accepts a digit on its inputs, such as a decimal or octal digit, and converts it to a coded output, such as binary or BCD. Encoders can also be devised to encode various symbols and alphabetic characters. This process of converting from familiar symbols or numbers to a coded format is called *encoding*.

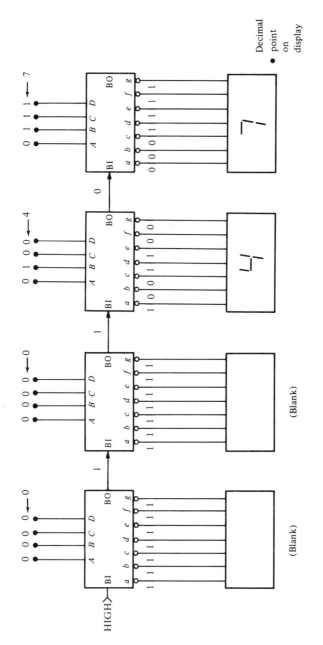

Figure 7–45. Four-Digit Decoder and Display Illustrating Leading Edge Zero Suppression.

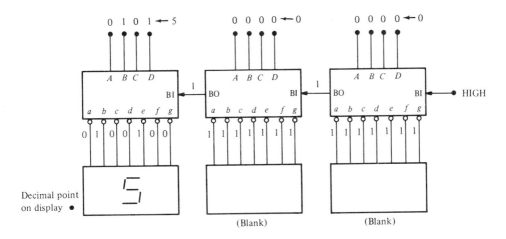

Figure 7–46.   Three-Digit Decoder and Display Illustrating Trailing Edge Zero Suppression.

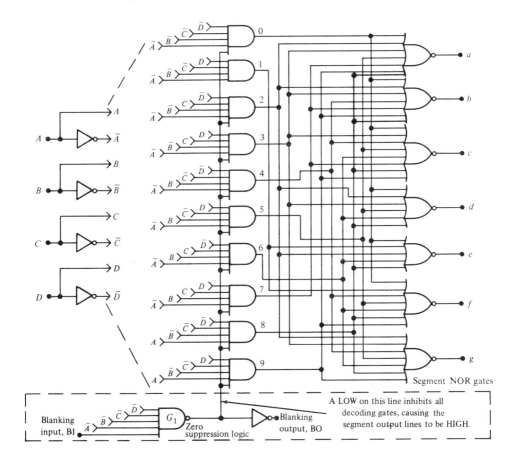

Figure 7–47.   Seven-Segment Decoder with Zero Suppression Logic.

The Decimal-to-BCD Encoder

This type of encoder has ten inputs—one for each decimal digit and four outputs corresponding to the BCD code, as shown in Figure 7-48.

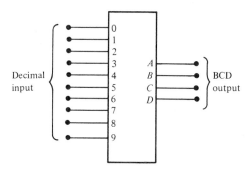

Figure 7–48.   Typical Decimal-to-BCD Encoder Block Diagram Symbol.

The BCD (8421) code is listed in Table 7–10, and from this we can determine the relationship between each BCD bit and the decimal digits. For instance, the most significant bit of the BCD code, A, is a 1 for decimal digit 8 or 9. The expression for bit A in terms of the decimal digits can therefore be written

$$A = 8 + 9$$

Bit B is a 1 for decimal digits 4, 5, 6, or 7, and can be expressed as follows:

$$B = 4 + 5 + 6 + 7$$

C is a 1 for decimal digits 2, 3, 6, or 7, and can be expressed as

$$C = 2 + 3 + 6 + 7$$

Finally, D is a 1 for digits 1, 3, 5, 7, or 9. The expression for D is

$$D = 1 + 3 + 5 + 7 + 9$$

TABLE 7–10.   BCD (8421)
Code.

Decimal Digit	BCD Code			
	A	B	C	D
0	0	0	0	0
1	0	0	0	1
2	0	0	1	0
3	0	0	1	1
4	0	1	0	0
5	0	1	0	1
6	0	1	1	0
7	0	1	1	1
8	1	0	0	0
9	1	0	0	1

Now we can implement the logic circuitry required for encoding each decimal digit to a BCD code by using the logic expressions just developed. It is simply a matter of ORing the appropriate decimal digit input lines to form each BCD output. The basic encoder logic resulting from these expressions is shown in Figure 7–49. This type of encoder is sometimes called a 10-line-to-4-line encoder.

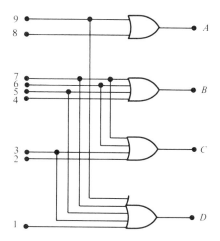

Figure 7–49.   Basic Logic for a Decimal-to-BCD Encoder.

The basic operation is as follows: when a HIGH appears on one of the decimal digit input lines, the appropriate levels occur on the four BCD output lines. For instance, if input line 9 is HIGH (assuming all other input lines are LOW), this will produce a HIGH on output A and output D and LOWs on outputs B and C, which is the BCD code (1001) for decimal 9.

Decimal-to-BCD Priority Encoder

This type of encoder performs the same basic encoding function as that previously discussed. It also offers additional flexibility in that it can be used in applications requiring priority detection. The *priority* function means that the encoder will produce a BCD output corresponding to the *highest order decimal digit* appearing on the inputs, and ignore all others. For instance, if the 6 and the 3 inputs are both HIGH, the BCD output is 0110 (which represents decimal 6).

Now let us look at the requirements for the priority logic. The purpose of this logic circuitry is to prevent a lower order digit input from disrupting the encoding of a higher order digit; this is accomplished by using inhibit gates. We will start by examining each BCD output (beginning with output D). Referring to Figure 7–49, note that D is HIGH when 1, 3, 5, 7, or 9 is HIGH. Digit input 1 is allowed to activate the D output if no higher order digits *other than those that also activate D* are HIGH. This can be stated as follows:

1. D is HIGH if 1 is HIGH and 2, 4, 6, and 8 are LOW. Similar statements can be made for the other digit inputs to the D OR gate.
2. D is HIGH if 3 is HIGH and 4, 6, and 8 are LOW.

3. D is HIGH if 5 is HIGH and 6 and 8 are LOW.

4. D is HIGH if 7 is HIGH and 8 is LOW.

5. D is HIGH if 9 is HIGH.

These five statements describe the priority of encoding for the BCD bit D. The D output is HIGH if any of the conditions listed occur; that is, D is true if statement 1, statement 2, statement 3, statement 4, or statement 5 is true. This can be expressed in the form of the following logic equation:

$$D = 1 \cdot \overline{2} \cdot \overline{4} \cdot \overline{6} \cdot \overline{8} + 3 \cdot \overline{4} \cdot \overline{6} \cdot \overline{8} + 5 \cdot \overline{6} \cdot \overline{8} + 7 \cdot \overline{8} + 9$$

From this expression the logic circuitry required for the D output with priority inhibits can be readily implemented, as shown in Figure 7–50.

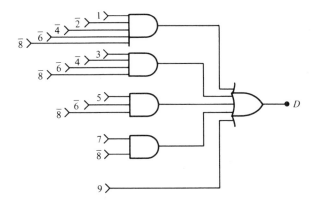

Figure 7–50. Logic for the D-Bit Output of a Decimal-to-BCD Priority Encoder.

The same reasoning process can be applied to output C, and the following logical statements can be made:

1. C is HIGH if 2 is HIGH and 4, 5, 8, and 9 are LOW.

2. C is HIGH if 3 is HIGH and 4, 5, 8, and 9 are LOW.

3. C is HIGH if 6 is HIGH and 8 and 9 are LOW.

4. C is HIGH if 7 is HIGH and 8 and 9 are LOW.

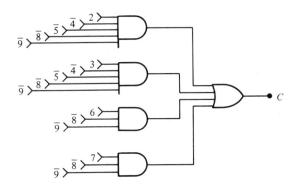

Figure 7–51. Logic for the C-Bit Output of a Decimal-to-BCD Priority Encoder.

These statements are summarized in the following equation and the logic implementation is shown in Figure 7–51:

$$C = 2 \cdot \overline{4} \cdot \overline{5} \cdot \overline{8} \cdot \overline{9} + 3 \cdot \overline{4} \cdot \overline{5} \cdot \overline{8} \cdot \overline{9} + 6 \cdot \overline{8} \cdot \overline{9} + 7 \cdot \overline{8} \cdot \overline{9}$$

Output B can be described as follows:

1. B is HIGH if 4 is HIGH and 8 and 9 are LOW.
2. B is HIGH if 5 is HIGH and 8 and 9 are LOW.
3. B is HIGH if 6 is HIGH and 8 and 9 are LOW.
4. B is HIGH if 7 is HIGH and 8 and 9 are LOW.

In equation form, output B is

$$B = 4 \cdot \overline{8} \cdot \overline{9} + 5 \cdot \overline{8} \cdot \overline{9} + 6 \cdot \overline{8} \cdot \overline{9} + 7 \cdot \overline{8} \cdot \overline{9}$$

The logic circuitry for the B output appears in Figure 7–52.

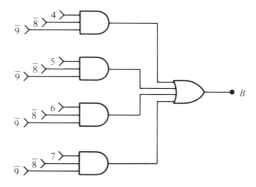

Figure 7–52. Logic for the B-Bit Output of a Decimal-to-BCD Priority Encoder.

Finally, for the A output,

A is HIGH if 8 is HIGH or if 9 is HIGH.

This statement appears in equation form as follows:

$$A = 8 + 9$$

The logic for this output is in Figure 7–53. No inhibits are required. We now have developed the basic logic for the Decimal-to-BCD Priority encoder. All of the complements of the input digit variables are realized by inverting the inputs. A complete logic diagram is shown in Figure 7-54.

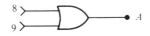

Figure 7–53. Logic for the A-Bit Output of a Decimal-to-BCD Priority Encoder.

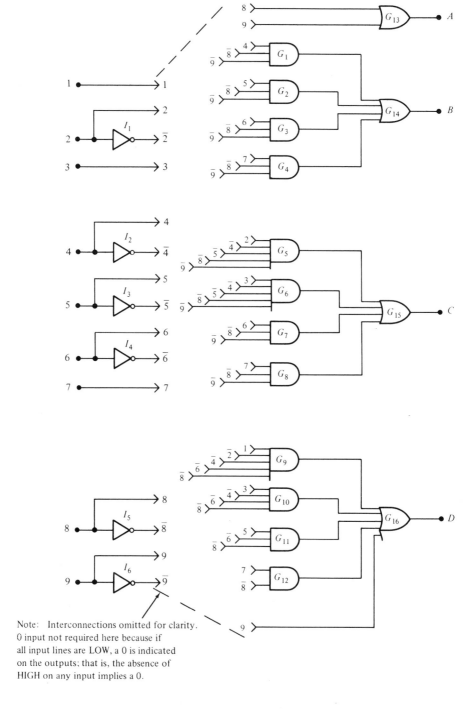

Note: Interconnections omitted for clarity. 0 input not required here because if all input lines are LOW, a 0 is indicated on the outputs; that is, the absence of HIGH on any input implies a 0.

Figure 7–54. Full Logic Diagram of a Decimal-to-BCD Priority Encoder.

Octal-to-Binary Priority Encoder

This commonly used type of encoder is often referred to as an 8-line to 3-line encoder. The same logic can be used in it as is used in the Decimal-to-BCD encoder, except that inputs 8 and 9 are omitted because there are only eight octal digits (0 through 7). Also, only three binary bits are required to represent the eight octal digits, so this type of encoder has only three output lines. A logic diagram for this encoder appears in Figure 7–55. Notice that the logic is similar to that for the first

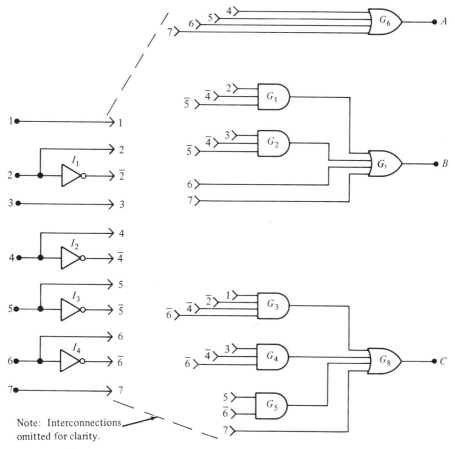

Figure 7–55. Logic Diagram for an Octal-to-Binary Encoder.

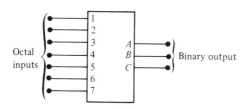

Figure 7–56. Typical Block Diagram Symbol for an Octal-to-Binary Encoder.

eight inputs and the three lower order outputs of the Decimal-to-BCD encoder, except that the decimal 8 or 9 inhibits are omitted. A block symbol is shown in Figure 7-56.

7–9 Multiplexers

A multiplexer is a device that allows digital information from several sources to be routed onto a common line for transmission over that line to a common destination. The basic multiplexer, then, has several input lines and a single output line. It also has control or selection inputs that permit digital data on any one of the inputs to be switched to the output line. A block diagram symbol for a four-input multiplexer is shown in Figure 7-57. Notice that there are two selection lines because with two selection bits, each of the four data input lines can be selected.

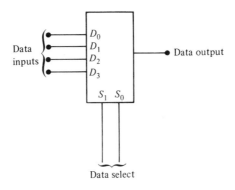

Figure 7–57. Four-Line Multiplexer Block Diagram.

In reference to Figure 7–57, a two-bit binary code on the data select inputs will allow the data on the corresponding data input to pass through to the data output. If a binary 0 ($S_1 = 0$ and $S_0 = 0$) is applied to the data select lines, the data on input D_0 appear on the data output line. If a binary 1 ($S_1 = 0$ and $S_0 = 1$) is applied to the data select lines, the data on input D_1 appear on the data output. If a binary 2 ($S_1 = 1$ and $S_0 = 0$) is applied, the data on D_2 appear on the output. If a binary 3 ($S_1 = 1$ and $S_0 = 1$) is applied, the data on D_3 are switched to the output line. A summary of this operation is given in Table 7–11.

TABLE 7–11.   Data Selection
for a Four-Input Multiplexer.

Data Select Inputs		Input Selected
S_1	S_0	
0	0	D_0
0	1	D_1
1	0	D_2
1	1	D_3

Now let us look at the logic circuitry required to perform this multiplexing operation. The data output is equal to the state of the *selected* data input. We should, therefore, be able to derive a logical expression for the output in terms of the data input and the select inputs. This can be done as follows:

The data output is equal to the data input D_0 if and only if $S_1 = 0$ and $S_0 = 0$:

$$\text{Data output} = D_0 \overline{S_1}\, \overline{S_0}$$

The data output is equal to D_1 if and only if $S_1 = 0$ and $S_0 = 1$:

$$\text{Data output} = D_1 \overline{S_1}\, S_0$$

The data output is equal to D_2 if and only if $S_1 = 1$ and $S_0 = 0$:

$$\text{Data output} = D_2 S_1 \overline{S_0}$$

The data output is equal to D_3 if and only if $S_1 = 1$ and $S_0 = 1$:

$$\text{Data output} = D_3 S_1 S_0$$

If these expressions are ORed, the total expression for the data output is

$$\text{Data output} = D_0 \overline{S_1}\,\overline{S_0} + D_1 \overline{S_1} S_0 + D_2 S_1 \overline{S_0} + D_3 S_1 S_0$$

The implementation of this equation requires four three-input AND gates, a four-input OR gate, and two inverters to generate the complements of S_1 and S_0, as shown in Figure 7–58.

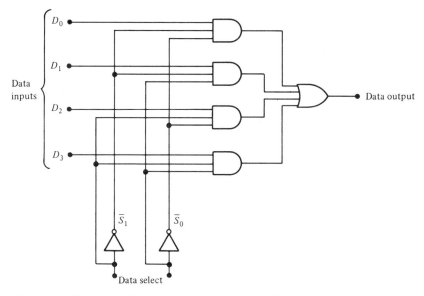

Figure 7–58.   Logic Diagram for a Four-Input Multiplexer.

Because data can be selected from any of the input lines, this circuit is sometimes referred to as a *data selector*. It can be implemented with any number of

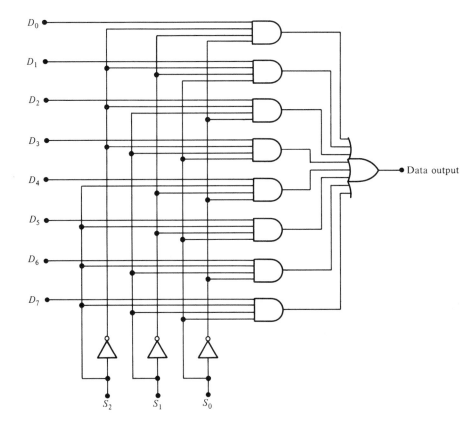

Figure 7–59.   Logic Diagram for an Eight-Input Multiplexer.

input lines. For example, an eight-input multiplexer is shown in Figure 7–59. Notice that three data select inputs are used because three bits are required to select each of eight inputs.

7–10   Code Converters

In this section we will examine some combinational circuits that convert from one code to another.

Binary-to-Gray Code Converter

The Gray code was discussed in Chapter 3. Table 3–3, showing the four-bit binary and corresponding Gray code, is repeated for convenience in Table 7–12. Close examination of this table reveals that the most significant bits in both the binary and the Gray codes are the same; and the sum (ignoring carry) of any two adjacent binary bits produces the Gray code bit located in the position corresponding to the lower ordered binary bit added.

TABLE 7–12.   Four-Bit
Gray and Binary Codes.

Decimal	Gray	Binary
0	0000	0000
1	0001	0001
2	0011	0010
3	0010	0011
4	0110	0100
5	0111	0101
6	0101	0110
7	0100	0111
8	1100	1000
9	1101	1001
10	1111	1010
11	1110	1011
12	1010	1100
13	1011	1101
14	1001	1110
15	1000	1111

Example 7–10

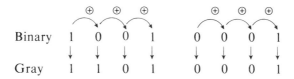

In order to convert from binary to Gray, the summing operations can be performed by exclusive-OR gates connected as shown in Figure 7–60.

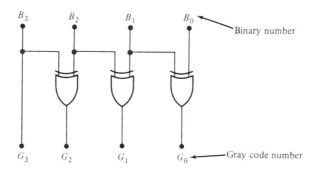

Figure 7–60.   Four-Bit Binary-to-Gray Converter.

Example 7–11   Convert the binary codes 0101, 00111, and 101011 to Gray code, using exclusive-OR converters.

Solutions:

Binary 0101 is 0111 Gray, 00111 is 00100 Gray, and 101011 is 111110 Gray. See Figure 7–61.

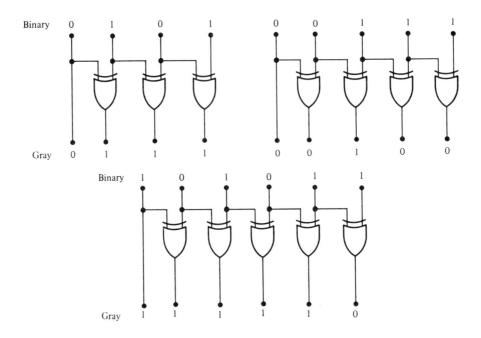

Figure 7-61.

Gray-to-Binary Code Converter

Conversion of a Gray code to binary involves the following procedure: each binary bit generated is added (no carry) to the next lower order Gray code bit. The most significant bits of both codes are always identical. Figure 7-62 shows a four-bit Gray-to-binary converter using exclusive-OR gates.

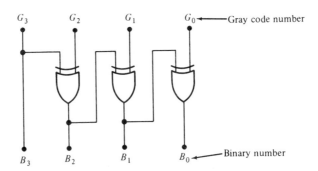

Figure 7-62.   Four-Bit Gray-to-Binary Converter.

Example 7-12   Convert the Gray code words 1011, 11000, and 1001011 to binary, using exclusive-OR converters.

Solutions:

Gray code 1011 is 1101 binary, 11000 is 10000, and 1001011 is 1110010. See Figure 7-63.

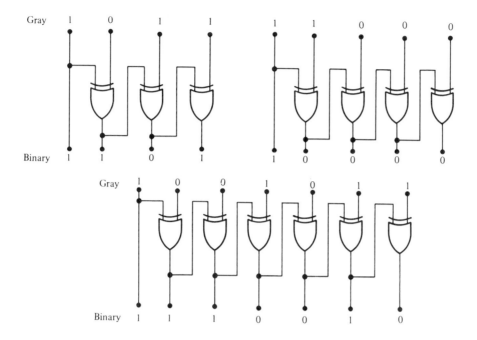

Figure 7–63.

BCD-to-Excess-3 Converter

Recall from Chapter 3 that the Excess-3 code for a decimal digit is formed by adding three to the BCD (8421) code for that digit. For example, given the BCD (8421) code word 1001, conversion to Excess-3 is as follows:

$$
\begin{array}{ll}
1001 & \text{8421 code for decimal 9} \\
+0011 & \text{add 3} \\ \hline
1100 & \text{Excess-3 code for decimal 9}
\end{array}
$$

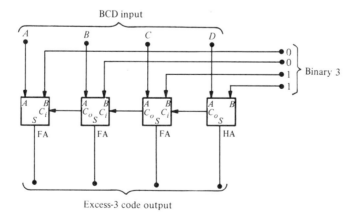

Figure 7–64.   BCD-to-Excess-3 Converter.

This conversion process is readily implemented by using four adders wired such that a binary 3 is added to the BCD code applied to the inputs. A logic diagram of this operation is shown in Figure 7–64.

9's Complement Generation

The conversion of a BCD number to its 9's complement is a form of code conversion that is useful in BCD arithmetic. The 9's complement of a decimal number was discussed in Chapter 2. Table 7–13 shows the binary representation of the 9's complement for each BCD digit.

TABLE 7–13.   Table of 9's Complements.

Decimal	9's Complement	BCD $ABCD$	9's Complement (Binary) $ABCD$
0	9	0000	1001
1	8	0001	1000
2	7	0010	0111
3	6	0011	0110
4	5	0100	0101
5	4	0101	0100
6	3	0110	0011
7	2	0111	0010
8	1	1000	0001
9	0	1001	0000

Notice in Table 7–13 that each D bit of the 9's complement code is the complement of the corresponding D bit of the BCD code. An inverter can be used to make this conversion.

Also notice that the C bits of both codes are identical, and that each B bit in the 9's complement code is the sum (exclusive-OR) of the B and C bits in the corresponding BCD code. Lastly, the A bit in the 9's complement code can be generated by NORing the A, B, and C bits of the corresponding BCD code. This implementation is illustrated in Figure 7–65.

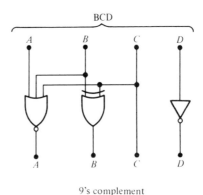

9's complement

Figure 7–65.   BCD-to-9's-Complement Converter.

BCD-to-Binary Converter

One method of BCD to binary code conversion involves the use of adder circuits. The basic conversion process is as follows:

1. The *value* of *each* bit in the BCD number is represented by a binary number.
2. All of the binary representations of bits that are 1s in the BCD number are added.
3. The result of this addition is the binary equivalent of the BCD number.

A more concise statement of this operation is that the binary numbers for the weights of the BCD bits are summed to produce the total binary number.

We will examine an eight-bit BCD code (one that represents a two-digit decimal number) in order to understand the relationship between BCD and binary. For instance, we already know the decimal number 87 can be expressed in BCD as

$$\underbrace{1000}_{8}\underbrace{0111}_{7}$$

The left-most four-bit group represents 80 and the right-most four-bit group represents 7; that is, the left-most group has a weight of 10 and the right-most has a weight of 1. Within each group, the binary weight of each bit is as follows:

	Tens Digit				Units Digit			
Weight:	80	40	20	10	8	4	2	1
Bit Designation:	A_1	B_1	C_1	D_1	A_0	B_0	C_0	D_0

The binary equivalent of each BCD bit is a binary number representing the *weight* of that bit within the total BCD number. This representation is given in Table 7-14.

TABLE 7-14.   Binary Representations of BCD Bit Weights.

BCD Bit	Weight	Binary Representation						
		64	32	16	8	4	2	1
D_0	1	0	0	0	0	0	0	1
C_0	2	0	0	0	0	0	1	0
B_0	4	0	0	0	0	1	0	0
A_0	8	0	0	0	1	0	0	0
D_1	10	0	0	0	1	0	1	0
C_1	20	0	0	1	0	1	0	0
B_1	40	0	1	0	1	0	0	0
A_1	80	1	0	1	0	0	0	0

If the binary representations for the weight of each 1 in the BCD number are added, the result is the binary number corresponding to the BCD number. An example will illustrate this.

Example 7–13 Convert the BCD numbers 00100111 (decimal 27) and 10011000 (decimal 98) to binary.

Solutions:
Write the binary for the weights of all 1s appearing in the numbers, and then add them together.

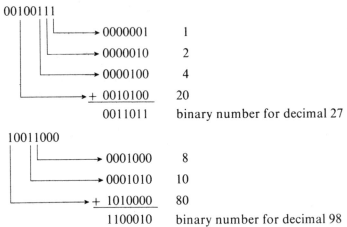

00100111	0000001	1
	0000010	2
	0000100	4
	+ 0010100	20
	0011011	binary number for decimal 27

10011000	0001000	8
	0001010	10
	+ 1010000	80
	1100010	binary number for decimal 98

With this basic procedure in mind, let us determine how the process can be implemented with logic circuits. Once the binary representation for each 1 in the BCD number is determined, adder circuits can be used to add the 1s in each column of the binary representation. The 1s occur in a given column only when the corresponding BCD bit is a 1. The occurrence of a BCD 1 can therefore be used to generate the proper binary 1 in the appropriate column of the adder structure. To handle a two-decimal digit BCD code, eight input lines and seven binary out-

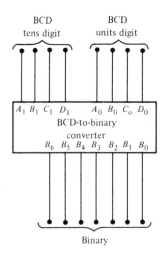

Figure 7–66. Typical Block Diagram Symbol for a BCD-to-Binary Converter (Two-Digit).

puts are required (it takes 7 binary bits to represent numbers up through 99). A block diagram is shown in Figure 7–66.

Referring to Table 7–14, notice that the "1" column of the binary representation has only a single 1 and no possibility of an input carry, so that a straight connection from the D_0 bit of the BCD input to the least significant binary output is sufficient. In the "2" column of the binary representation, the possible occurrence of the two 1s can be accommodated by adding the C_0 bit and the D_1 bit of the BCD number. In the 4 column of the binary representation, the possible occurrence of the two 1s is handled by adding the B_0 bit and the C_1 bit of the BCD number. In the "8" column of the binary representation, the possibility of the three 1s is handled by adding the A_0, D_1, and B_1 bits of the BCD number. In the "16" column, the C_1 and the A_1 bits are added. In the "32" column, only a single 1 is possible, so the B_1 bit is added to the carry from the "16" column. In the "64" column, only a single 1 can occur, so the A_1 bit is added only to the carry from the "32" column. A method of implementing these requirements is shown in Figure 7–67.

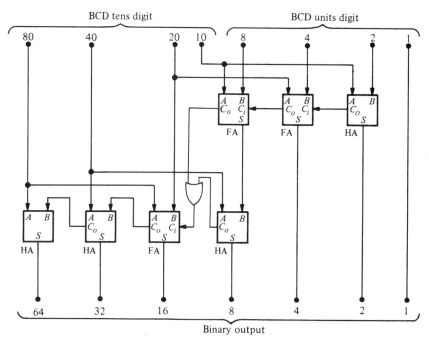

Figure 7–67.   Two-Digit BCD-to-Binary Converter.

7–11   Parity Detection Logic

Recall from Chapter 3 that the *parity* of a digital code indicates if the total number of 1s in the code is even or odd and provides a means of single-bit error detection. A parity bit is normally attached to the beginning or end of a digital code word. For an *even parity* code, the total number of 1s, *including* the parity bit, is always

(a) Summing of Two Bits

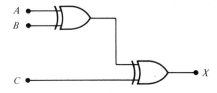

(b) Summing of Three Bits

Figure 7–68.

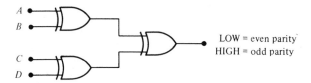

(a) Four-Bit Parity Checker

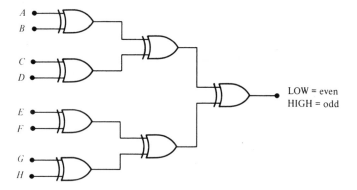

(b) Eight-Bit Parity Checker

Figure 7–69.

even. For an *odd parity* code, the total number of 1s, *including* the parity bit, is always odd.

In order to check for the proper parity in a given code word, a very basic principle can be used. *The sum* (disregarding carries) *of an even number of 1s is always 0, and the sum of an odd number of 1s is always 1.* Therefore, in order to determine if a given code word is even or odd parity, all of the bits in that code word are summed. As we know, the sum of two bits can be generated by an exclusive-OR gate, as shown in Figure 7–68(a); the sum of three bits can be formed by two exclusive-OR gates connected as shown in Figure 7–68(b); and so on.

The parity detection logic for a four-bit code (including parity) is shown in Figure 7–69(a), and that for an eight-bit code is in Figure 7–69(b).

Example 7–14   Verify that the parity detector circuit of Figure 7–69(a) produces the proper indication for the code words 1010 and 1101.

Solutions:

First we will take the first code word listed and follow the logic states of each gate through to the output, as shown in Figure 7–70. A 0 output results which indicates even parity. This is the proper output since there are two 1s in the code word. Next, the second code word is applied to the inputs, as shown in Figure 7–71. A 1 on the output indicates odd parity, which is correct.

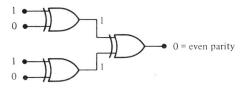

Figure 7–70.

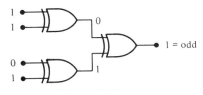

Figure 7–71.

7–12   A Multiplier

Arithmetic multiplication can be achieved with combinational logic, although there are other ways. To understand how this function can be implemented, let us examine the multiplication process using a four-bit number multiplied by a two-bit number. The general format of this multiplication is as follows:

$$A_3\ A_2\ A_1\ A_0 \qquad \text{multiplicand}$$
$$\underline{B_1\ B_0} \qquad\qquad \text{multiplier}$$
$$P_3\ P_2\ P_1\ P_0 \qquad \text{first partial product}$$
$$\underline{+ P_3\ P_2\ P_1\ P_0} \qquad \text{second partial product}$$
$$P_5\ P_4\ P_3\ P_2\ P_1\ P_0 \qquad \text{final product}$$

A specific example will illustrate the process:

$$
\begin{array}{r}
1111 \\
\times\quad 10 \\
\hline
0000 \\
+1111 \\
\hline
11110
\end{array}
$$

The first partial product is formed by multiplying each of the multiplicand bits by 0 (B_0); the second partial product is formed by multiplying each of the multiplicand bits by 1 (B_1); and the final product is found by adding the two partial products.

Recall that multiplication of two binary bits is accomplished by ANDing the two bits. Therefore, each of the partial products is generated by ANDing the multiplier bit with each multiplicand bit. The first partial product can be realized with logic gates, as illustrated in Figure 7–72(a), and the second partial product implementation is as shown in Figure 7–72(b).

Now, each bit of the second partial product is added to the *next higher order* bit of the first partial product. This effectively creates a one-bit shift to the left of the second partial product. The total implementation of a four-bit-by-two-bit multiplier is shown in Figure 7–73.

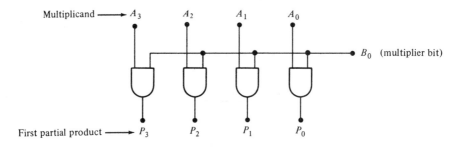

(a) First Partial Product Logic

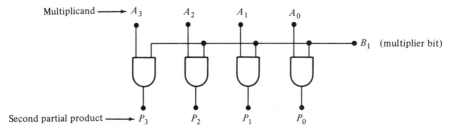

(b) Second Partial Product Logic

Figure 7–72.

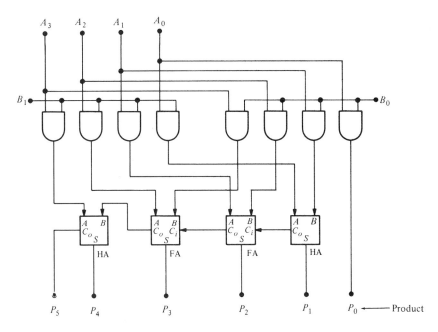

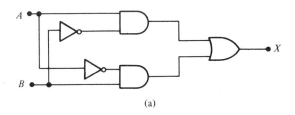

Figure 7–73.   A Four-Bit-by-Two-Bit Multiplier.

PROBLEMS

7–1   Show that the two circuits in Figure 7–74 are equivalent, and identify their logic function.

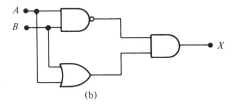

Figure 7–74.

7-2   For the full-adder of Figure 7–6, determine the logic state (1 or 0) at each point in the circuit for the following inputs:

  (a) $A = 1, B = 1, C_i = 1$ (b) $A = 0, B = 1, C_i = 1$
  (c) $A = 0, B = 1, C_i = 0$

7-3   Repeat Problem 7–2 for the following inputs:

  (a) $A = 0, B = 0, C_i = 0$ (b) $A = 1, B = 0, C_i = 0$
  (c) $A = 1, B = 0, C_i = 1$

7-4   Simplify, if possible, the full-adder circuit of Example 7–1, using Karnaugh map simplification.

7-5   For the parallel adder in Figure 7–75, determine the sum by analysis of the logical operation of the circuit. Verify your result by long hand addition of the two input numbers.

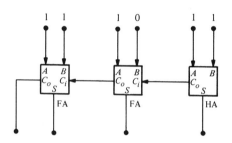

Figure 7–75.

7-6   Repeat Problem 7–5 for the circuit and input conditions in Figure 7–76.

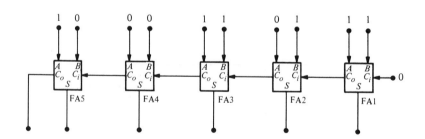

Figure 7–76.

7-7   In Problem 7–6, determine the carry out of:

  (a) FA1   (b) FA2   (c) FA3   (d) FA4   (e) FA5

7-8   For the full-subtractor of Figure 7–17(b), show the logic states at each point in the circuit for all possible input conditions.

7-9   If the parallel subtractor in Figure 7–77 is used to subtract 1011 from 1100, determine the difference based on the logical operation of the device.

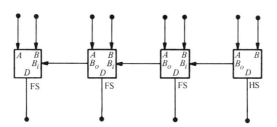

Figure 7–77.

7–10 Repeat Problem 7–9 for the subtraction of 0101 from 1000.

7–11 The waveforms in Figure 7–78 are applied to the comparator as shown. Determine the output $(A = B)$ waveform. Assume a HIGH on the output indicates equality of the inputs.

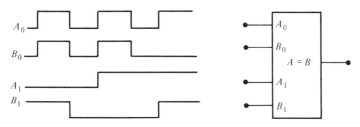

Figure 7–78.

7–12 For the four-bit comparator in Figure 7–79, plot each output waveform for the inputs shown. The outputs are active HIGH.

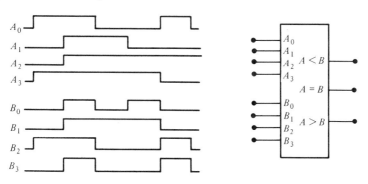

Figure 7–79.

7–13 For each following set of binary numbers, determine the logic states at each point in the comparator circuit of Figure 7–26, and verify that the output indications are correct:

(a) $A_3 \, A_2 \, A_1 \, A_0 \equiv 1100$ (b) $A_3 \, A_2 \, A_1 \, A_0 \equiv 1000$
$B_3 \, B_2 \, B_1 \, B_0 \equiv 1001$ $B_3 \, B_2 \, B_1 \, B_0 \equiv 1011$

(c) $A_3 \, A_2 \, A_1 \, A_0 \equiv 0100$
$B_3 \, B_2 \, B_1 \, B_0 \equiv 0100$

7-14   When a HIGH is on the output of each of the decoding gates in Figure 7–80, what is the binary code word appearing on the inputs? A is the MSB.

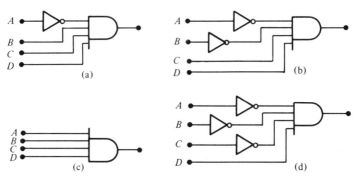

Figure 7–80.

7-15   Show the decoding logic for each of the following code words if an active HIGH indication is required:

(a)   1101   (b)   1000   (c)   11011   (d)   11100
(e)   101010   (f)   111110   (g)   000101   (h)   1110110

7-16   Repeat Problem 7–15 given that an active LOW output is required.

7-17   We wish to detect the presence of the code words 1010, 1100, 0001, and 1011. An active HIGH output is required to indicate their presence. Develop the complete decoding logic with a single output that will tell us when any one of these codes is on the inputs.

7-18   If the input waveforms are applied to the decoding logic as indicated in Figure 7–81, sketch the output waveform in proper relation to the inputs.

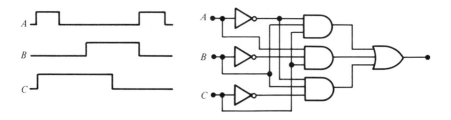

Figure 7–81.

7-19   BCD numbers are applied sequentially to the BCD-to-Decimal decoder in Figure 7–82. Draw the output waveforms, showing each in the proper relationship to the others and to the inputs.

7-20   A seven-segment decoder drives the display in Figure 7–83. If the waveforms are applied as indicated, determine the sequence of digits that appears on the display.

7-21   In the Decimal-to-BCD encoder of Figure 7–49, assume the 9 input and the 3 input are both HIGH. What is the output code? Is it a valid BCD (8421) code word?

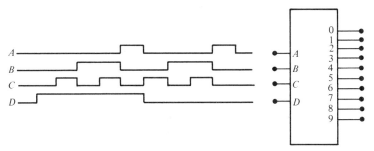

Figure 7–82.

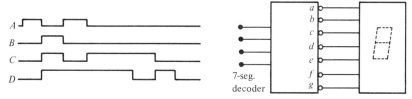

Figure 7–83.

7-22 For the priority encoder of Figure 7–54, analyze the operation and verify that the proper output occurs when a HIGH is applied to both the 2 and the 7 inputs.

7-23 Repeat Problem 7–22 for inputs 4, 6, and 9 being HIGH at the same time.

7-24 Explain why an input line for the decimal digit 0 is not required in Figure 7–54.

7-25 Explain how the octal-to-binary encoder of Figure 7–55 produces the proper output for a HIGH on the 3 input.

7-26 For the four-input multiplexer in Figure 7–84, determine the output state. Refer to logic diagram in Figure 7–58 if necessary.

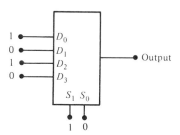

Figure 7–84.

7-27 For the multiplexer of Problem 7–26, determine the output for the following input states: $D_0 = 0, D_1 = 1, D_2 = 1, D_3 = 0, S_0 = 1, S_1 = 0$.

7-28 If the data select inputs to the multiplexer of Problem 7–26 are sequenced as shown by the waveforms in Figure 7–85, determine the output waveform with the data inputs as shown in Problem 7–26.

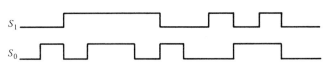

Figure 7–85.

7-29 Show the logic required to convert a ten-bit binary number to Gray code, and use that logic to convert the following binary code words to Gray code:

    (a) 1010101010   (b) 1111100000
    (c) 0000001110   (d) 1111111111

7-30 Show the logic required to convert a ten-bit Gray code to binary, and use that logic to convert the following Gray code words to binary:

    (a) 1010000000   (b) 0011001100
    (c) 1111000111   (d) 0000000001

7-31 Convert the following decimal numbers first to BCD and then, using the BCD-to-Binary converter of Figure 7–67, convert the BCD to binary. Verify the result in each case.

    (a) 2   (b) 8   (c) 13   (d) 26   (e) 33
    (f) 45   (g) 61   (h) 70   (i) 84   (j) 99

7-32 The waveforms in Figure 7–86 are applied to the four-bit parity checker. Determine the output waveform in proper relation to the inputs. How many times does even parity occur?

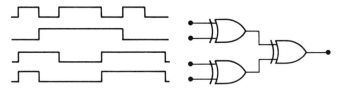

Figure 7–86.

7-33 Perform the following multiplications using the multiplier circuit shown in Figure 7–73 based on its logical operation. Verify each result by long hand multiplication.

    (a) 01 × 0010   (b) 10 × 0101   (c) 01 × 1011
    (d) 11 × 1100   (e) 11 × 1111   (f) 1000 × 10

CHAPTER 8

Flip-Flops

The flip-flop belongs to a category of digital circuits known as multivibrators. There are three basic types of multivibrators in common use: the bistable multivibrator, the monostable multivibrator, and the astable multivibrator. The bistable multivibrator is commonly called a flip-flop, and will be discussed in this chapter. The other two types are covered in Chapter 13.

The flip-flop has two stable states. It is capable of being in either a HIGH state (logic 1) or a LOW state (logic 0) *indefinitely*. Since it can reside in either state for an extended period of time, it is useful as a *storage* or *memory* device. The flip-flop finds wide application in digital systems as a "building block" for counters, registers, memories, control logic, and other functions. It is the most widely applied type of multivibrator and therefore, in general, ranks highest in importance. In this chapter we will study several important types of flip-flops.

The emphasis will be on the functional operation and characteristics of the flip-flop as a logic device rather than on the detailed circuit configuration. The advent of integrated circuits and the large number of ways in which a given type of circuit can be implemented have made the internal circuitry less important *from an application point of view* than the characteristics associated with the inputs, outputs, and functional or logical operation. For application purposes, it is generally more important to know what logic function the circuit performs, and what the limitations are on this performance, than it is to know the details of the circuit design.

8–1 The Basic RS Flip-Flop

In this section we examine the operation of a basic type of RS (reset-set) flip-flop and its function as a logic element. The basic flip-flop, as illustrated in Figure 8–1, is formed by two cross-coupled NOR gates. It may also be implemented in the same manner with other types of logic gates such as the NAND, as is shown in Figure 8–2.

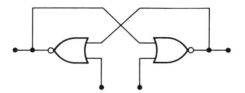

Figure 8–1.   NOR Gates Connected as a Basic Flip-Flop.

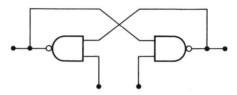

Figure 8–2.   NAND Gates Connected as a Basic Flip-Flop.

Figure 8–3 is the logic diagram of Figure 8–1 arranged in a more convenient manner (inputs on the left and outputs on the right), as might normally be seen in logic schematics.

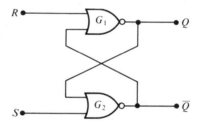

Figure 8–3.   Logic Diagram for a Basic RS Flip-Flop Implemented with NOR Gates.

The RS flip-flop has two outputs, Q and $\overline{Q}$, and two controlling inputs, R (RESET) and S (SET). The operation of the circuit in Figure 8–3 can be understood by initially assuming that the Q output, R input, and S input are all LOW. Since the Q output is connected back into one input of gate G_2, and the S input to gate G_2 is LOW, the output of the gate is HIGH. This HIGH is coupled back into one input of gate G_1, insuring that its output is LOW. When the Q output is LOW, the flip-flop is said to be in the RESET state. It will remain indefinitely in this state until a HIGH is applied to the S input. With a HIGH on the S input, the output of gate G_2 becomes LOW. This LOW on the $\overline{Q}$ output is coupled back into one input of gate G_1 and, since the R input is LOW, the output of gate G_1 becomes HIGH. The HIGH on the Q output is coupled back into one input on gate G_2, insuring that the $\overline{Q}$ output remains LOW even when the HIGH on the S input is removed. When the Q output is HIGH the flip-flop is said to be in the SET state. Now the flip-flop will remain indefinitely in the SET state until a HIGH is applied to the R input, causing the flip-flop to RESET. Notice that the outputs are always complements of each other; that is, when Q is HIGH, $\overline{Q}$ is LOW, and when Q is LOW, $\overline{Q}$ is HIGH.

A condition commonly referred to as indeterminate, unallowed, or illegal occurs when a HIGH is applied to both the R and the S inputs simultaneously. A HIGH on S forces the $\overline{Q}$ output LOW, and a HIGH on R forces the Q output LOW also. This in itself is an unallowed condition for a flip-flop, since Q and $\overline{Q}$ cannot be in the same state at the same time without violating the definition of flip-flop operation. Now, if the HIGH levels on R and S are simultaneously removed, both of the outputs will attempt to go HIGH. Since there is always some minute difference in the propagation delay of the gates, one of the gates will dominate in its rise toward the HIGH state. This, in turn, forces the output of the slower gate to remain LOW. An unpredictable mode of operation exists because of the race condition, and we cannot determine in which state the flip-flop will finally come to rest.

The operation of the basic RS flip-flop can be summarized as follows:

1. If Q is HIGH, and a HIGH is applied to the R input with the S input LOW, Q will go LOW. This is the RESET condition.
2. If Q is LOW, and a HIGH is applied to the R input with the S input LOW, Q will remain LOW. The flip-flop remains RESET.
3. If Q is LOW, and a HIGH is applied to the S input with the R input LOW, Q will go HIGH. This is the SET condition.
4. If Q is HIGH, and a HIGH is applied to the S input with the R input LOW, Q will remain HIGH. The flip-flop remains SET.
5. If both S and R are LOW, the flip-flop will not change state. Q remains in its present state. This is a NO CHANGE condition.
6. If HIGHs are applied to both S and R simultaneously, it cannot be predicted what the flip-flop will do when the HIGHs are removed.

Figure 8–4 illustrates the flip-flop operation for each of the four possible combinations of R and S. Table 8–1 is a summary of the logical operation in truth table form.

In many cases, a simplified symbol is used to represent a flip-flop. Figure 8–5 shows a typical logic symbol that represents the basic RS flip-flop of Figure 8–3.

The following example will illustrate how a basic RS flip-flop responds to conditions on its inputs. We will pulse HIGH levels on each input in a certain sequence and observe the resulting Q output waveform. The indeterminate condi-

TABLE 8–1.   Truth Table for Basic RS Flip-Flop.

Inputs		Outputs		
S	R	Q	$\overline{Q}$	Comments
LOW	LOW	NC	NC	No change. Flip-flop remains in present state.
LOW	HIGH	LOW	HIGH	Flip-flop RESETS.
HIGH	LOW	HIGH	LOW	Flip-flop SETS.
HIGH	HIGH	?	?	Next state is not predictable.

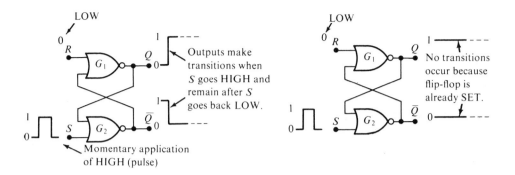

Flip-flop starts out RESET.

Flip-flop starts out SET.

(a) Two possibilities for the SET operation

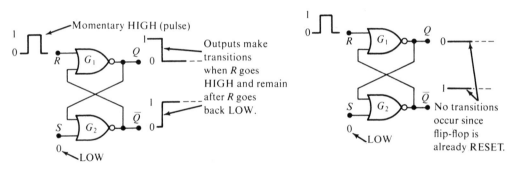

Flip-flop starts out SET.

Flip-flop starts out RESET.

(b) Two possibilities for the RESET operation

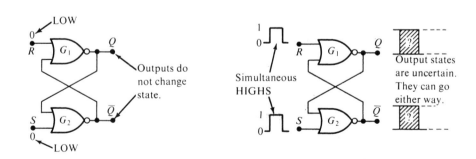

(c) NO CHANGE condition

(d) Indeterminate condition

Figure 8-4. The Four Modes of Basic RS Flip-Flop Operation.

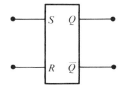

Figure 8–5.   Logic Symbol for the Basic RS Flip-Flop.

tion is avoided because it results in an undesirable mode of operation and is a major drawback of the RS type of flip-flop.

Example 8–1   If the waveforms in Figure 8–6(a) are applied to the inputs of the flip-flop of Figure 8–5, determine the waveform that would be observed on the Q output. Assume that Q is initially LOW.

Solution:
See Figure 8–6(b).

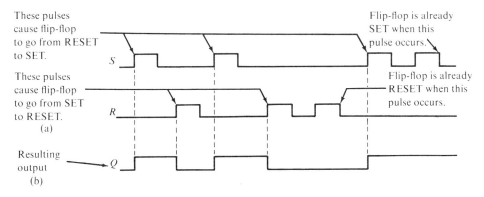

Figure 8–6.

8–2   The Edge-Triggered RS Flip-Flop

The edge-triggered RS flip-flop is a type of *clocked* flip-flop. It differs from the basic RS flip-flop previously discussed in that an additional input is required for its operation. This third input is commonly called the clock or trigger. Figure 8–7 shows a typical logic implementation using NAND gates. The NAND gates in the flip-flop portion are shown as negative OR gates because a LOW on any input activates the flip-flop.

Notice that the circuit in Figure 8–7 is partitioned into two sections, one labeled "steering gates" and the other "basic flip-flop." The steering gates direct or "steer" the clock pulse (CP) to either the input to gate G_3 or the input to gate G_4, depending on the state of the S and R inputs. In order to understand the operation of this flip-flop, let us begin with the assumption that it is in the RESET state ($Q = 0$) and the S, R, and clock inputs are all LOW. For this condition, the outputs of gate G_1 and gate G_2 are both HIGH. The LOW on the Q output is

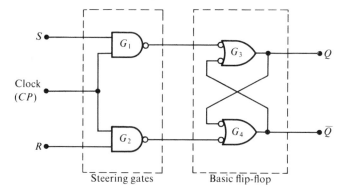

Figure 8-7.   Logic Diagram for an Edge-Triggered RS Flip-Flop.

coupled back into one input of gate G_4, making the $\overline{Q}$ output HIGH. Because $\overline{Q}$ is HIGH, both inputs to gate G_3 are HIGH (remember, the output of gate G_1 is HIGH), holding the Q output LOW. If a HIGH is now momentarily applied to the clock input, the outputs of gates G_1 and G_2 remain HIGH because they are disabled by the LOWs on the S input and the R input; therefore, there is no change in the state of the flip-flop—it remains RESET. We will now make S HIGH, leave R LOW, and apply a clock pulse. Because the S input to gate G_1 is now HIGH, the output of gate G_1 goes LOW when the clock input goes HIGH, causing the Q output to go HIGH. Both inputs to gate G_4 are now HIGH (remember, gate G_2 output is HIGH because R is LOW), forcing the $\overline{Q}$ output LOW. This LOW on $\overline{Q}$ is coupled back into one input of gate G_3, insuring that the Q output will remain HIGH after the clock pulse goes back LOW. The flip-flop is now in the SET state. Figure 8-8 illustrates the logic level transitions that take place within the flip-flop for this condition.

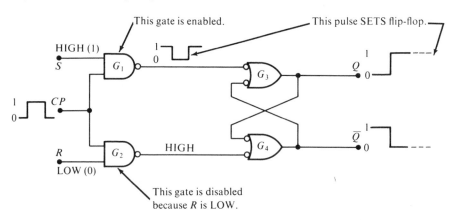

Figure 8-8.   Flip-Flop Making Transition from the RESET State to the SET State.

Next, we will make S LOW, R HIGH, and apply a clock pulse. Because the R input is now HIGH, a HIGH on the clock input produces a LOW on the output of gate G_2, causing the $\overline{Q}$ output to go HIGH. Because of this HIGH on $\overline{Q}$, both inputs to gate G_3 are now HIGH (remember, the output of gate G_1 is HIGH because of the LOW on S), forcing the Q output to go LOW. This LOW on Q is coupled back into one input of gate G_4, insuring that $\overline{Q}$ will remain HIGH after the clock pulse goes back LOW. The flip-flop is now in the RESET state. Figure 8–9 illustrates the logic level transitions that occur within the flip-flop for this condition.

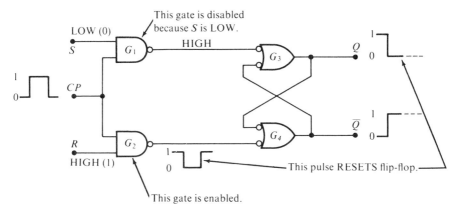

Figure 8–9. Flip-Flop Making Transition from the SET State to the RESET State.

As with the basic flip-flop, an indeterminate condition exists when both S and R are HIGH at the same time. Because we are trying to SET and RESET the flip-flop simultaneously, a race condition occurs when the clock pulse is removed, and the final state depends on the delays through the gates.

To summarize, the edge-triggered RS flip-flop is controlled by the logic states of the S and R inputs when a clock pulse is applied. A change in the state of the flip-flop can occur only when the leading edge of the clock pulse is applied. In the flip-flop that we have examined, the leading edge of the clock pulse is a positive-going transition (LOW to HIGH), so the flip-flop is said to be positive edge-triggered. Notice that as long as the clock input is LOW, the S and R inputs can be changed without affecting the state of the flip-flop. The only time that the states of the S and R inputs are critical is when a clock pulse occurs. Notice also that the flip-flop will assume the state determined by the inputs, and the Q and $\overline{Q}$ outputs will reflect this state after the triggering edge of the clock pulse. The four possible input conditions for the edge-triggered RS flip-flop are summarized below:

1. If S is LOW and R is LOW when the triggering edge of the clock pulse occurs, the flip-flop will not change from its present state.
2. If S is LOW and R is HIGH when the clock edge occurs, the flip-flop will RESET ($Q = 0$). If Q is already LOW, it will remain LOW.
3. If S is HIGH and R is LOW when the triggering edge of the clock pulse occurs, the flip-flop will SET ($Q = 1$). If Q is already HIGH, it will remain HIGH.

4. If S is HIGH and R is HIGH, when the clock pulse is removed (goes back LOW) the next state of the flip-flop is unpredictable. This imposes a restriction on the use of the RS flip-flop and is an undesirable feature.

Table 8–2 provides a more concise description of the logical operation. The fact that the flip-flop is clocked does not alter the basic truth table from that of Table 8–1. However, in order to distinguish this flip-flop as being a clocked type, a refinement in notation is introduced. We will use the notation Q_n to indicate the Q output *before* the clock pulse is applied and Q_{n+1} to indicate the Q output after the clock pulse is applied.

TABLE 8–2.   Truth Table for the Clocked RS Flip-Flop.

Inputs		Output	
S	R	Q_{n+1}	Comments
LOW	LOW	Q_n	NO CHANGE. The output remains in its previous state.
LOW	HIGH	LOW	RESET condition.
HIGH	LOW	HIGH	SET condition.
HIGH	HIGH	?	Indeterminate.

A typical logic symbol for the clocked RS flip-flop is shown in Figure 8–10; symbols like this are normally used in schematic drawings to represent flip-flops rather than the more involved diagram of Figure 8–7. Once the internal logical operation is understood, the "block" symbol conveys all of the information necessary and serves to greatly simplify logic drawings.

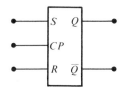

Figure 8–10.   Logic Symbol for a Clocked RS Flip-Flop.

The following example will illustrate the edge-triggered RS flip-flop operation.

Example 8–2   Given the waveforms for the S, R, and CP inputs in Figure 8–11(a), to the flip-flop of Figure 8–10, determine the Q and $\bar{Q}$ output waveforms. Assume the flip-flop is initially RESET.

Solutions:

1. At clock pulse 1, S is LOW and R is LOW. Q does not change.
2. At clock pulse 2, S is LOW and R is HIGH. Q remains LOW (RESET).
3. At clock pulse 3, S is HIGH and R is LOW. Q goes HIGH (SET).
4. At clock pulse 4, S is LOW and R is HIGH. Q goes LOW (RESET).

5.   At clock pulse 5, S is HIGH and R is LOW. Q goes HIGH (SET).

6.   At clock pulse 6, S is HIGH and R is LOW. Q stays HIGH.

Once Q is determined, $\overline{Q}$ is easily found since it is simply the complement of Q. The resulting waveforms for Q and $\overline{Q}$ are shown in Figure 8–11(b).

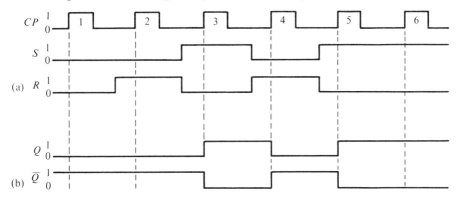

Figure 8–11.

8-3   The Master-Slave RS Flip-Flop

Another important type of clocked RS flip-flop is commonly referred to as a "master-slave" type and is shown in Figure 8–12. The truth table operation is identical to that of an edge-triggered RS flip-flop. The primary difference in operation is the manner in which the flip-flop is clocked.

This type of flip-flop is composed of two sections, the *master* section and the *slave* section. The master section is basically an edge-triggered RS flip-flop, and the slave section is the same except that it is triggered or clocked on the inverted clock pulse and is controlled by the outputs of the master section rather than by external inputs.

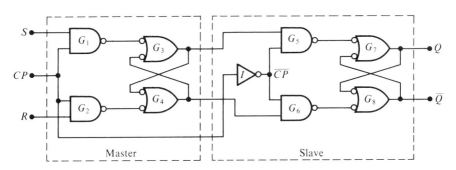

Figure 8–12.   Logic Diagram for a Master-Slave RS Flip-Flop.

The master section will assume the state determined by the R and S inputs at the leading edge of the clock pulse. The state of the master section is then transferred into the slave section on the trailing edge of the clock pulse because the outputs of the master are applied to the inputs of the slave, and the clock pulse to the slave is inverted. The state of the slave then immediately appears on the Q and $\overline{Q}$ outputs.

Figure 8–13 illustrates the master-slave timing operation, which is described as follows. The clock pulse shown is nonideal since it has a rise time and fall time other than 0 (as is always the situation, in practice). At point 1 on the rising edge, the voltage level on the inverted clock pulse reaches the threshold point below which the G_5 and G_6 gates of Figure 8–12 are disabled. At point 2, the voltage level on the true (noninverted) clock enables gates G_1 and G_2, allowing information to be entered into the master section from the R and S inputs. At point 3, the voltage level on the true clock goes below the threshold and disables gates G_1 and G_2; at this point, the S and R inputs are essentially "locked out." At point 4, the inverted clock pulse ($\overline{CP}$) reaches the threshold level and enables gates G_5 and G_6, allowing the output levels of the master section to be transferred to the slave section and appear on the Q and $\overline{Q}$ outputs.

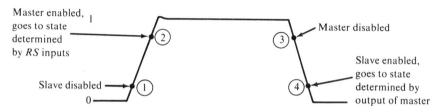

Figure 8–13.   True Clock Pulse Showing Points of Triggering for Master-Slave Operation.

Let us now further examine the operation of this circuit. To begin, we will assume that the flip-flop is RESET and that S is HIGH and R is LOW. Figure 8–14 shows the logic level transitions within the flip-flop when it goes from the RESET state to the SET state. The circled number associated with each transition indicates when it occurs with respect to the clock pulse. A ① corresponds to the leading edge of the clock pulse and a ② corresponds to the trailing edge of the clock pulse. The following statements describe what happens on the leading edge of the clock pulse:

1.   The output of gate G_1 goes from a HIGH to a LOW because both of its inputs are HIGH.

2.   The output of gate G_3 goes from a LOW to a HIGH, and the output of gate G_4 goes from a HIGH to a LOW because of the LOW on the output of gate G_1.

3.   The inverted clock ($\overline{CP}$) input to gates G_5 and G_6 goes LOW because the true clock goes HIGH; this disables the G_5 and G_6 gates and forces their outputs HIGH.

Let us review what has taken place on the leading edge of the clock pulse. The master section has been SET because the S input is HIGH and the R input

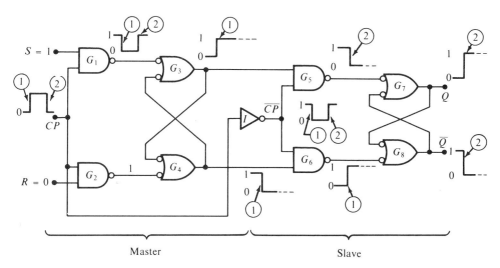

Figure 8–14.   Flip-Flop Making Transition from a RESET State to a SET State.

is LOW. The slave section has not changed state, and therefore the Q and $\overline{Q}$ outputs remain in the initial RESET state.

Now let us look at what happens on the trailing edge of the clock pulse.

1.   The master section remains in the SET condition.
2.   The output of gate G_5 goes from a HIGH to a LOW because both of its inputs are now HIGH.
3.   The output of gate G_7 goes from a LOW to a HIGH, and the output of gate G_8 goes from a HIGH to a LOW because the output of gate G_5 is LOW.

The flip-flop is now in the SET state, and therefore the Q output is HIGH. It did not become SET until the trailing edge of the clock pulse, although the master section was SET on the leading edge. Figure 8–15(a) is a timing diagram illustrating this action.

You should verify the operations for the RESET and NO CHANGE conditions. In Figure 8–15, the timing relationship between the clock pulse and the Q output for each legitimate condition on the R and S inputs is illustrated.

The logic symbol for the master-slave RS flip-flop is the same as for an edge-triggered type. The distinction is usually made by simply stating the type on the data sheet that describes the flip-flop; this is especially true in the case of integrated circuits. In this text we will use the designation shown in Figure 8–16 to indicate a master-slave flip-flop. The circle, or "bubble," on the clock input indicates that the output changes on the negative-going edge of the clock pulse and is often referred to as an "active low" input.

Example 8–3   If the waveforms in Figure 8–17(a) are applied to the inputs of a master-slave RS flip-flop as indicated, determine the Q output. Assume the flip-flop is initially RESET.

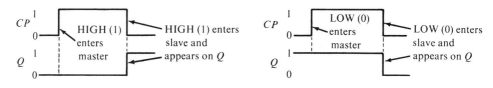

(a) SET condition ($S = 1, R = 0$),
    Q initially LOW (0).

(b) RESET condition ($S = 0, R = 1$),
    Q initially HIGH (1).

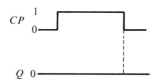

(c) NO CHANGE condition ($S = 0, R = 0$),
    Q initially LOW.

Figure 8–15.   Master-Slave RS Flip-Flop Clock vs. Output Timing for Each Input
Condition.

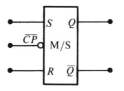

Figure 8–16.   Logic Symbol for a Master-Slave Type of RS Flip-Flop.

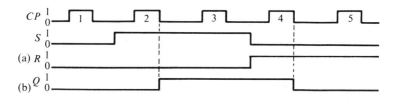

Figure 8–17.

Solution:

1. At clock pulse 1, both S and R are LOW. Q does not change state.
2. At clock pulse 2, S is HIGH and R is LOW, making a SET condition.
    Q goes HIGH.
3. At clock pulse 3, a SET condition still exists on the inputs. Q remains
    HIGH.
4. At clock pulse 4, S is LOW and R is HIGH, creating a RESET condition.
    Q goes LOW.

5. At clock pulse 5, the RESET condition still exists on the inputs. Q remains LOW.

In each case, the Q output does not change until the trailing edge of the clock pulse because of the master-slave operation. The waveform for the Q output is as shown in Figure 8–17(b).

8–4   The Toggle Flip-Flop

A simple modification of the master-slave type of clocked RS flip-flop produces what is commonly referred to as a toggle or T flip-flop. This type of flip-flop changes state on *each* clock pulse, and there are no controlling inputs other than the clock. Figure 8–18(a) shows a clocked master-slave RS flip-flop connected as a toggle flip-flop. A more common representation is shown in Figure 8–18(b). These logic symbols are equivalent and represent the same functional circuit.

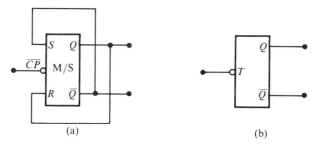

Figure 8–18.   Logic Representations of a Toggle Flip-Flop.

Notice that in Figure 8–18(a), the Q output is cross-coupled back into the R input and the $\overline{Q}$ output is cross-coupled back into the S input. We have already examined the operation of a master-slave RS flip-flop, so all that is required here is an examination of the effect of the cross-coupling. Let us begin by assuming that the flip-flop starts out RESET, so that the Q output is LOW and the $\overline{Q}$ output is HIGH. *Since Q is connected to the R input and $\overline{Q}$ is connected to the S input, the flip-flop will SET when the trailing edge of a clock pulse is applied because S is HIGH and R is LOW.* If another clock pulse is applied, the flip-flop will RESET because S is now LOW and R is HIGH. When a flip-flop changes state on each clock pulse, it is said to be "toggling"—hence the name toggle flip-flop. The clock input is usually designated with a T, as shown in Figure 8–18(b). Toggle operation is illustrated by the timing diagram in Figure 8–19.

There is a particularly interesting feature of the toggle flip-flop operation that should be noted. Notice in Figure 8–19 that the output has a period twice that of the clock. This makes the frequency of the output one-half the frequency of the input, as indicated by the following relationships:

$$T_Q = 2T_{CP} \quad \text{and} \quad f_Q = \frac{1}{T_Q}$$

$$f_Q = \frac{1}{2T_{CP}} = \frac{f_{CP}}{2}$$

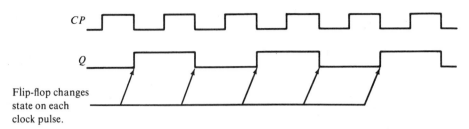

Flip-flop changes
state on each
clock pulse.

Figure 8–19.   Timing Diagram for Toggle Operation.

where T_Q and f_Q are the period and frequency of the Q output, and T_{CP} and f_{CP} are the period and frequency of the clock. The toggle flip-flop is, therefore, a frequency-dividing device because it divides its input (clock) frequency by two; this makes it useful in digital counters and frequency-control applications.

8–5   The Master-Slave JK Flip-Flop

The JK flip-flop is very versatile and is perhaps the most widely used type of flip-flop. The J and K designations for the inputs have no known significance except that they are adjacent letters in the alphabet.

The functioning of the JK flip-flop is identical to that of the RS flip-flop in SET, RESET, and NO CHANGE conditions of operation. The difference is that the JK flip-flop has no indeterminate state as does the RS flip-flop. When both the J and K inputs are HIGH, it becomes essentially a *toggle* flip-flop. Therefore, the JK flip-flop combines the features of both the RS and toggle types and, for this reason, it is a very flexible device that finds wide application in digital systems.

Two types of clocked JK flip-flops will be considered: the master-slave type in this section and the edge-triggered type in the next section. The truth table operation for both of these types is identical; the primary difference is the clocking operation. For an edge-triggered type, the data on the inputs are entered into the flip-flop and appear on the outputs on the same edge of the clock pulse. For a master-slave type, the data on the J and K inputs are entered on the leading edge of the clock but do not appear on the outputs until the trailing edge of the clock.

Figure 8–20 is the logic diagram for a master-slave JK flip-flop. A brief inspection of this diagram shows that this circuit is very similar to the master-slave RS flip-flop. Notice the two main differences: first, the Q output is connected back into the input of gate G_2 and the $\overline{Q}$ output is connected back into the input of gate G_1; and second, one input is now designated J and the other input is now designated K. We will now examine the operation of this circuit in detail.

To begin, let us assume that the flip-flop is RESET. We will also let J be HIGH and K be LOW, creating a SET condition on the inputs. Figure 8–21 shows the transitions for the flip-flop going from the RESET state to the SET state upon application of a clock pulse. The circled number associated with each transition indicates when that transition occurs with respect to the clock pulse. A ① corresponds to the leading edge of the clock pulse and a ② corresponds to the trail-

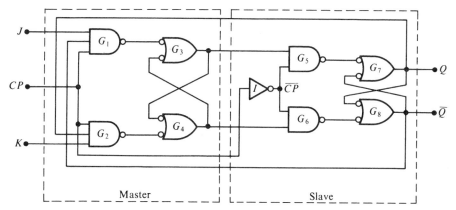

Figure 8–20.   Logic Diagram for a Master-Slave JK Flip-Flop.

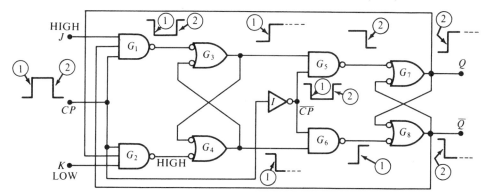

Figure 8–21.   Flip-Flop Making Transition from the RESET State to the SET State.

ing edge. The following statements describe what happens on the leading edge of the clock pulse:

1. The output of gate G_1 goes from HIGH to LOW because all of its input are HIGH.
2. The output of gate G_3 goes from LOW to HIGH and the output of gate G_4 goes from HIGH to LOW because of the LOW on the output of gate G_1.
3. The inverted clock ($\overline{CP}$) goes LOW, disabling both the G_5 and G_6 gates. This insures that their outputs remain HIGH.

Let us review what has taken place on the leading edge of the clock pulse. The master section has been SET because the J input is HIGH and the K input is LOW. The slave section has not changed state, and therefore the Q and $\overline{Q}$ outputs remain in the RESET state. The net result is that the flip-flop is still RESET, as indicated by the Q and $\overline{Q}$ outputs.

The following statements describe what happens on the trailing edge of the clock pulse:

1. The master section remains in the SET state.
2. The output of gate G_5 goes from HIGH to LOW because both of its inputs are now HIGH.
3. The output of gate G_7 goes from LOW to HIGH and the output of gate G_8 goes from HIGH to LOW because of the LOW on the output of gate G_5.

The flip-flop is now in the SET state because the Q output is HIGH. It did not become SET until the trailing edge of the clock pulse, although the master section was set on the leading edge. Figure 8–22 is a timing diagram illustrating the relationship between the clock pulse and the Q output for each condition on the J and K inputs.

The transitions for the SET condition have been examined. You should verify the operation for the other conditions in a similar manner.

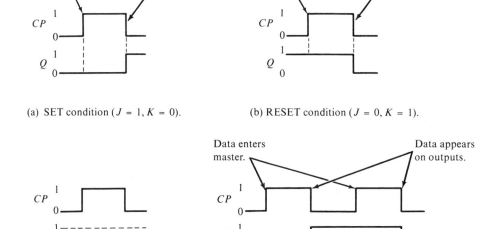

(a) SET condition ($J = 1, K = 0$).   (b) RESET condition ($J = 0, K = 1$).

(b) NO CHANGE condition   (c) TOGGLE condition ($J = 1, K = 1$).
($J = 0, K = 0$).

Figure 8–22. Master-Slave JK Timing for Each Input Condition.

The logic symbol for the master-slave JK flip-flop is shown in Figure 8–23. As was stated for the master-slave RS flip-flop, the distinction is usually made by simply stating the type (master-slave or edge-triggered) on the data sheet normally available for the device. In this text we will use the designation shown in Figure 8–23 to indicate a master-slave type of flip-flop, unless otherwise stated.

The truth table for the master-slave JK flip-flop is shown in Table 8–3. Note that the SET, RESET, and NO CHANGE conditions are the same as for an RS flip-flop. *The difference is that the condition where J and K are both HIGH produces a toggle operation rather than an indeterminate condition.*

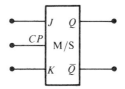

Figure 8–23.   Logic Symbol for a Master-Slave JK Flip-Flop.

TABLE 8–3.   Truth Table for the JK Flip-Flop.

Inputs		Output	
S	R	Q_{n+1}	Comments
LOW	LOW	Q_n	NO CHANGE
LOW	HIGH	LOW	RESET
HIGH	LOW	HIGH	SET
HIGH	HIGH	$\overline{Q}_n$	TOGGLE

In reference to Table 8–3, recall that the designation Q_{n+1} means the state of the Q output *after* a clock pulse and Q_n means the state of the Q output *before* a clock pulse. Therefore, for the NO CHANGE condition, the output after the clock pulse, Q_{n+1}, is equal to the output before the clock pulse, Q_n. For the toggle condition, the output after the occurrence of a clock pulse is equal to the complement of the output before the clock, $\overline{Q}_n$; that is, the flip-flop changes state.

Example 8–4   The waveforms in Figure 8–24(a) are applied to the J, K, and clock inputs as indicated. Determine the Q output, assuming the flip-flop of Figure 8–23 is initially RESET.

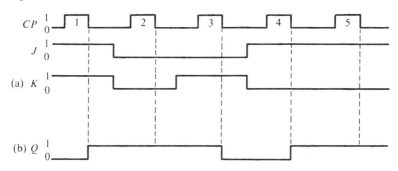

Figure 8–24.

Solution:

1. First, since this is a master-slave flip-flop, we know that the Q output will change only on the trailing edge of the clock pulse.
2. At the first clock pulse both J and K are HIGH, and because this is a toggle condition, Q goes HIGH.
3. At clock pulse 2, a NO CHANGE condition exists on the inputs, keeping Q at a HIGH level.

4. When clock pulse 3 occurs, *J* is LOW and *K* is HIGH, resulting in a RESET condition. *Q* goes LOW.

5. At clock pulse 4, *J* is HIGH and *K* is LOW, resulting in a SET condition. *Q* goes HIGH.

6. A SET condition still exists on *J* and *K* when clock pulse 5 occurs. *Q* will remain HIGH.

The resulting *Q* waveform is indicated in Figure 8–24(b).

8–6 The Edge-Triggered JK Flip-Flop

The logical operation of the edge-triggered JK flip-flop is identical to the operation of the master-slave type. The only difference is the manner in which it is triggered or clocked. In this section we will limit ourselves to discussing this difference and not go into a detailed analysis of the internal operation.

The information (data) on the *J* and *K* inputs of an edge-triggered type of flip-flop enters the flip-flop and appears at the outputs on the *same* edge of the clock pulse. The edge-triggered timing for each input condition is shown in Figure 8–25.

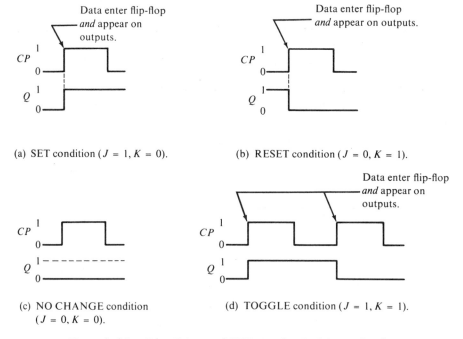

(a) SET condition (*J* = 1, *K* = 0).

(b) RESET condition (*J* = 0, *K* = 1).

(c) NO CHANGE condition (*J* = 0, *K* = 0).

(d) TOGGLE condition (*J* = 1, *K* = 1).

Figure 8–25. Edge-Triggered JK Timing for Each Input Condition.

Notice the difference between this operation and that of the master-slave version in Figure 8–22. Here, the data (states of the *J* and *K* inputs) go into the flip-flop and appear on the outputs on the *positive* edge of the clock pulse. It should be noted here that a *negative* edge-triggered JK flip-flop is also possible.

The logic symbol for an edge-triggered JK flip-flop is shown in Figure 8–26. It is the same as for the master-slave, except that in this text we will use the designation M/S to mean master-slave and no special mark implies an edge-triggered type.

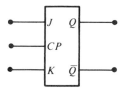

Figure 8–26.   Logic Symbol for an Edge-Triggered JK Flip-Flop.

An example will serve to illustrate the operation of this flip-flop.

Example 8–5   The waveforms in Figure 8–27(a) are applied to the *J*, *K*, and clock inputs of an edge-triggered flip-flop. Determine the *Q* output waveform, assuming that there is positive edge-triggering and that the flip-flop starts out RESET.

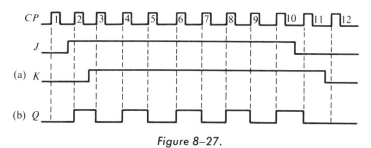

Figure 8–27.

Solution:

1. At clock pulse 1, *J* is LOW and *K* is LOW, so *Q* does not change.
2. At clock pulse 2, *J* is HIGH and *K* is LOW, so the flip-flop SETS.
3. For clock pulses 3 through 10, a toggle condition exists, so *Q* will change state on each pulse.
4. At clock pulse 11, *J* is LOW and *K* is HIGH, so a RESET occurs.
5. At clock pulse 12, *J* is LOW and *K* is LOW, creating a NO CHANGE condition.

The resulting *Q* waveform is shown in Figure 8–27(b). Note that the changes occur only on positive edges of the clock because positive edge-triggering was specified.

8–7   The D Flip-Flop

The D flip-flop is very useful in cases where only a data bit (1 or 0) is to be stored. A simple addition to a clocked RS flip-flop creates a basic D flip-flop, as shown in Figure 8–28.

Notice that this flip-flop has only one input in addition to the clock. This is called the *D* input. If there is a HIGH on the *D* input when a clock pulse is applied, the flip-flop will SET because gate G_1 is enabled by the HIGH on the D input and gate G_2 is disabled by the LOW on its input produced by inverter I. The HIGH on the D input is "stored" by the flip-flop on the leading edge of the clock pulse. If there is a LOW on the D input when the clock pulse is applied, the flip-flop will RESET because gate G_1 is disabled by the LOW on the D input and gate G_2 is enabled by the HIGH produced by the inverter I. The LOW on the D input is thus stored by the flip-flop on the leading edge of the clock pulse. In the SET state, the flip-flop can be said to be storing a logic 1, and in the RESET state it is storing a logic 0.

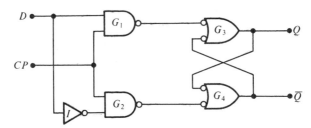

Figure 8–28.   **Logic Diagram for a D Flip-Flop.**

As we have seen, the state of the D input (HIGH or LOW) is transferred to the Q output on the leading edge of the clock pulse (negative edge-triggering is also possible). Figure 8–29 illustrates the transitions within the flip-flop when a HIGH is entered on the D input, followed by a LOW. The HIGH is transferred to the Q output on the first clock pulse, and the LOW is transferred to the Q output on the second clock pulse.

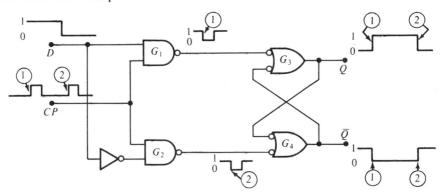

Figure 8–29.   **Transitions within a D Flip-Flop When a HIGH Followed by a LOW is Clocked into the Flip-Flop.**

A typical logic symbol for the D flip-flop appears in Figure 8–30, and the truth table is given in Table 8–4.

Example 8–6   Given the waveforms in Figure 8–31(a) for the D input and the clock, determine the Q output waveform if the flip-flop starts out RESET.

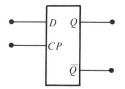

Figure 8–30.   Logic Symbol for a D Flip-Flop.

TABLE 8–4.   Truth Table for a D Flip-Flop.

D	Q_{n+1}	Comments
LOW	LOW	The Q output assumes the
HIGH	HIGH	state of D at each CP.

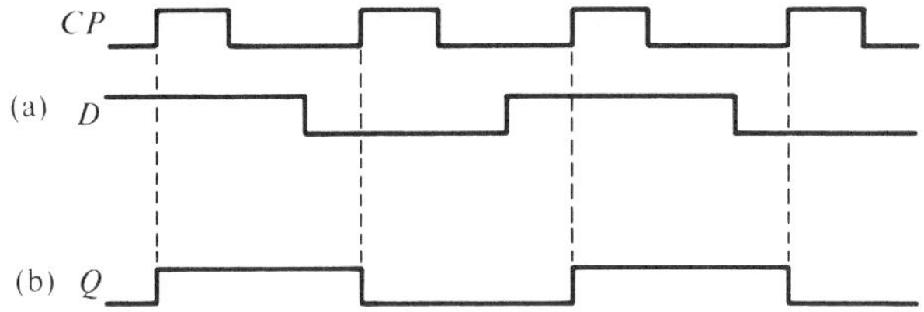

Figure 8–31.

Solution:

The Q output assumes the state of the D input at the time of the leading clock edge. The resultant output is shown in Figure 8–31(b).

8–8   PRESET and CLEAR Functions

The flip-flops that we have studied up to this point are all of a basic form in that they have only the primary inputs (S, R or J, K or D, and CP) that are required to produce the proper truth table operation. These are often called the "synchronous" inputs because they control the state of the flip-flop only when a clock pulse occurs; that is, changes can occur only in synchronization with the clock.

There are additional functions in the form of secondary inputs that can be added to a flip-flop to increase its flexibility and usefulness in digital applications or to alter the operation somewhat from what we have discussed.

We will examine two of these secondary functions, PRESET and CLEAR. The CLEAR function is sometimes called direct RESET, and is shown incorporated on the edge-triggered RS flip-flop in Figure 8-32(a). This type of flip-flop will be used to illustrate the special functions discussed in this section (the application of each function can be extended to the other types of flip-flop in a similar manner).

Notice that the CLEAR input goes into the $\overline{Q}$ side of the flip-flop, so that when a LOW is applied to this input, the flip-flop will RESET regardless of what the other inputs (S, R, and CP) are. Therefore, this CLEAR input RESETS the flip-flop directly and is not dependent on the clock pulse or the S and R inputs. Essentially, it "overrides" these other inputs and, because it has to be LOW to

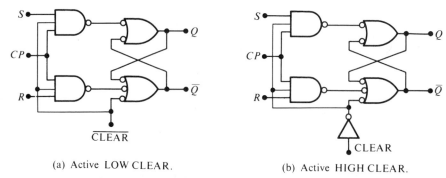

(a) Active LOW CLEAR.   (b) Active HIGH CLEAR.

Figure 8–32.   Logic Diagrams of RS Flip-Flops with CLEAR Inputs.

cause the flip-flop to RESET, it is sometimes referred to as an "active low" type of input. An "active high" CLEAR can be implemented by adding an inverter, as shown in Figure 8–32(b). Here a HIGH level on the CLEAR input produces a RESET of the flip-flop.

The next function to be presented is commonly called the PRESET or direct SET. Figure 8–33(a) shows this input added to the logic diagram of Figure 8–32(a). Notice that this input is connected to the Q side of the flip-flop, so that when a LOW is applied, the flip-flop will be SET regardless of what the S, R, and CP inputs are. Therefore, the PRESET input SETS the flip-flop directly, and is not dependent on the clock pulse or the S and R inputs. Again, by addition of an inverter as shown in Figure 8–33(b), an "active high" PRESET can be created.

Figure 8–34(a) is the logic symbol, including the CLEAR (C) and PRESET (PS) inputs. The circle, or "bubble," on both of these inputs indicates that they are "active low." Figure 8–34(b) shows the logic symbol with "active high" CLEAR and PRESET inputs (no bubble).

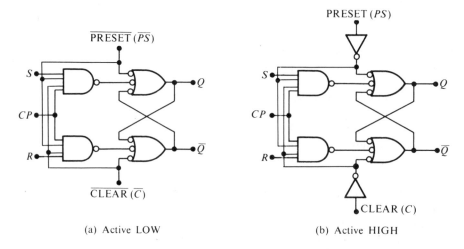

(a) Active LOW   (b) Active HIGH

Figure 8–33.   Logic Diagram for an RS Flip-Flop with both PRESET and CLEAR Inputs.

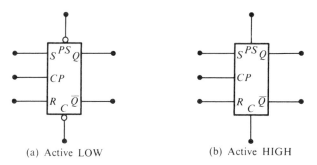

(a) Active LOW   (b) Active HIGH

Figure 8–34.   Logic Symbols for an RS Flip-Flop with PRESET and CLEAR Inputs.

Example 8–7   If the signals having the waveforms in Figure 8–35(a) are applied to the inputs of the flip-flop shown, what is the waveform of the Q output when the flip–flop starts in the SET state?

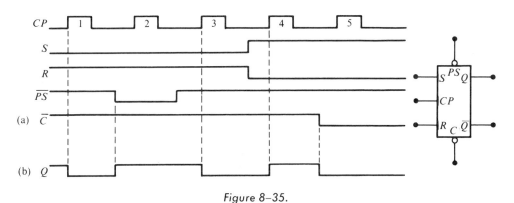

Figure 8–35.

Solution:

1. At clock pulse 1, the flip-flop is in a RESET condition because S is LOW and R is HIGH. The PRESET and CLEAR inputs are HIGH and have no effect because they are "active low." Therefore, the flip-flop RESETS when the clock pulse occurs (positive edge-triggered).

2. When the PRESET input goes LOW, the flip-flop immediately SETS, regardless of the other synchronous inputs.

3. Clock pulse 2 has no effect, since the PRESET input remains LOW during this time and overrides the clock.

4. Clock pulse 3 causes the flip-flop to RESET because S is still LOW and R is still HIGH.

5. Clock pulse 4 causes the flip-flop to SET because S is HIGH and R is LOW.

6. Clock pulse 5 has no effect because the CLEAR input is LOW, forcing the flip-flop to RESET. The resulting Q output is shown in Figure 8-35(b).

Example 8–8   For the edge-triggered JK flip-flop with PRESET and CLEAR inputs in Figure 8–36(a), determine the Q output for the inputs shown in the timing diagram. Q is initially LOW.

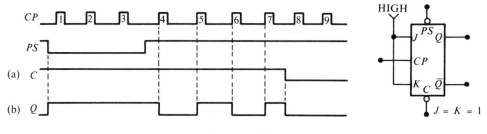

Figure 8–36.

Solution:

1. During clock pulses 1, 2, and 3, the PRESET (*PS*) is LOW, keeping the flip-flop SET.
2. For clock pulses 4, 5, 6, and 7, toggle operation occurs because J is HIGH, K is HIGH, and both PRESET and CLEAR are HIGH.
3. For clock pulses 8 and 9, the CLEAR (*C*) input is LOW, keeping the flip-flop RESET regardless of the synchronous inputs. The resulting Q output is shown in Figure 8–36(b).

8–9   Operating Characteristics

Several operating characteristics or parameters important in the application of flip-flops specify the performance, operating requirements, and operating limitations of the circuit. They are typically found in the data sheets for integrated circuits, and are applicable to all flip-flops regardless of the particular form of the circuit.

Propagation Delays

A propagation delay is the interval of time required after the input signal has been applied for the resulting output change to occur.

There are several categories of propagation delay that are important in the operation of a flip-flop.

1. The *turn-off delay* is measured from the triggering edge of the clock pulse to the *LOW-to-HIGH transition* of the output. This delay is illustrated in Figure 8–37.
2. The *turn-on delay* is measured from the triggering edge of the clock pulse to the *HIGH-to-LOW transition* of the output. This delay is illustrated in Figure 8–38.
3. The *turn-off delay* is measured from the PRESET input to the *LOW-to-HIGH transition* of the output. This delay is illustrated in Figure 8–39 for an "active low" PRESET.

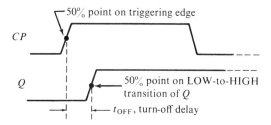

Figure 8–37. Turn-Off Delay, Clock to Output.

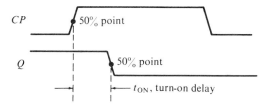

Figure 8–38. Turn-On Delay, Clock to Output.

4. The *turn-on delay* is measured from the CLEAR input to the *HIGH-to-LOW transition* of the output. This delay is illustrated in Figure 8–40 for an "active low" CLEAR.

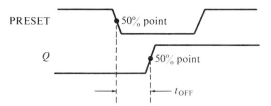

Figure 8–39. Turn-Off Delay, PRESET to Output.

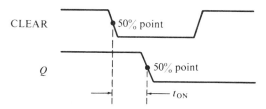

Figure 8–40. Turn-On Delay, Clear to Output.

Set-Up Time

The *set-up time* is the interval required for the control levels to be on the inputs (*J* and *K*, or *S* and *R*, or *D*) *prior* to the triggering edge of the clock pulse in order for the levels to be reliably clocked into the flip-flop. This is illustrated in Figure 8–41 for a D flip-flop.

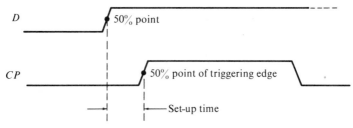

Figure 8–41. Set-Up Time.

Hold Time

The *hold time* is the interval required for the control levels to remain on the inputs *after* the triggering edge of the clock pulse in order for the levels to be reliably clocked into the flip-flop. This is illustrated in Figure 8–42 for a D flip-flop.

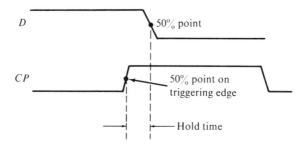

Figure 8–42. Hold Time.

Maximum Clock Frequency

The maximum clock frequency is the highest rate at which a flip-flop can be reliably triggered. At clock frequencies above the maximum, the flip-flop would be unable to respond quickly enough and its operation would be impaired.

Power Dissipation

The power dissipation of a flip-flop is the total power consumption of the device. For example, if the flip-flop operates on a + 5 V DC source and draws 50 mA of current, the power dissipation is

$$P = 5\,V \times 50\,mA = 250\,mW$$

The power dissipation is very important in most applications where the capacity of the DC supply is a concern. As an example, let us assume we have a digital system that requires a total of ten flip-flops and each flip-flop dissipates 250 mW of power. The total power requirement is

$$P_{TOT} = 10 \times 250\,mW = 2500\,mW = 2.5\,W$$

This tells us the type of DC supply that is required as far as output capacity is concerned. If the flip-flops operate on +5 V DC, then the amount of current that the supply must provide is as follows:

$$I = \frac{2.5\,\text{W}}{5\,\text{V}} = 0.5\,\text{A}$$

We must use a $+5\,\text{V}$ DC supply that is capable of providing at least $0.5\,\text{A}$ of current.

Other Characteristics

Many characteristics discussed in Chapter 4 in relation to gates—such as fan-out, input voltages, output voltages, and noise margin—apply equally to flip-flops and are not repeated here. Since integrated circuits are widely used in digital systems today, a variety of flip-flops are available in IC form; these range from single (one-per-package) flip-flops with various types of gated inputs and different operating characteristics to dual (two-per-package) flip-flops in several varieties.

8–10   Summary of Flip-Flop Operation

1. Master-slave operation:
   The state of the control inputs (sometimes called data or information) is entered into the flip-flop on the leading edge of the clock pulse, but does not appear on the Q and $\overline{Q}$ outputs until the trailing edge of the clock pulse.

2. Edge-triggered operation:
   The state of the control inputs is entered into the flip-flop *and* appears on the Q and $\overline{Q}$ outputs on the *same* edge of the clock pulse (triggering edge).

3. Synchronous operation:
   The synchronous operation of a flip-flop refers to the clocked operation. The control inputs (S, R or J, K or D) are synchronous inputs because they determine the next state of the flip-flop only on the occurrence of a clock pulse. In other words, the output changes are synchronized with the clock.

4. Asynchronous operation:
   The asynchronous operation of a flip-flop refers to the unclocked operation. The PRESET and CLEAR inputs are asynchronous because they are independent of the clock and control inputs. In other words, they can cause a change in the output without the occurrence of a clock pulse.

5. The RS flip-flop operation:
   Logic Symbol:

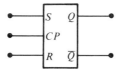

Figure 8-43.

Truth Table:

	S	R	Q_{n+1}
NO CHANGE	LOW	LOW	Q_n
RESET	LOW	HIGH	LOW
SET	HIGH	LOW	HIGH
INDETERMINATE	HIGH	HIGH	?

6. The JK flip-flop operation:
    Logic Symbol:

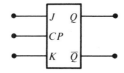

Figure 8–44.

Truth Table:

	J	K	Q_{n+1}
NO CHANGE	LOW	LOW	Q_n
RESET	LOW	HIGH	LOW
SET	HIGH	LOW	HIGH
TOGGLE	HIGH	HIGH	$\overline{Q}_n$

7. The toggle flip-flop operation:
    Logic Symbol:

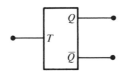

Figure 8–45.

Truth Table: $\dfrac{Q_{n+1}}{\overline{Q}_n}$

Q goes to opposite state on each clock pulse.

8. The D flip-flop operation:
    Logic Symbol:

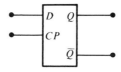

Figure 8–46.

Truth Table:

D	Q_{n+1}
LOW	LOW
HIGH	HIGH

The output follows the input on each clock pulse

PROBLEMS

8-1 If the waveforms in Figure 8–47 are applied to an unclocked RS flip-flop, sketch the resulting Q output waveform in relation to the inputs. Assume Q starts LOW.

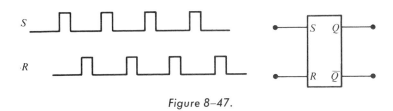

Figure 8–47.

8-2 Repeat Problem 8–1 for the input waveforms in Figure 8–48.

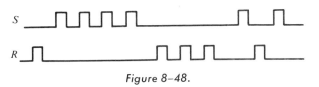

Figure 8–48.

8-3 Repeat Problem 8–1 for the input waveforms in Figure 8–49.

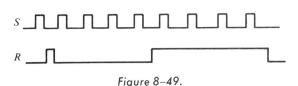

Figure 8–49.

8-4 For a positive edge-triggered RS flip-flop, determine the Q and $\overline{Q}$ outputs for the given inputs in Figure 8–50. Show them in proper relation to the clock. Assume Q starts LOW.

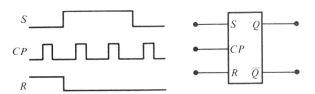

Figure 8–50.

8–5   Repeat Problem 8–4 for the inputs in Figure 8–51.

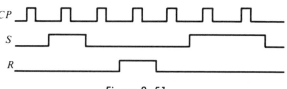

Figure 8–51.

8–6   For a master-slave RS flip-flop, determine the Q and $\overline{Q}$ outputs in proper relation to the clock for the waveforms given in Problem 8–5.

8–7   The Q output of a clocked RS flip-flop in Figure 8–52 is shown in relation to the clock signal. Determine the input waveforms on the R and S inputs that are required to produce this output if the flip-flop is a positive edge-triggered type.

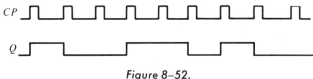

Figure 8–52.

8–8   The clock signal applied to a toggle flip-flop has a period of 0.5 microsecond. What is the frequency of the output waveform?

8–9   For Problem 8–8 sketch the relationship of the clock and output waveforms for a square wave clock and a T flip-flop such as that shown in Figure 8–18.

8–10   For the flip-flop of Figure 8–12, show the transitions that occur at all points within the circuit when it is clocked from the SET state to the RESET state.

8–11   For a positive edge-triggered JK flip-flop with inputs in Figure 8–53, determine the Q output. Assume Q starts LOW.

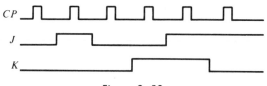

Figure 8–53.

8–12   Repeat Problem 8–11 for the inputs in Figure 8–54.

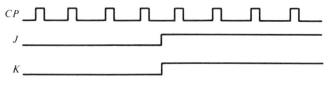

Figure 8–54.

8–13   For a master-slave JK flip-flop with the inputs in Figure 8–55, sketch the Q output waveform. Assume Q is initially LOW.

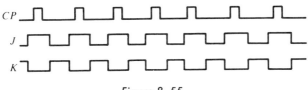

Figure 8–55.

8–14 Determine the J and K inputs required to generate the waveform in Figure 8–56 on the Q output of a master-slave flip-flop.

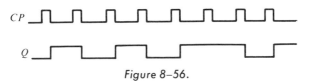

Figure 8–56.

8–15 For the flip-flop of Figure 8–20, show the transitions that occur at all points within the circuit when it is clocked from the SET state to the RESET state.

8–16 Repeat Problem 8–14 for the output in Figure 8–57.

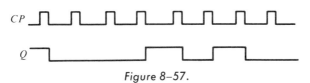

Figure 8–57.

8–17 The following serial data stream is to be generated using a JK positive edge-triggered flip-flop. Determine the inputs required. The data are to be in NRZ format.

100101100011010000111101   Left bit is first out.

8–18 Determine the Q waveform if the signals shown in Figure 8–58 are applied to the inputs of the edge-triggered JK flip-flop. Q is initially LOW.

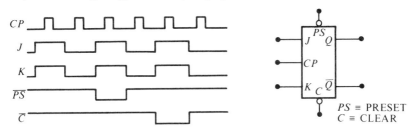

Figure 8–58.

8–19 Sketch the Q output for a D flip-flop, with the inputs as shown in Figure 8–59. Assume positive edge-triggering and Q initially LOW.

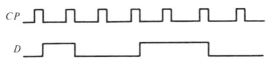

Figure 8–59.

8-20 Repeat Problem 8-19 for the inputs in Figure 8-60.

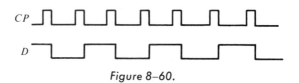

Figure 8-60.

8-21 The DC current required by a particular flip-flop that operates on +5 V DC is found to be 10 mA. A certain digital system uses 15 of these flip-flops. Determine the current capacity required for the +5 V DC supply and the total power dissipation of the system.

8-22 For the circuit in Figure 8-61, determine the maximum frequency of the clock signal for reliable operation if the set-up time for each flip-flop is 20 nanoseconds and the propagation delays from clock to output are 50 nanoseconds for each flip-flop. Assume positive edge-triggering.

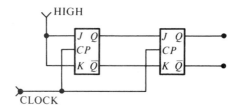

Figure 8-61.

8-23 The following serial data are applied to the flip-flop as indicated in Figure 8-62. Determine the resulting serial data that appear on the Q output. There is one clock pulse for each bit time. Assume Q is initially 0.

$$J_1:\quad 1010011$$
$$J_2:\quad 0111010$$
$$J_3:\quad 1111000$$
$$K_1:\quad 0001110$$
$$K_2:\quad 1101100$$
$$K_3:\quad 1010101$$

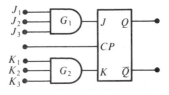

Figure 8-62.

8-24 Sketch the Q output of flip-flop B in Figure 8-63 in proper relation to the clock.

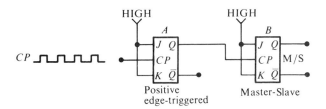

Figure 8–63.

8-25 Sketch the Q output of flip-flop B in Figure 8–64 in proper relation to the clock.

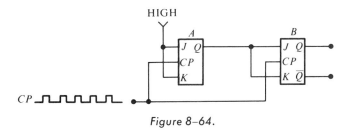

Figure 8–64.

8-26 The flip-flop in Figure 8–65 is initially RESET. Show the relation between the Q output and the clock pulse if the propagation delay is 8 nanosecond.

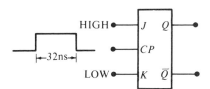

Figure 8–65.

CHAPTER 9

Sequential Logic: Counters

When we count a sequence of numbers, two basic factors are involved: we must know where we are in the sequence at any given time, and we must know the next number in the sequence. A digital *counter* is a logic circuit that can progress through a sequence of numbers or states when activated by an input normally called the clock. The outputs of a counter indicate the binary number contained within the counter at any given time. Counters are all characterized by a storage or memory capability because they must be able to retain their present state until the clock forces a change to the next state in the sequence. The clock signal is normally a series of pulses (pulse train) of a specified frequency, but it can also be random pulses, depending on the particular application.

Digital counters have many wide-ranging applications in digital systems, but all have one thing in common—they utilize the counter's ability to progress through a certain sequence of states. The number of states through which a counter progresses before it goes back to the original state (recycles) is called the *modulus* of the counter.

9-1 Using Flip-Flops to Count

Since the flip-flop has memory or storage capability, it is the primary logic element in digital counters. In fact, a single flip-flop can be considered a simple counter. Let's look at Figure 9-1 and see how this works. First, assume the flip-flop is RESET ($Q = 0$) and a clock pulse occurs. The flip-flop is connected for toggle operation ($J = 1$, $K = 1$), so the first clock pulse (CP_1) causes it to SET ($Q = 1$). The fact that Q has gone from a 0 to a 1 state tells us that a single clock pulse has occurred—that is, the flip-flop has essentially "counted" one clock pulse. At CP_2, the flip-flop RESETS. If we know that the flip-flop has been SET only *once* and is then RESET, we know that two pulses have occurred. The flip-flop has essentially counted two pulses.

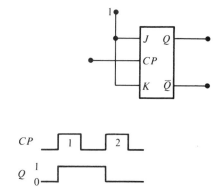

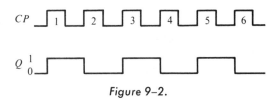

Figure 9–1.   The Flip-Flop as a Simple Counter.

A limitation of the "counting" ability of a single flip-flop should be apparent at this time. Assuming RESET is the beginning state, a series of clock pulses will produce the output indicated in Figure 9–2.

Figure 9–2.

Notice that the Q output is HIGH (1) after each odd clock pulse, and it is LOW (0) initially and after each even clock pulse. Therefore, if $Q = 0$ at any time, either no clock pulses or an even number of clock pulses have occurred. If $Q = 1$ at any time, this tells us that an odd number of clock pulses have occurred, but not exactly how many.

If there is a situation where either *none* or *only one* clock pulse can possibly occur, then the single flip-flop counter can give us a precise indication. Assuming that it is initially RESET, then $Q = 0$ indicates that no clock pulses have occurred and $Q = 1$ indicates that one has occurred. As you can see, the single flip-flop is limited to counting two occurrences—none or one.

In order to increase the counting capacity, additional flip-flops are used. For instance, let us look at the counting capacity of two flip-flops. At this point, we are not concerned with the actual connections of the flip-flops. As indicated in Figure 9–3, the possible combinations of states of two flip-flops together are as follows:

1. FFA RESET and FFB RESET
2. FFA RESET and FFB SET
3. FFA SET and FFB RESET
4. FFA SET and FFB SET

The two flip-flops together have a total of four possible states and can therefore be used to count the occurrence of none, one, two, or three clock pulses when connected properly.

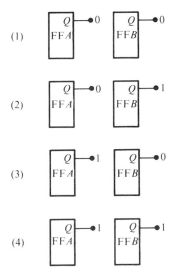

Figure 9–3.   All Possible State Combinations of Two Flip-Flops.

When three flip-flops are used together, the possible combination of states are as follows:

1. FF*A* RESET, FF*B* RESET, FF*C* RESET
2. FF*A* RESET, FF*B* RESET, FF*C* SET
3. FF*A* RESET, FF*B* SET, FF*C* RESET
4. FF*A* RESET, FF*B* SET, FF*C* SET
5. FF*A* SET, FF*B* RESET, FF*C* RESET
6. FF*A* SET, FF*B* RESET, FF*C* SET
7. FF*A* SET, FF*B* SET, FF*C* RESET
8. FF*A* SET, FF*B* SET, FF*C* SET

These possible combinations are shown in table form in Table 9–1 as logic states of the Q outputs of each flip-flop.

TABLE 9–1.   All Possible
State Combinations of a
Three Flip-Flop Counter.

Q_A	Q_B	Q_C
0	0	0
0	0	1
0	1	0
0	1	1
1	0	0
1	0	1
1	1	0
1	1	1

As you can see, the three flip-flops together exhibit a total of eight possible states and can be used to count zero to seven clock pulses when connected properly. It becomes apparent that the maximum number of binary states in which a counter can exist is dependent on the number of flip-flops used to construct the counter; this can be expressed by the following relationship, where N = maximum number of counter states and n = number of flip-flops in the counter:

$$N = 2^n \tag{9-1}$$

Example 9–1 Determine the maximum number of binary states that a counter can have if it is constructed with four, five, and six flip-fiops.

Solutions:
Four flip-flops: $2^4 = 16$ states
Five flip-flops: $2^5 = 32$ states
Six flip-flops: $2^6 = 64$ states

9–2   Asynchronous Binary Counters

Basic binary counters can be placed into two categories relating to the method by which they are *clocked,* that is, the way in which the clock pulses are used to sequence the counter. One category is called *asynchronous* and is discussed in this section. The second category is covered in Section 9–3.

The term *asynchronous* refers to events that do not occur at the same time. With respect to counter operation, asynchronous means that the flip-flops within the counter are not made to change states at exactly the same time (simultaneously); this is because the clock pulses are not connected directly to the clock or trigger input of each flip-flop in the counter. Figure 9–4 shows a two-stage counter connected for asynchronous operation. Each flip-flop in a counter is commonly referred to as a *stage* of the counter.

Notice in Figure 9–4 that the clock line is connected to the clock (CP) input of only the first stage, FFA. The second stage, FFB, is triggered by the $\overline{Q}_A$ output of FFA. FFA changes state when a clock pulse occurs, but FFB changes only when triggered by a transition of the $\overline{Q}_A$ output of FFA. Because of the inherent propagation delay through a flip-flop, a transition of the input clock pulse and a transition of the $\overline{Q}_A$ output of FFA can never occur at exactly the same time. Therefore, the two flip-flops are *never simultaneously triggered,* which results in *asynchronous* counter operation.

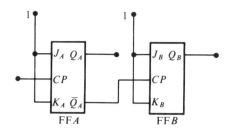

Figure 9–4.   A Two-Stage Asynchronous Binary Counter.

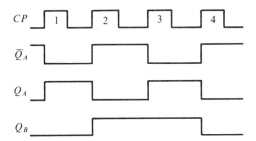

Figure 9-5. Timing Diagram for the Counter of Figure 9-4.

Let us examine the basic operation of the counter of Figure 9-4 by applying four clock pulses to FFA and observing the Q output of each flip-flop; Figure 9-5 illustrates the changes in the state of the flip-flop outputs in response to the clock pulses. Positive edge-triggering is assumed for each flip-flop. Also, both flip-flops are connected for toggle operation ($J = 1, K = 1$) and are initially RESET.

The positive-going edge of CP_1 (clock pulse 1) causes the Q_A output of FFA to go HIGH. The $\overline{Q}_A$ output at the same time goes LOW, but has no effect on FFB because a *positive-going* transition must occur to trigger the flip-flop. After the leading edge of CP_1, $Q_A = 1$ and $Q_B = 0$. The positive-going edge of CP_2 causes Q_A to go LOW. $\overline{Q}_A$ goes HIGH and triggers FFB, causing Q_B to go HIGH. After the leading edge of CP_2, $Q_A = 0$ and $Q_B = 1$. The positive-going edge of CP_3 causes Q_A to go HIGH again. $\overline{Q}_A$ goes LOW and has no effect on FFB. Thus, after the leading edge of CP_3, $Q_A = 1$ and $Q_B = 1$. The positive-going edge of CP_4 causes Q_A to go LOW. $\overline{Q}_A$ goes HIGH and triggers FFB, causing Q_B to go LOW. After the leading edge of CP_4, $Q_A = 0$ and $Q_B = 0$. The counter is back in its original state (both flip-flops RESET).

The waveforms of the Q_A and Q_B outputs are shown relative to the clock pulses in Figure 9-5. This graphic waveform relationship is sometimes called a *timing diagram.* It should be pointed out that, for simplicity, the transitions of Q_A, Q_B, and the clock pulses are shown simultaneous even though this is an asynchronous counter. There is, of course, some small delay between the Q_A and the Q_B transitions.

A summary of the sequence of states through which the counter progresses as the clock pulses occur follows:

Initially	$Q_A = 0,$	$Q_B = 0$
After CP_1	$Q_A = 1,$	$Q_B = 0$
After CP_2	$Q_A = 0,$	$Q_B = 1$
After CP_3	$Q_A = 1,$	$Q_B = 1$
After CP_4	$Q_A = 0,$	$Q_B = 0$ (original state)

Notice that the counter exhibits four different states, as you would expect with two flip-flops ($N = 2^2 = 4$). Also, notice that if Q_A represents the least significant bit (LSB) and Q_B represents the most significant bit (MSB), the sequence of counter states is actually a sequence of binary numbers as shown in Table 9-2, which is a state table or truth table for a two-stage binary counter.

TABLE 9–2.

Clock Pulse	Q_B	Q_A
0	0	0
1	0	1
2	1	0
3	1	1

Since it goes through a binary sequence, the counter in Figure 9–4 is a form of *binary counter*. It actually counts the number of clock pulses up to three, and on the fourth pulse it recycles to its original state ($Q_A = 0$, $Q_B = 0$). The term *recycle* is commonly applied to counter operation, and refers to the transition of the counter from its final state back to its original state.

Sometimes it is convenient to show the sequence of counter states with a *state diagram,* as illustrated for the two-stage counter in Figure 9–6. This state diagram tells us that the counter starts in the binary 0 state and progresses to the binary 1 state at CP_1, goes to the binary 2 state at CP_2, to the binary 3 state at CP_3, and then recycles to the binary 0 state at CP_4, thus completing its entire cycle.

A three-stage asynchronous binary counter is shown in Figure 9–7(a). The basic operation, of course, is the same as that of the two-stage counter just discussed, except that it has *eight* states due to its three stages. A timing diagram appears in Figure 9–7(b) for eight clock pulses.

Notice that the counter progresses through a binary count of 0 to 7 and then recycles to the 0 state. This counter sequence is presented in the state diagram of Figure 9–8 and the state table of Table 9–3.

Asynchronous counters are commonly referred to as *ripple* counters for the following reason. The effect of the input clock pulse is first "felt" by FFA. This effect cannot get to FFB immediately due to propagation delay through FFA. Then there is the propagation delay through FFB before FFC can be triggered. Thus, the effect of an input clock pulse "ripples" through the counter, taking some time due to propagation delays to reach the last flip-flop. To illustrate, all three flip-flops in the counter of Figure 9–7 change state as a result of CP_4. The HIGH-to-LOW transition of Q_A occurs one delay time after the positive-going transition of the clock pulse. The HIGH-to-LOW transition of Q_B occurs one delay time

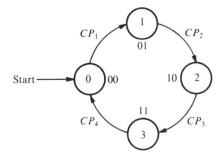

Figure 9–6.   State Diagram for a Two-Stage Binary Counter.

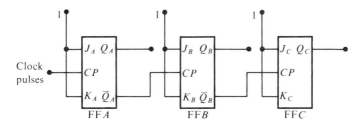

(a) Three-Stage Binary Counter (Asynchronous or Ripple Type)

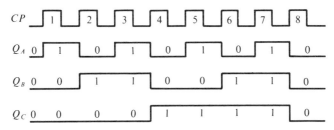

(b) Timing Diagram for the Three-Stage Binary Counter

Figure 9–7.

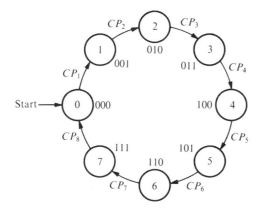

Figure 9–8.   State Diagram for a Three-Stage Binary Counter.

TABLE 9–3.   State Table for a
Three-Stage Binary Counter.

Clock Pulse	Q_C	Q_B	Q_A
0	0	0	0
1	0	0	1
2	0	1	0
3	0	1	1
4	1	0	0
5	1	0	1
6	1	1	0
7	1	1	1

after the positive-going transition of $\overline{Q}_A$. The LOW-to-HIGH transition of Q_C occurs one delay time after the positive-going transition of $\overline{Q}_B$. As you can see, FFC is not triggered until two delay times after the positive-going edge of the clock pulse, CP_4. In other words, it takes two flip-flop delay times for the effect of the clock pulse to "ripple" through the counter and trigger FFC.

This cumulative delay of an asynchronous counter is a major disadvantage in many applications because it limits the rate at which the counter can be clocked and creates decoding problems.

Example 9–2 A four-stage asynchronous binary counter is shown in Figure 9–9(a). Each flip-flop is positive edge-triggered and has a propagation delay of 10 nanoseconds (ns). Draw a timing diagram showing the Q output of each stage, and determine the total delay time from the leading edge of a clock pulse until a corresponding change can occur in the state of Q_D.

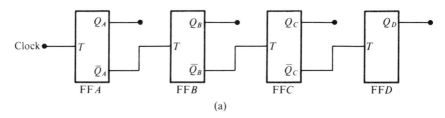

(a)

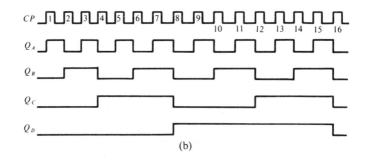

(b)

Figure 9–9.

Solutions:

The timing diagram is as shown in Figure 9–9(b). To determine the total delay time, the effect of CP_8 or CP_{16} must propagate through four flip-flops before Q_D changes.

$$t_D = 4 \times 10 \text{ ns} = 40 \text{ ns total delay}$$

9–3   Synchronous Binary Counters

The term *synchronous* as applied to counter operation means that the counter is clocked such that each flip-flop in the counter is *triggered at the same time*. This is accomplished by connecting the clock line to *each* stage of the counter, as shown

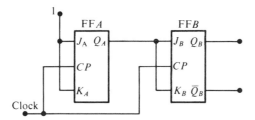

Figure 9-10. A Two-Stage Synchronous Binary Counter.

in Figure 9-10 for a two-stage counter. The synchronous counter is also called a *parallel* counter because the clock line is connected in parallel to each flip-flop.

Notice that an arrangement different from that for the asynchronous counter must be used for the J and K inputs of FFB in order to achieve a binary sequence. The operation of this counter is as follows. First, we will assume the counter is

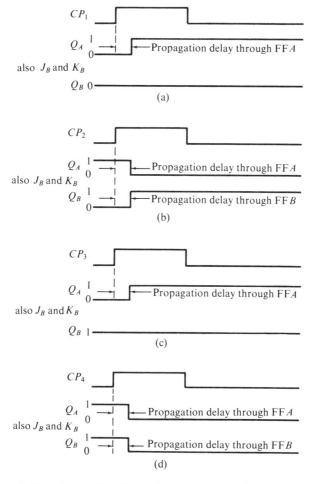

Figure 9-11. Timing Details for the Synchronous Counter Operation.

initially in the binary 0 state—both flip-flops RESET. When the positive edge of the first clock pulse is applied, FFA will toggle, and Q_A will therefore go HIGH. What happens to FFB at the positive-going edge of CP_1? To find out, let us look at the input conditions of FFB. J_B and K_B are both LOW because Q_A, to which they are connected, has not yet gone HIGH. Remember, there is a propagation delay from the triggering edge of the clock pulse until the Q output actually makes a transition. So, $J_B = 0$ and $K_B = 0$ when the leading edge of the first clock pulse is applied. This is a NO CHANGE condition, and therefore FFB does not change state. A timing detail of this portion of the counter operation is given in Figure 9–11(a).

After CP_1, $Q_A = 1$ and $Q_B = 0$ (which is the binary 1 state). At the leading edge of CP_2, FFA will toggle, and Q will go LOW. Since FFB "sees" a HIGH on its J and K inputs when the triggering edge of this clock pulse occurs, Q_B goes HIGH. Thus, after CP_2, $Q_A = 0$ and $Q_B = 1$ (which is a binary 2 state). The timing detail for this condition is given in Figure 9–11(b). At the leading edge of CP_3, FFA again toggles to the SET state ($Q_A = 1$) and FFB remains SET ($Q_B = 1$) because J_B and K_B are both LOW. After this triggering edge, $Q_A = 1$ and $Q_B = 1$ (which is a binary 3 state). The timing detail is shown in Figure 9–11(c). Finally, at the leading edge of CP_4, Q_A and Q_B go LOW because they both have a toggle condition on their J and K inputs. The timing detail is shown in Figure 9–11(d). The counter has now recycled back to its original state, binary 0. The complete timing diagram is shown in Figure 9–12.

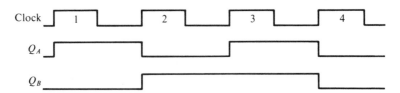

Figure 9–12.   Timing Diagram for the Counter of Figure 9–10.

Notice that all of the waveform transitions appear coincident; that is, the delays are not indicated. Although the delays are a very important factor in the counter operation, as we have seen in the preceding discussion, in an overall timing diagram they are normally omitted for simplicity. Major waveform relationships resulting from the logical operation of a circuit can be conveyed completely without showing minute delay and timing differences.

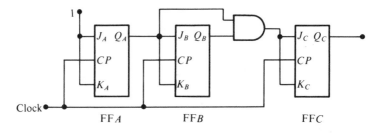

Figure 9–13.   A Three-Stage Synchronous Binary Counter.

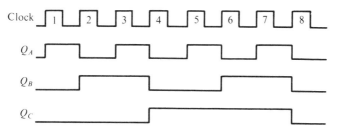

Figure 9–14.   Timing Diagram for the Counter of Figure 9–13.

A three-stage synchronous binary counter is shown in Figure 9–13 and its timing diagram in Figure 9–14. An understanding of this counter can be achieved by a careful examination of its sequence of states as shown in Table 9–4.

TABLE 9–4.   State Table for a
Three-Stage Binary Counter.

CP	Q_C	Q_B	Q_A
0	0	0	0
1	0	0	1
2	0	1	0
3	0	1	1
4	1	0	0
5	1	0	1
6	1	1	0
7	1	1	1

First, let us look at Q_A. Notice that Q_A changes on each clock pulse as we progress from its original state to its final state and then back to its original state. To produce this operation FFA must be held in the toggle mode by constant HIGHs on its J and K inputs. Now let us see what Q_B does. Notice that it goes to the opposite state following each time Q_A is a 1. This occurs at CP_2, CP_4, CP_6, and the clock pulse after CP_7 that causes the counter to recycle. To produce this operation, Q_A is connected to the J and K inputs of FFB. When Q_A is a 1 and a clock pulse occurs, FFB is in the toggle mode and will change state. The other times when Q_A is a 0, FFB is in the NO CHANGE mode and remains in its state. Next, let us see how FFC is made to change at the proper times according to the binary sequence. Notice that both times Q_C changes state, it is preceded by the unique condition of both Q_A and Q_B being 1s. This condition is detected by the AND gate and applied to the J and K inputs of FFC. Whenever both Q_A and Q_B are 1s, the output of the AND gate makes J_C and K_C HIGH and FFC toggles on the following clock pulse. At all other times J_C and K_C are held LOW by the AND gate output, and FFC does not change state.

Because the clock is applied to all flip-flops simultaneously in a synchronous counter, the clocking rate is not limited by the cumulative delays of all the flip-flops (as is the case with the asynchronous counter). For this reason the synchron-

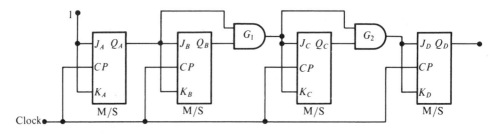

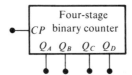

Figure 9–15. A Four-Stage Synchronous Binary Counter and Its Typical Logic Block Symbol.

ous type of counter can be operated at a much faster rate (higher clock frequency) than the asynchronous type.

We will look at one more synchronous binary counter before we move on. Figure 9–15 shows a four-stage binary counter and its equivalent logic symbol. Figure 9–16 shows the timing diagram for this counter. This particular counter is implemented with master-slave flip-flops rather than the edge-triggered types, and the counter makes its transitions on the trailing edge of the clock pulses. The reasoning behind the J and K input control for the first three flip-flops is the same as presented previously for the three-stage counter. The fourth stage, FFD, changes only twice in the sequence. Notice that both of these transitions occur following the times Q_A, Q_B, and Q_C are all 1s. This condition is detected by AND gate G_2, so that when a clock pulse occurs, FFD will change state. For all other times FFD's J and K inputs are LOW, and it is in a NO CHANGE condition.

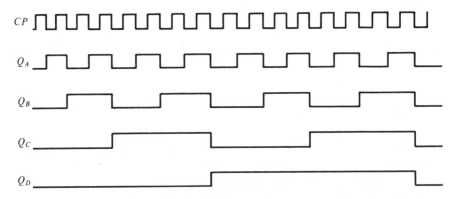

Figure 9–16. Timing Diagram for the Counter of Figure 9–15.

9–4 Frequency Division

The frequency division characteristic is inherent in the sequential operation of all counters. It was pointed out in Chapter 8 that the toggle flip-flop is basically a divide-by-two device because the Q or $\overline{Q}$ output is one-half the frequency of the clock input. This is shown in Figure 9–17.

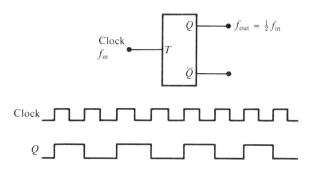

Figure 9–17. The Toggle Flip-Flop as a Divide-by-Two Device.

The two-stage binary counter is a divide-by-four device. The output of the second stage, Q_B, is one-fourth the frequency of the input clock, as illustrated in Figure 9–18.

In general, a binary counter with n stages will divide the clock frequency by a factor of 2^n. For example, a three-stage binary counter is a divide-by-eight counter, a four-stage binary counter divides by sixteen, a five-stage counter divides by thirty-two, and so on. Note that the maximum number of states through which a counter progresses (the modulus of a counter) is the same as the division factor.

A single counter can be used to divide a given clock frequency into a number of lower-frequency waveforms by taking the outputs off various stages of the counter. This is illustrated for a five-stage binary counter in Figure 9–19. The Q_A output of the first stage is one-half the frequency of the clock; the Q_B output of the second stage is one-fourth the frequency of the clock; the Q_C output of the

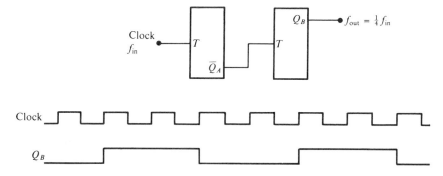

Figure 9–18. A Two-Stage Binary Counter as a Divide-by-Four Device.

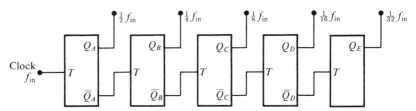

Figure 9–19.   Frequency Division with a Five-Stage Binary Counter.

third stage is one-eighth of the clock frequency; the output of the fourth stage is one-sixteenth of the clock frequency; and the output of the fifth stage is one-thirty-second of the clock frequency.

9–5   Modified Modulus Counters

Each of the binary counters discussed so far has had a modulus equal to 2^n, where n is the number of stages. This is the *maximum* modulus a counter can have, but it can be modified to have a *lesser* modulus without changing the number of stages. This is necessary when a non-power-of-two modulus is required. For example, if a counter with a modulus of five is required, three flip-flops must be used because a two-stage has a maximum modulus of four, which is insufficient. To produce a counter with a modulus of twelve, four stages are required because a three-stage counter can produce a maximum modulus of only eight.

In order to achieve a less-than-maximum modulus in a counter, some of the "natural" states must be skipped over. In other words, the counter is modified so that it does not go through all of the possible states. For instance, in a three-stage binary counter, the sequence of states is as follows:

CP	Q_C	Q_B	Q_A	
0	0	0	0	
1	0	0	1	
2	0	1	0	Recycles after state 4
3	0	1	1	for a total of five states.
4	1	0	0	
5	1	0	1	
6	1	1	0	
7	1	1	1	

If the counter is made to recycle after, say, the binary 4 state rather than after the binary 7 state, a modulus of five is achieved. How can this be done? Obviously the counter has a "natural" modulus of seven, so changes in the circuitry are necessary to modify it to a modulus-5 operation. When the counter reaches the binary 4 state, it must recycle to the binary 0 state on the next clock pulse. Notice that several things have to occur at this time: FFA is RESET and must remain RESET in going from state 4 to state 0; FFB is RESET and must remain RESET; and FF*C* is SET and must RESET. To summarize, when the counter is in the binary 4 state and a clock pulse occurs, the following conditions must exist on the *J* and *K* inputs of the flip-flops:

1.   FF*A*, NO CHANGE
2.   FF*B*, NO CHANGE
3.   FF*C*, TOGGLE

Since Q_C is a 1 for the first time in the sequence when the counter enters the binary 4 state, it can be used to set up the conditions listed, as shown in Figure 9–20. Notice that FF*A* is in a toggle mode ($J = 1$, $K = 1$) for all states prior to state 4. When the counter reaches state 4, $\overline{Q}_C$ goes to a 0, making $J_A = 0$ and $K_A = 0$, and holding FF*A* in the NO CHANGE condition. Since FF*A* is RESET, FF*B* is in the *NO CHANGE* condition also. The *J* and *K* inputs to FF*C* are both 1s because Q_C is connected to an input of the OR gate. These conditions are indicated in Figure 9–20.

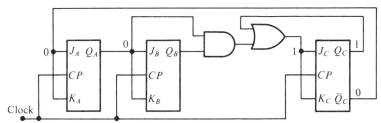

Figure 9–20.   A Modulus-5 Binary Counter.

From the specific case we have just discussed, a general approach to modifying the modulus of a counter can be stated.

1.   Determine the state at which the counter should recycle in order to produce the desired modulus.
2.   Detect the occurrence of this state so that a *unique* indication is available.
3.   Use this unique indication (1 or 0) to set up the proper levels on the inputs to each flip-flop so that they will recycle properly at a clock pulse.

To better understand this procedure, let us relate it to what was done to implement the modulus-5 counter previously discussed.

1.   The state at which the counter should recycle was determined to be the binary 4 state. This gives the counter a total of five states in its modified sequence.
2.   The occurrence of the binary 4 state was detected uniquely by looking for a HIGH on the Q_C output.
3.   The $Q_C = 1$ condition was inverted ($\overline{Q}_C$) and applied to J_A and K_A to keep FF*A* from changing state on the next clock pulse. The $Q_C = 1$ condition was applied to J_C and K_C through the OR gate to make FF*C* toggle on the next clock pulse.

Example 9–3   Implement a modulus-7 binary counter that recycles after state 6.

Solution:

The counter will have the following sequence:

	Q_C	Q_B	Q_A
0	0	0	0
1	0	0	1
2	0	1	0
3	0	1	1
4	1	0	0
5	1	0	1
6	1	1	0

A unique indication of the binary 6 state is

$$Q_B = 1 \quad \text{and} \quad Q_C = 1$$

In order for the counter to change from state 6 back to state 0, the following conditions are required:

FFA: NO CHANGE	$J_A = K_A = 0$
FFB: TOGGLE	$J_B = K_B = 1$
FFC: TOGGLE	$J_C = K_C = 1$

An implementation and the logic levels for the counter in state 6 are indicated in Figure 9–21.

Example 9–4   Implement a modulus-12 binary counter that recycles after state 11.

Solution:

The counter will have the following sequence:

	Q_D	Q_C	Q_B	Q_A
0	0	0	0	0
1	0	0	0	1
2	0	0	1	0
3	0	0	1	1
4	0	1	0	0
5	0	1	0	1
6	0	1	1	0
7	0	1	1	1
8	1	0	0	0
9	1	0	0	1
10	1	0	1	0
11	1	0	1	1

A unique indication of the binary 11 state is

$$Q_A = 1 \quad \text{and} \quad Q_B = 1 \quad \text{and} \quad Q_D = 1$$

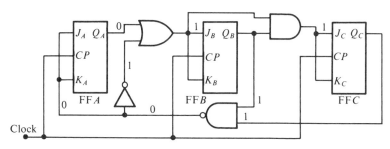

Figure 9–21.

In order for the counter to change from state 11 to state 0, the following conditions are required:

FF*A*: TOGGLE	$J_A = K_A = 1$	
FF*B*: TOGGLE	$J_B = K_B = 1$	
FF*C*: NO CHANGE	$J_C = K_C = 0$	
FF*D*: TOGGLE	$J_D = K_D = 1$	

A counter implementation is shown in Figure 9–22.

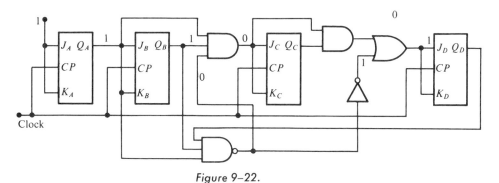

Figure 9–22.

9–6   Decade Counters

Decade counters are a very important category of digital counter because of their wide application. A decade counter has *ten* states in its sequence—that is, it has a *modulus of ten*. A logic symbol for a typical decade counter is shown in Figure 9–23. It consists of four stages and can have any given sequence of states as long as there are ten. A very common type of decade counter is the BCD (8421) coun-

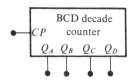

Figure 9–23.   A Typical Logic Block Symbol for a BCD Decade Counter.

TABLE 9–5.   States of a Decade
BCD Counter.

CP	Q_D	Q_C	Q_B	Q_A
0	0	0	0	0
1	0	0	0	1
2	0	0	1	0
3	0	0	1	1
4	0	1	0	0
5	0	1	0	1
6	0	1	1	0
7	0	1	1	1
8	1	0	0	0
9	1	0	0	1

ter, which exhibits a Binary-Coded-Decimal sequence as shown in Table 9–5. A state diagram is shown in Figure 9–24.

As you can see, the BCD decade counter goes through a straight binary sequence through the binary 9 state. Rather than going to the binary 10 state, it recycles to the 0 state. A synchronous BCD decade counter is shown in Figure 9–25.

The counter operation can be understood by examining the sequence of states in Table 9–5. First, notice that the FFA toggles on each clock pulse, so the logic equation for its J and K inputs is

$$J_A = K_A = 1$$

This is implemented by connecting these inputs to a constant HIGH level. Next

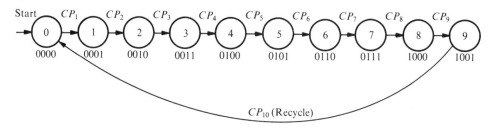

Figure 9–24.   State Diagram for a BCD Decade Counter.

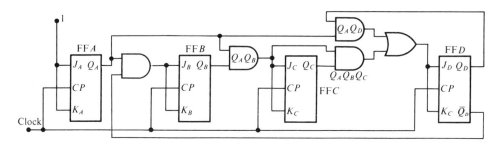

Figure 9–25.   A Synchronous BCD Decade Counter.

notice that FFB changes on the next clock pulse each time $Q_A = 1$, so the logic equation for its J and K inputs is

$$J_B = K_B = Q_A$$

This is implemented by connecting Q_A to the J_B and K_B inputs. FFC changes on the next clock pulse each time both $Q_A = 1$ and $Q_B = 1$. This requires an input logic equation as follows:

$$J_C = K_C = Q_A Q_B$$

The control is implemented by ANDing Q_A and Q_B and connecting the gate output to J_C and K_C. Finally, FFD changes to the opposite state on the next clock pulse each time $Q_A = 1$, $Q_B = 1$, and $Q_C = 1$ (state 7), or when $Q_A = 1$ and $Q_D = 1$ (state 9). The equation for this is as follows:

$$J_D = K_D = Q_A Q_B Q_C + Q_A Q_D$$

This function is implemented with the AND/OR logic as shown in the logic diagram in Figure 9–25. Notice that the only difference between this decade counter and a four-stage binary counter is the $Q_A Q_D$ AND gate and the OR gate; this essentially detects the occurrence of the 9 state and causes the counter to recycle properly on the next clock pulse. The timing diagram for the decade counter is given in Figure 9–26.

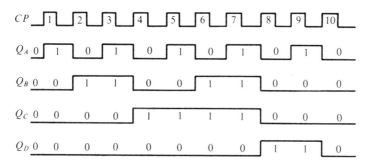

Figure 9–26. Timing Diagram for the BCD Decade Counter.

Example 9–5 Determine the sequence of the decade counter in Figure 9–27.

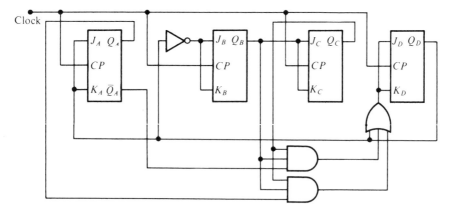

Figure 9–27.

Solution:

We will begin by assuming the counter is in the all 0s state, and evaluate the J and K inputs to each flip-flop at each clock pulse to determine what each stage does.

At CP_1: $J_A = K_A = 0$ Q_A remains 0 ⎫
                $J_B = K_B = 1$ Q_B goes to 1 ⎪ 0010
                $J_C = K_C = 0$ Q_C remains 0 ⎪
                $J_D = K_D = 0$ Q_D remains 0 ⎭

At CP_2: $J_A = K_A = 0$ Q_A remains 0 ⎫
                $J_B = K_B = 1$ Q_B goes to 0 ⎪ 0100
                $J_C = K_C = 1$ Q_C goes to 1 ⎪
                $J_D = K_D = 0$ Q_D remains 0 ⎭

At CP_3: $J_A = K_A = 0$ Q_A remains 0 ⎫
                $J_B = K_B = 1$ Q_B goes to 1 ⎪ 0110
                $J_C = K_C = 0$ Q_C remains 1 ⎪
                $J_D = K_D = 0$ Q_D remains 0 ⎭

At CP_4: $J_A = K_A = 0$ Q_A remains 0 ⎫
                $J_B = K_B = 1$ Q_B goes to 0 ⎪ 1000
                $J_C = K_C = 1$ Q_C goes to 0 ⎪
                $J_D = K_D = 1$ Q_D goes to 1 ⎭

At CP_5: $J_A = K_A = 1$ Q_A goes to 1 ⎫
                $J_B = K_B = 0$ Q_B remains 0 ⎪ 0001
                $J_C = K_C = 0$ Q_C remains 0 ⎪
                $J_D = K_D = 1$ Q_D goes to 0 ⎭

At CP_6: $J_A = K_A = 0$ Q_A remains 1 ⎫
                $J_B = K_B = 1$ Q_B goes to 1 ⎪ 0011
                $J_C = K_C = 0$ Q_C remains 0 ⎪
                $J_D = K_D = 0$ Q_D remains 0 ⎭

At CP_7: $J_A = K_A = 0$ Q_A remains 1 ⎫
                $J_B = K_B = 1$ Q_B goes to 0 ⎪ 0101
                $J_C = K_C = 1$ Q_C goes to 1 ⎪
                $J_D = K_D = 0$ Q_D remains 0 ⎭

At CP_8: $J_A = K_A = 0$ Q_A remains 1 ⎫
                $J_B = K_B = 1$ Q_B goes to 1 ⎪ 0111
                $J_C = K_C = 0$ Q_C remains 1 ⎪
                $J_D = K_D = 0$ Q_D remains 0 ⎭

At CP_9: $J_A = K_A = 0$ Q_A remains 1 ⎫
                $J_B = K_B = 1$ Q_B goes to 0 ⎪ 1001
                $J_C = K_C = 1$ Q_C goes to 0 ⎪
                $J_D = K_D = 1$ Q_D goes to 1 ⎭

At CP_{10}: $J_A = K_A = 1$ Q_A goes to 0 ⎫
                $J_B = K_B = 0$ Q_B remains 0 ⎪ 0000
                $J_C = K_C = 0$ Q_C remains 0 ⎪
                $J_D = K_D = 1$ Q_D goes to 0 ⎭

Sequence: 0, 2, 4, 6, 8, 1, 3, 5, 7, 9, and back to 0.

9–7   Up-Down Counters

An up-down counter is one that is capable of progressing in *either direction* through a certain sequence. An up-down counter is sometimes called a bidirectional counter and can have any specified sequence of states. For example, a BCD decade counter that advances upwards through its sequence (0, 1, 2, 3, 4, 5, 6, 7, 8, 9) and then can be reversed so it goes through the sequence in the opposite direction (9, 8, 7, 6, 5, 4, 3, 2, 1, 0) is an illustration of up-down sequential operation.

In general, most up-down counters can be reversed at any point in their sequence. For instance, the BCD decade counter mentioned can be made to go through the following sequence:

$$\overbrace{}^{\text{up}} \qquad \overbrace{}^{\text{up}}$$
$$0, 1, 2, 3, 4, \underbrace{5, 4, 3, 2}_{\text{down}}, 3, 4, 5, 6, 7, 8, \underbrace{7, 6, 5}_{\text{down}}, \text{etc.}$$

This is an illustration of just one possible sequence. A block diagram symbol for an up-down four-stage counter is shown in Figure 9–28.

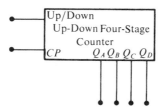

Figure 9–28.   Typical Logic Symbol for a Four-Stage Up-Down Counter.

Notice that in addition to the Q outputs and the clock pulse input, there is an *up-down control* input. The counter progresses in its forward (up) sequence when one level, say, a HIGH, is on the up-down control line and clock pulses are applied. It moves in the reverse (down) sequence when the opposite level is on the up-down control line and clock pulses are applied.

The following illustrates both the up and down sequences for a BCD up-down counter:

Q_D	Q_C	Q_B	Q_A
0	0	0	0
0	0	0	1
0	0	1	0
0	0	1	1
0	1	0	0
0	1	0	1
0	1	1	0
0	1	1	1
1	0	0	0
1	0	0	1

up   down

Keep in mind that, in general, an up-down counter can be characterized by any sequence or any number of stages. We are using the BCD type of counter just as an example of up-down counter operation. The arrows indicate the state-to-

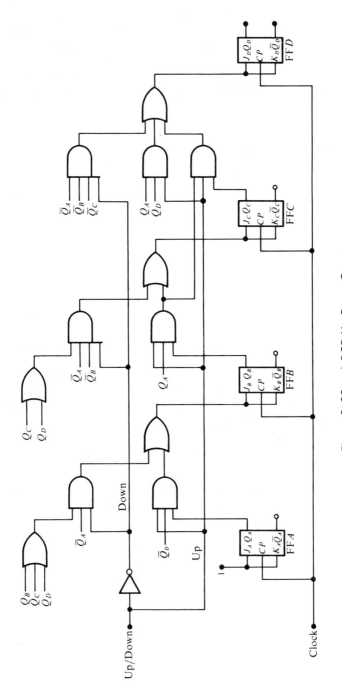

Figure 9-29. A BCD Up-Down Counter.

state movement of the counter for both its up and down modes of operation. (In Section 9–6, the logic required for the decade counter up sequence was developed. Obviously, some additional controls must be provided to enable the counter to operate in a down sequence. This requirement can best be understood by analyzing the down sequence as was done in the case of the forward BCD sequence. The up-down counter logic diagram is shown in Figure 9–29.)

Notice in the preceding sequence that Q_A changes state each time a clock pulse occurs for the down sequence. So, FFA operates continuously in the toggle mode, which means its J and K inputs always remain HIGH.

$$J_A = K_A = 1$$

Next, notice that Q_B changes state each time Q_A is a 0 *except* between the 0 state and the 9 state. The logic equation for the J and K inputs of FFB is therefore

$$J_B = K_B = \bar{Q}_A(Q_B + Q_C + Q_D)$$

Q_C makes a change to the opposite state only when both $Q_A = 0$ and $Q_B = 0$ *except* when the counter recycles from the 0 state to the 9 state. In order to make FFC toggle at these times, the equation for its J and K inputs is

$$J_C = K_C = \bar{Q}_A\bar{Q}_B(Q_C + Q_D)$$

This equation says that J_C and K_C are 1s when Q_A and Q_B are 0 *and* Q_C or Q_D are 1. If both Q_C and Q_D are 0, then $J_C = K_C = 0$, and the counter is in the NO CHANGE condition. This occurs only when the counter is in the 0 state. Q_D changes state when $Q_A = 0$ and $Q_B = 0$ and $Q_C = 0$. Notice that this occurs twice in the down sequence. The equation for J and K logic of FFD is therefore

$$J_D = K_D = \bar{Q}_A\bar{Q}_B\bar{Q}_C$$

Each of the conditions we have discussed simply produces a toggle condition or a NO CHANGE condition on the J and K inputs of each flip-flop at the appropriate time in the counter sequence.

Remember, these logic functions for the J and K inputs to each flip-flop are enabled only when the counter is in the down mode of operation (LOW on the up-down control line). The functions required for the up mode must be enabled when the counter is required to count in a forward sequence. This is accomplished by ANDing the up-down input with these functions, as illustrated in Figure 9–29.

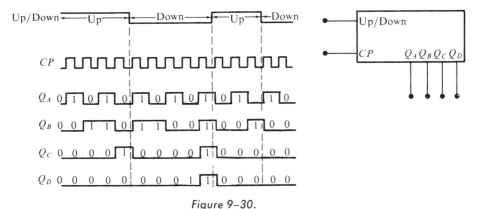

Figure 9–30.

A HIGH on the up-down control line activates the up sequence for this particular implementation.

Example 9–6   Determine the sequence of the four-stage binary up-down counter if the clock and up-down control inputs have waveforms as shown in Figure 9–30. Counter starts in all 0s state.

Solution:

From the waveforms, we can see that the counter sequence is as follows:

Q_D	Q_C	Q_B	Q_A	
0	0	0	0	
0	0	0	1	
0	0	1	0	up
0	0	1	1	
0	1	0	0	
0	0	1	1	
0	0	1	0	
0	0	0	1	down
1	1	1	1	
1	1	1	0	
1	1	1	1	
0	0	0	0	up
0	0	0	1	
0	0	0	0	down
1	1	1	1	

9–8   The Johnson Counter

The Johnson counter is sometimes known as a *shift counter* and produces a special sequence; the sequence for a four-stage counter is shown in Table 9–6(a) and that for a five-stage counter appears in Table 9–6(b).

TABLE 9–6.

(a) Four-Bit Johnson Sequence.					(b) Five-Bit Johnson Sequence.					
CP	Q_0	Q_1	Q_2	Q_3	CP	Q_0	Q_1	Q_2	Q_3	Q_4
0	0	0	0	0	0	0	0	0	0	0
1	1	0	0	0	1	1	0	0	0	0
2	1	1	0	0	2	1	1	0	0	0
3	1	1	1	0	3	1	1	1	0	0
4	1	1	1	1	4	1	1	1	1	0
5	0	1	1	1	5	1	1	1	1	1
6	0	0	1	1	6	0	1	1	1	1
7	0	0	0	1	7	0	0	1	1	1
	(back to 0)				8	0	0	0	1	1
					9	0	0	0	0	1
						(back to 0)				

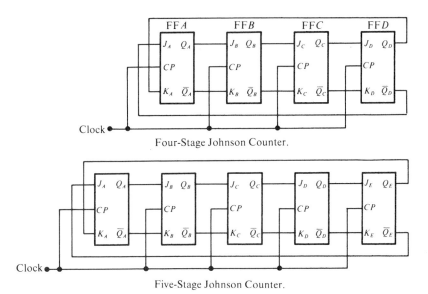

Four-Stage Johnson Counter.

Five-Stage Johnson Counter.

Figure 9–31.

Notice that the four-bit sequence has a total of *eight* states and that the five-bit sequence has a total of *ten* states. In general, an n-stage Johnson counter will produce a modulus of $2n$, where n is the number of stages in the counter. The implementations for the four- and five-stage Johnson counters are shown in Figure 9–31.

The implementation of a Johnson counter is very straightforward and is the same regardless of the number of stages. The Q output of each stage is connected to the J input of the next stage (assuming JK flip-flops are used), and the $\overline{Q}$ output of each stage is connected to the K input of the following stage. The single exception is that the last stage Q output is connected back to the K input of the first stage, and the $\overline{Q}$ output of the last stage is connected back to the J input of the first stage. As the sequences in Table 9–6 show, the counter will "fill up" with 1s from left to right, and then it will "fill up" with 0s again. One advantage of this type of sequence is that it is readily decoded with two-input AND gates. Diagrams of the timing operations of both the four- and five-stage counters are shown in Figures 9–32 and 9–33, respectively.

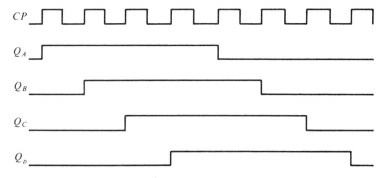

Figure 9–32. Timing Sequence for a Four-Stage Johnson Counter.

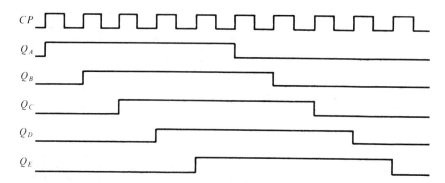

Figure 9-33.   Timing Sequence for a Five-Stage Johnson Counter.

Example 9-7 Show a modulus-12 Johnson counter and the timing diagram for its complete cycle.

Solutions:

See Figures 9-34 and 9-35.

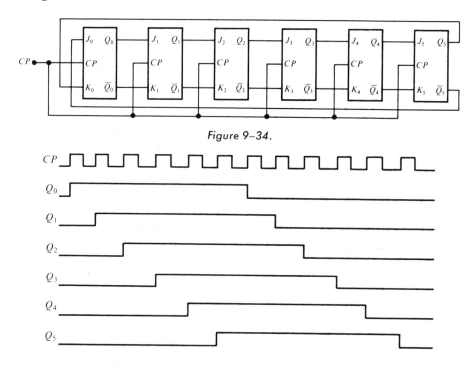

Figure 9-34.

Figure 9-35.

9-9 The Ring Counter

The ring counter utilizes one flip-flop for each state in its sequence, and is therefore the most wasteful of flip-flops. It does have the advantage that decoding is not required for decimal conversion. A logic diagram for a ten-stage ring counter is shown in Figure 9–36. The sequence for this ring counter is given in Table 9–7.

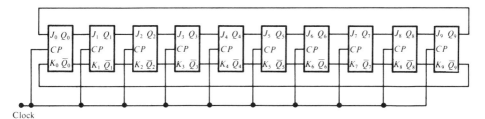

Figure 9–36.   A Ten-Stage Ring Counter.

TABLE 9–7.   Ring Counter Sequence (Ten Bits).

CP	Q_0	Q_1	Q_2	Q_3	Q_4	Q_5	Q_6	Q_7	Q_8	Q_9
0	1	0	0	0	0	0	0	0	0	0
1	0	1	0	0	0	0	0	0	0	0
2	0	0	1	0	0	0	0	0	0	0
3	0	0	0	1	0	0	0	0	0	0
4	0	0	0	0	1	0	0	0	0	0
5	0	0	0	0	0	1	0	0	0	0
6	0	0	0	0	0	0	1	0	0	0
7	0	0	0	0	0	0	0	1	0	0
8	0	0	0	0	0	0	0	0	1	0
9	0	0	0	0	0	0	0	0	0	1
					(back to 0)					

Actually, the ring counter can be thought of as a form of shift register since the 1 is simply shifted from one stage to the next. However, since a particular sequence is characteristic of this device, it is normally classified as a counter. Notice that the interstage connections are the same as for a Johnson, except that the feedback connections from the last stage to the first are not reversed. The ten outputs of the counter indicate directly the decimal count of the clock pulse. For instance, a 1 on Q_0 is a zero, a 1 on Q_1 is a one, a 1 on Q_2 is a two, a 1 on Q_3 is a three, and so on. You should verify for yourself that the 1 is always retained in the counter and simply shifted "around the ring," advancing one stage for each clock pulse.

Modified sequences can be achieved by having more than a single 1 in the counter, as will be illustrated in the following example.

Example 9–8   If the ten-stage ring counter of Figure 9–36 has the initial state 1010000000, determine the waveforms for each of the Q outputs.

Solutions:

See Figure 9–37.

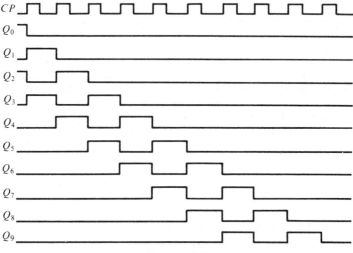

Figure 9–37.

9–10   Map Method for Counter Analysis and Design

A general, systematic method can be applied to implement *any* specified counter sequence using the Karnaugh map as the basic tool. This involves "plotting" on the maps the conditions that will cause each flip-flop in the counter to go to the proper next state from any state in the specified sequence. The method is best illustrated by example.

Example 9–9   Develop a three-bit counter using JK flip-flops to produce the following sequence of states:

Q_A	Q_B	Q_C
0	0	1
1	0	0
0	1	0
1	0	1
1	1	0
1	1	1
0	1	1

Solution:

Step 1.   Construct a Karnaugh map for three variables.

Step 2.   Plot the sequence of states on the map. We start with the 001 cell, go to the 100 cell, and then to the 010 cell, etc., as the sequence in Figure 9–38 indicates.

Step 3.   Contruct a Karnaugh map for the *J* input and a Karnaugh map for the *K* input of *each* flip-flop.

Step 4.   In each of the *J* and *K* maps, enter a 1 or a "*don't care*" as follows (an X is used for the *don't care*):

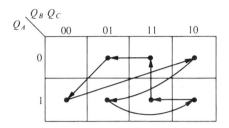

Figure 9–38.

(a)   Start with the cell corresponding to the first state in the sequence, 001. For FFA, Q_A changes from a 0 to a 1 in going to the next state. For this to occur, J_A must be a 1, so we enter a 1 into this cell of the J_A map. K_A can be either a 1 or a 0 in this case—we "don't care." Enter a *don't care* symbol in this cell of the K_A map.

(b)   For this same state we observe what Q_B does when it goes to the next state. As you can see, it remains a 0. In this case, J_B must be a 0, so nothing is entered into this cell of the J_B map. K_B can be either a 1 or a 0, and FFB will stay in the same state. Enter a *don't care* symbol in this cell of the K_B map.

(c)   Next we observe that Q_C goes from a 1 to a 0 from the first to the second state in the sequence. For this to occur, we don't care what J_C is as long as K_C is a 1. Enter a *don't care* symbol in this cell of the J_C map, and a 1 in this cell of the K_C map.

(d)   The procedure is repeated for each state in the sequence, with proper entries in each appropriate cell of the J and K maps as shown in Figure 9–39(a).

(e)   For all states that are not in the counter sequence, enter a *don't care* in the corresponding cell of each J and K map. This is done because the counter will never get into this un-allowed state in its normal operation, so we really don't care what the J and K inputs are for all of the flip-flops. In this case we enter a *don't care* symbol in the 000 cells.

Step 5.   After filling in the maps, we can now factor by grouping the 1s and Xs (*don't cares*) to determine the logic required for the J and K inputs to each of the flip-flops. The Xs can be counted as either a 1 or a 0 for the purpose of factoring the maps. This means that they can be included in a grouping to get a minimum logic expression or they can be omitted and not factored. In other words, the Xs do not have to be factored unless they can be included in a grouping to minimize the resulting expression. All of the procedures that have been de-scribed are done on the maps in Figure 9–39(a).

Step 6.   When the maps have been properly factored and we have a resulting logic expression for all of the J and K inputs, the counter in Figure 9-39(b) can be implemented to produce the required sequence.

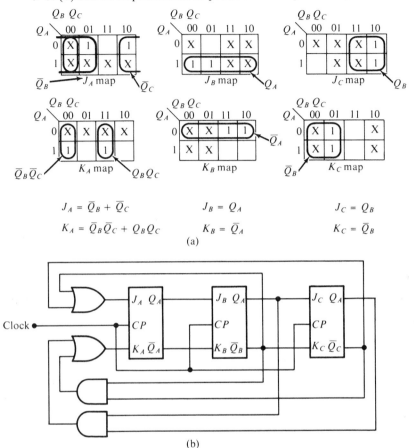

$$J_A = \overline{Q}_B + \overline{Q}_C \qquad\qquad J_B = Q_A \qquad\qquad J_C = Q_B$$

$$K_A = \overline{Q}_B\overline{Q}_C + Q_BQ_C \qquad K_B = \overline{Q}_A \qquad\qquad K_C = \overline{Q}_B$$

(a)

(b)

Figure 9-39.

With the previous example in mind, some general rules for implementing any counter sequence can be outlined as follows:

1. For a specified sequence requiring an n-stage counter, construct a Karnaugh map for n variables.

2. Plot the sequence of counter states on this map by starting with the cell corresponding to the first state in the sequence and drawing an arrow from that cell to the cell representing the next state in the sequence. Continue connecting each succeeding cell until the total specified counter sequence is represented on the map. This map is for convenience in plotting the J and K maps to follow.

3. Contruct an n-variable Karnaugh map for the J input and one for the K input of *each* flip-flop in the counter.

4. Beginning with the first state in the sequence, and for each succeeding state, determine what each of the n flip-flops must do in order to get to the

next state in the sequence. Does the flip-flop have to change its state or does it remain in the same state? This determines what is to be entered on the J and K maps for that particular flip-flop as follows:

    (a)   If the Q output of the flip-flop is to go from a 0 in its present state to a 1 in its next state, J must be a 1. Enter a 1 in the cell of the J map corresponding to the present state of the sequence. K can be either a 1 or a 0 (if you do not see why, refer to the JK flip-flop truth table). Enter an X in the cell of the K map for the present state of the sequence.

    (b)   If the Q output of the flip-flop is to go from a 1 in its present state to a 0 in its next state, J can be either a 1 or a 0 and K must be a 1. Enter an X in the present state cell of the J map and a 1 in the present state cell of the K map.

    (c)   If the Q output of the flip-flop is a 1 in its present state and does not change in its next state, J can be either a 1 or a 0 and K must be a 0. Enter an X in the present state cell of the J map and leave the present state cell of the K map blank.

    (d)   If the Q output of the flip-flop is a 0 in its present state and does not change in its next, J must be a 0 and K can be either a 1 or a 0. Leave the present state cell of the J map blank and enter an X in the present state cell of the K map.

    (e)   Enter an X in any cell of all J and K maps that corresponds to an unallowed state in the counter sequence.

5.   Factor each J and K map for a minimum Boolean expression. Use the Xs as either 1s or 0s. If an X can be included to create a larger grouping, use it. If it cannot, disregard it.

6.   When each of the J and K maps has been factored, you have a Boolean expression for the logic required for each J and each K input. The counter is then implemented by connecting each logic function to the corresponding J or K input. A similar procedure can be applied with RS flip-flops by taking into account the truth table differences.

Another example will serve to further illustrate this technique as applied to a four-variable sequence.

Example 9–10 Develop a four-stage counter using JK flip-flops to produce the following sequence

	Q_A	Q_B	Q_C	Q_D			Q_A	Q_B	Q_C	Q_D
1.	0	0	0	0		9.	0	1	1	1
2.	1	0	1	0		10.	1	1	0	1
3.	1	0	0	1		11.	0	1	0	1
4.	1	0	0	0		12.	1	1	0	0
5.	0	1	0	0		13.	0	1	1	0
6.	1	1	1	0		14.	0	0	1	0
7.	1	1	1	1		15.	0	0	1	1
8.	0	0	0	1		16.	1	0	1	1

Solution:
See Figure 9–40.

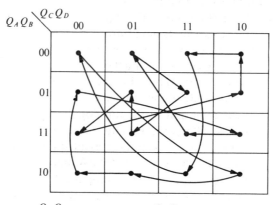

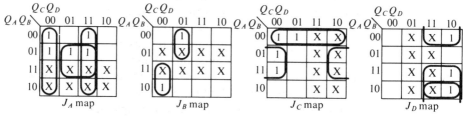

$$J_A = \bar{Q}_C\bar{Q}_D + Q_BQ_D + Q_C Q_D$$

$$J_B = Q_A\bar{Q}_C\bar{Q}_D + \bar{Q}_A\bar{Q}_C Q_D$$

$$J_C = \bar{Q}_A\bar{Q}_B + Q_B\bar{Q}_D$$

$$J_D = \bar{Q}_B Q_C + Q_A Q_C$$

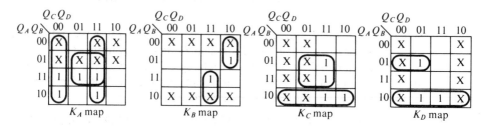

$$K_A = \bar{Q}_C\bar{Q}_D + Q_BQ_D + Q_CQ_D$$

$$K_B = Q_AQ_CQ_D + \bar{Q}_AQ_C\bar{Q}_D$$

$$K_C = Q_A\bar{Q}_B + Q_B Q_D$$

$$K_D = Q_A\bar{Q}_B + \bar{Q}_A Q_B\bar{Q}_C$$

Implementation of stage A

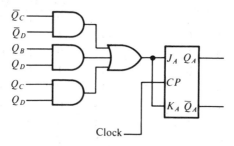

Figure 9–40.

Implementation of stage *B*

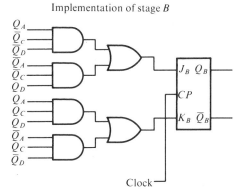

Implementation of stage *C*

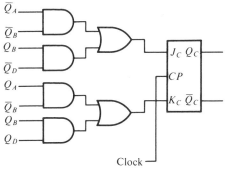

Implementation of stage *D*

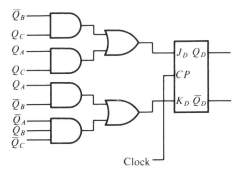

Figure 9–40, cont.

9–11 Special Counter Functions

Many counters incorporate extra logic functions that make them more versatile and useful. In this section we will discuss some of the most important of these functions.

The CLEAR Function

The purpose of the *clear* function is to initialize the counter to an all 0s state. This is normally done by connecting the *direct reset* or *clear* (C_D) of each flip-flop in the counter to a common line, as shown in Figure 9–41. In this particular case, the *clear* input is "active LOW" so that when a LOW level is applied to this line, each flip-flop in the counter is RESET and the counter is said to be cleared of its contents or initialized. Of course, "active HIGH" clear inputs can also be implemented as shown in Figure 9–42. This function is sometimes called *master reset*.

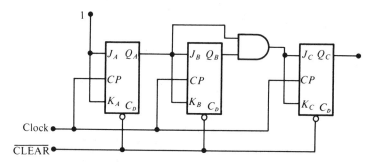

Figure 9–41.   Counter with a Clear Input Function (Active LOW).

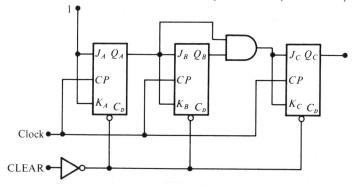

Figure 9–42.   Counter with a Clear Input Function (Active HIGH).

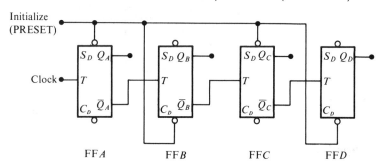

Figure 9–43.   A Counter That Can Be Initialized to the 1010 State.

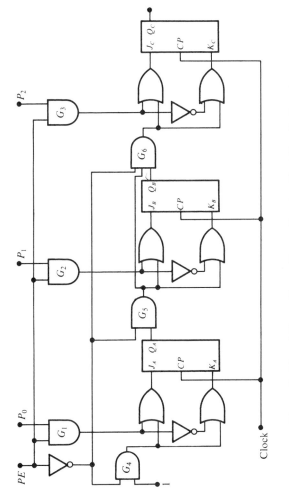

Figure 9–44. A Three-Stage Binary Counter That Can Be Preset.

301

Nonzero Initialization of a Counter

In some applications it is necessary to start the counter in a state other than all 0s. If a *single* nonzero initial state is required, it can be achieved by either setting or resetting the individual flip-flops in the counter as required. For instance, suppose a four-stage counter is to be initialized to the 1010 state. This can be done as shown in Figure 9–43 by causing FFA to SET, FFB to RESET, FFC to SET, and FFD to RESET by connecting a common line to the appropriate direct set (S_D) or clear input of each flip-flop in the counter.

As you can see, by proper connections to the direct set or clear of each counter stage, a counter can be initialized to a given state. A more versatile technique allows a counter to be initialized to any selected state as determined by inputs to the counter and synchronous with the clock. A configuration for accomplishing the initialization of a three-stage counter to any desired state is shown in Figure 9–44. The operation is as follows. The three parallel input lines, P_0, P_1, and P_2, are for applying the desired initial state. The parallel enable line, PE, when HIGH, enables AND gates G_1 through G_3, and allows the 1 or 0 appearing on each parallel input line to be gated through the OR gates onto the J or K input of the flip-flop. Notice that AND gates G_4 through G_6 are used to disable the normal J or K function required to sequence the counter during the time the counter is being initialized. During the normal operation of the counter when the PE line is LOW, the parallel initialization logic is disabled and the counter can advance through its specified sequence. The particular counter in Figure 9–44 is a three-stage binary counter, but the same technique can be applied to any type of digital counter. The following example will illustrate the counter being initialized to a specific state.

Example 9–11 Show the logic levels within the counter in Figure 9–45 for initialization to the binary 5 state, and describe the operation of the counter at this time.

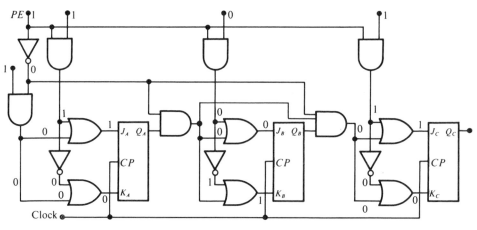

Figure 9–45.

Solutions:

When a clock pulse is applied, the counter is initialized to 101 because

$$J_A = 1, \quad K_A = 0$$
$$J_B = 0, \quad K_B = 1$$
$$J_C = 1, \quad K_C = 0$$

Count Enable

The count enable function can be implemented in several ways and is basically a counter control that allows the counter to be stopped or started at any time with the clock pulses still running. One way to implement this function is shown in Figure 9–46 for a three-stage binary counter. The count enable line must be HIGH for the counter to operate normally. If this line is LOW, the *J* and *K* inputs of each counter stage are held LOW, which is a NO CHANGE condition.

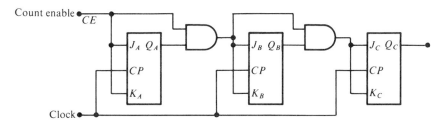

Figure 9–46.   Counter with a Count Enable Function.

Terminal Count Decode

Many counters utilize an extra gate to decode the last or *terminal* state in the sequence. This function is useful in many applications where several counters are connected and one counter must go through its entire sequence before the next counter can be advanced. A more thorough discussion of this is presented in Section 9-12 on cascaded counters. Figure 9–47 shows the terminal count output for a three-stage counter. In this particular case the terminal count is a binary 7 (111), which is decoded by AND gate G_1.

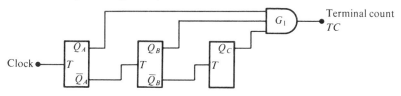

Figure 9–47.   Counter with a Terminal Count Function.

9–12   Cascaded Counters

Counters can be connected in series or *cascaded* in order to achieve higher modulus operation. In essence, cascading means that the last stage of one counter drives the input of the next counter. An example of two counters connected in cascade is shown for a two-stage and a three-stage ripple counter in Figure 9–48. The timing diagram is in Figure 9–49.

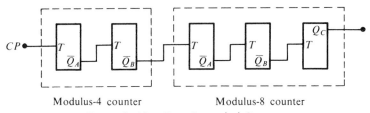

Modulus-4 counter   Modulus-8 counter

Figure 9–48. Two Cascaded Counters.

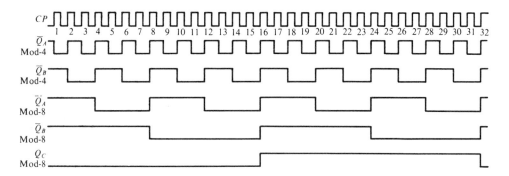

Figure 9–49. Timing Diagram for the Counter Configuration of Figure 9–48.

Notice in the timing diagram of Figure 9–49 that the final output of the modulus-8 counter Q_C occurs once for every 32 input clock pulses applied to the modulus-4 counter. The overall modulus of the cascaded counters is 32—that is, it acts as a divide-by-32 counter.

In general, the overall modulus of cascaded counters is equal to the product of each individual modulus. For instance, for the counter in Figure 9–48, 4 × 8 = 32.

When operating synchronous counters in a cascaded configuration, it is convenient to use the *count enable* and the *terminal count* functions to achieve higher modulus operation. Figure 9–50 shows two decade counters connected in cascade. The terminal count output of counter 1 is connected to the count enable input of counter 2. Counter 2 is inhibited by the LOW on its *CE* input until counter 1 reaches its last or terminal state and its terminal count output goes HIGH. This HIGH now enables counter 2, so that on the first clock pulse after counter 1 reaches its terminal count (CP_{10}), counter 2 goes from its initial state to its second state. Upon completion of the entire second cycle of counter 1 (when counter 1 reaches terminal count the second time), counter 2 is again enabled and advances to its next state. This sequence continues. Since these are decade counters, counter 1 must go through ten complete cycles before counter 2 completes its first cycle. In other words, for every ten cycles of counter 1, counter 2 goes through one cycle. This means that counter 2 will complete one cycle after 100 clock pulses. The overall modulus of these two cascaded counters is 10 × 10 = 100.

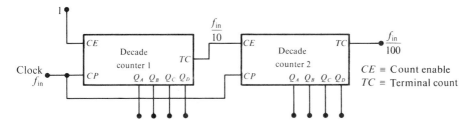

Figure 9–50.   A Modulus-100 Cascaded Counter.

When viewed as a frequency divider, the circuit of Figure 9–50 divides the input clock frequency by 100. Cascaded counters are often used to divide a high clock signal to obtain highly accurate pulse frequencies. Cascaded counter configurations used for such purposes are sometimes called *countdown chains*.

For example, suppose we have a basic clock frequency of 1 MHz and we wish to obtain 100 kHz, 10 kHz, and 1 kHz; a series of cascaded counters can be used. If the 1-MHz signal is divided by ten, we get 100 kHz. Then if the 100-kHz signal is divided by ten, we get 10 kHz. Another division by ten yields the 1-kHz frequency. The implementation of this countdown chain is shown in Figure 9–51.

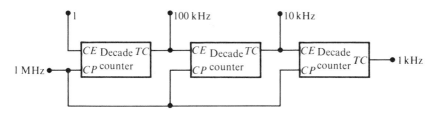

Figure 9–51.

Example 9–12   Determine the overall modulus of the two cascaded counter configurations in Figure 9–52.

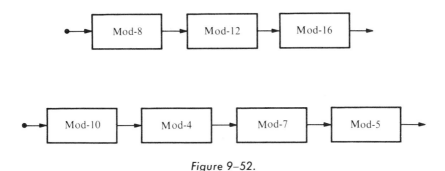

Figure 9–52.

Solutions:
The overall modulus for the three-counter configuration is

$$8 \times 12 \times 16 = 1536$$

The modulus for four counters is

$$10 \times 4 \times 7 \times 5 = 1400$$

Example 9-13   Assume that the following modulus counters are available: mod-5, mod-8, mod-10, and single flip-flops. Determine the appropriate cascaded configurations to obtain the following higher modulus counters: mod-4, mod-16, mod-25, mod-32, mod-40, and mod-50.

Solutions:
See Figure 9-53.

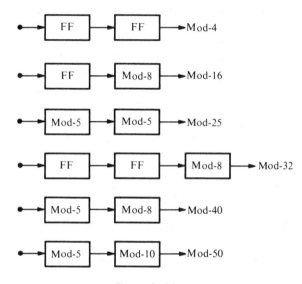

Figure 9-53.

9-13   Decoding a Counter

In many digital applications it is necessary that some or all of the counter states be decoded. The decoding of a counter involves using decoding logic gates to determine when the counter is in a certain state or states in its sequence. For instance, the terminal count function previously discussed is a case of a single state (the last state) in the counter sequence being decoded. Any state in a counter sequence can be decoded in a similar manner.

For example, let us say we wish to decode the binary 6 state of a three-stage binary counter. This can be done as shown in Figure 9-54. When $Q_C = 1$, $Q_B = 1$, and $Q_A = 0$, a HIGH appears on the output of the decoding gate indicating that the counter is in state 6.

Example 9-14   Implement the decoding of the binary 2 and the binary 7 state of a three-stage synchronous counter. Show the entire counter timing and the output waveforms of the decoding gates. Binary 2 = $\bar{Q}_C Q_B \bar{Q}_A$ and binary 7 = $Q_C Q_B Q_A$.

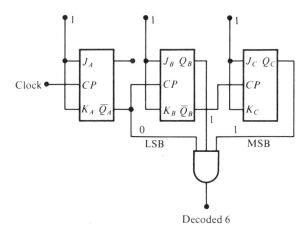

Figure 9–54. Decoding of Count 6.

Solutions:
See Figure 9–55.

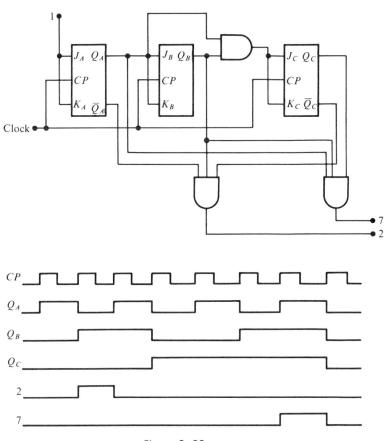

Figure 9–55.

PROBLEMS

9-1   Assuming a toggle flip-flop starts out in the RESET state, in what state will it be after
     (a)   two clock pulses   (b)   five clock pulses
     (c)   ten clock pulses   (d)   twenty-one clock pulses

9-2   What is the maximum modulus for a counter with each of the following numbers of .
flip-flops?
     (a)   2   (b)   4   (c)   5   (d)   6
     (e)   7   (f)   8   (g)   9   (h)   10

9-3   Determine the number of flip-flops required to implement each of the following
counters:
     (a)   modulus-4   (b)   modulus-3   (c)   modulus-5
     (d)   modulus-9   (e)   modulus-12   (f)   modulus-17
     (g)   modulus-36   (h)   modulus-65

9-4   Repeat Problem 9-3 for the following:
     (a)   modulus-32   (b)   modulus-39   (c)   modulus-50
     (d)   modulus-64   (e)   modulus-75   (f)   modulus-144
     (g)   modulus-257   (h)   modulus-512

9-5   For the ripple counter shown in Figure 9-56, draw the complete timing diagram for
eight clock pulses showing the clock, Q_A, and Q_B waveforms.

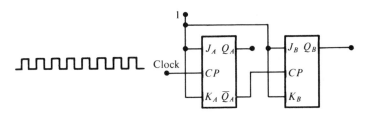

Figure 9-56.

9-6   For the ripple counter in Figure 9-57, draw the complete timing diagram for 16 clock
pulses. Show the clock, Q_A, Q_B, and Q_C waveforms.

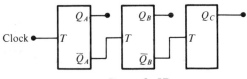

Figure 9-57.

9-7   Repeat Problem 9-6 using master-slave type of flip-flops that trigger on a negative-
going clock edge.

9-8   In the counter of Problem 9-6, assume each flip-flop has a propagation delay from
the triggering edge of the clock to a change in the Q output of 12 nanoseconds. Determine the worst (longest) delay time from a clock pulse to the arrival of the counter
in a given state. Specify the state or states for which this worst delay occurs.

9-9   If the counter of Problem 9–8 were synchronous rather than asynchronous, what would be the longest delay time?

9-10   Draw the complete timing diagram for the five-stage synchronous binary counter in Figure 9–58. Verify that the waveforms of the Q outputs represent the proper binary number after each clock pulse.

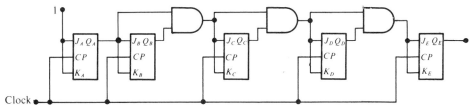

Figure 9–58.

9-11   Prove that the decade counter in Figure 9–59 progresses through a BCD sequence by analyzing the J and K inputs to each flip-flop prior to each clock pulse. Explain how these conditions in each case cause the counter to go to the proper next state.

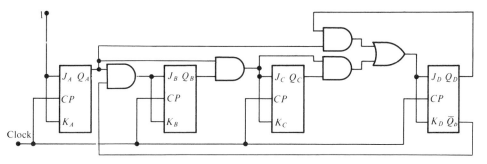

Figure 9–59.

9-12   Explain how the BCD decade counter acts as a divide-by-10 device, and show this with the proper waveforms.

9-13   List the unallowed or illegal states of the BCD decade counter.

9-14   What happens to the BCD decade counter if it accidentally gets into any of its unallowed states? Will it get back into its normal sequence or will it "lock up" in an illegal sequence?

9-15   Show how you would obtain the following frequencies from a 10-MHz clock signal if you have available single flip-flops, modulus-5 counters, and decade counters:

   (a) 5 MHz   (b) 2.5 MHz   (c) 2 MHz
   (d) 1 MHz   (e) 500 kHz   (f) 250 kHz
   (g) 62.5 kHz   (h) 40 kHz   (i) 10 kHz
   (j) 1 kHz

9-16   Draw a complete timing diagram for an up-down counter that goes through the following sequence. Indicate when the counter is in the up mode and when it is in the down mode.

$$0, 1, 2, 3, 2, 1, 2, 3, 4, 5, 6, 5, 4, 3, 2, 1, 0$$

9–17 The state diagram for a counter is shown in Figure 9–60. Sketch the corresponding timing diagram.

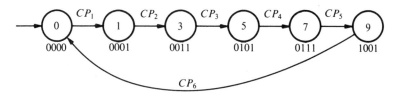

Figure 9–60.

9–18 Repeat Problem 9–17 for the state diagram in Figure 9–61.

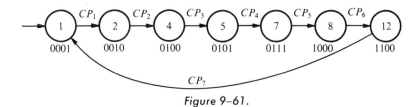

Figure 9–61.

9–19 The logic symbol for a BCD decade counter is shown in Figure 9–62. The waveforms are applied to the clock and clear inputs as indicated. Determine the waveforms for each of the counter outputs (Q_A, Q_B, Q_C, and Q_D). Changes in the outputs occur on the LOW-to-HIGH transitions of the clock. The clear is active LOW.

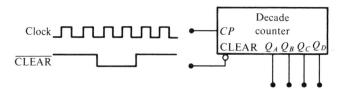

Figure 9–62.

9–20 The waveforms in Figure 9–63 are applied to the count enable (CE) and clock inputs (CP) as indicated. Draw the counter output waveforms in proper relation to these inputs. Changes in the outputs occur on the LOW-to-HIGH transitions of the clock. Count enable is active HIGH.

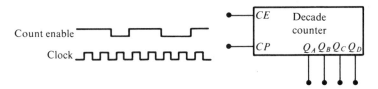

Figure 9–63.

9–21 The waveforms in Figure 9–64 are applied to the count enable, clear, and clock inputs as indicated. Sketch the counter output waveforms in proper relation to these inputs. Changes in the outputs occur on the LOW-to-HIGH transition of the clock. Count enable is active HIGH and clear is active LOW.

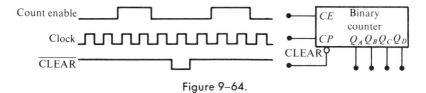

Figure 9–64.

9–22 Repeat Problem 9–21 including a terminal count output, and draw its waveform relative to the clock.

9–23 For the counter logic in Figure 9–65, properly connect the direct set and clear inputs of each stage so that the counter may be initialized to 10110 (left-most bit = Q_A).

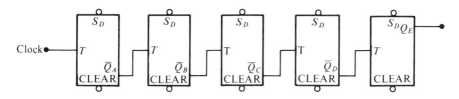

Figure 9–65.

9–24 Determine the sequence of the counter in Figure 9–66.

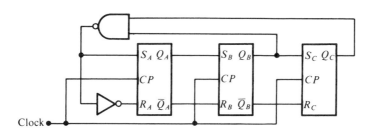

Figure 9–66.

9–25 Determine the sequence of the counter in Figure 9–67.

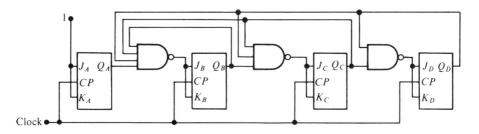

Figure 9–67.

9-26   Implement a counter to produce the following sequence. Use JK flip-flops.

CP	Q_A	Q_B
0	0	0
1	1	0
2	0	1
3	1	1

9-27   Implement a counter to produce the following sequence. Use JK flip-flops.

CP	Q_A	Q_B	Q_C
0	0	0	1
1	1	0	0
2	0	1	1
3	1	0	1
4	1	1	1
5	1	1	0
6	0	1	0

9-28   Implement a counter to produce the following sequence. Use JK flip-flops.

CP	Q_A	Q_B	Q_C	Q_D
0	0	0	0	0
1	1	0	0	1
2	0	0	0	1
3	1	0	0	0
4	0	0	1	0
5	0	1	1	1
6	0	0	1	1
7	0	1	1	0
8	0	1	0	0
9	0	1	0	1

9-29   Repeat Problem 9-27 using RS flip-flops.

9-30   How many flip-flops are required to implement each of the following in a Johnson counter configuration?

     (a)   divide-by-6   (b)   divide-by-10   (c)   divide-by-14
     (d)   divide-by-16   (e)   divide-by-20   (f)   divide-by-24
     (g)   divide-by-36

9-31   Draw the logic diagram for a divide-by-18 Johnson counter. Sketch the timing diagram and write the sequence in tabular form.

9-32   For the ring counter in Figure 9-68, draw the waveforms for each flip-flop output with respect to the clock. Assume FF0 is initially SET and the rest RESET. Show at least ten clock pulses.

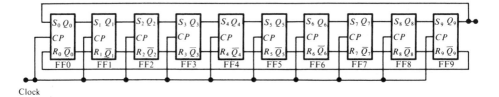

Figure 9-68.

9-33 The waveform pattern in Figure 9-69 is required. Show a ring counter and indicate how it can be preset to produce this waveform.

Figure 9-69.

9-34 For each of the cascaded counter configurations in Figure 9-70, determine the frequency of the waveform at each point indicated by circled numbers and the overall modulus.

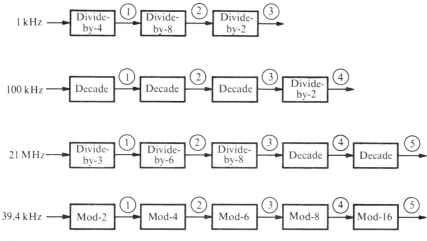

Figure 9-70.

9-35 Given a BCD decade counter, show the decoding logic required to decode each of the following states and how it should be connected to the counter. A HIGH output indication is required for each decoded state. MSB is to the left.

(a) 0001   (b) 0011   (c) 0101   (d) 0111   (e) 1000

9-36 For the four-stage binary counter connected to the 1-of-10 decoder in Figure 9-71, determine each of the decoder output waveforms in relation to the clock pulses.

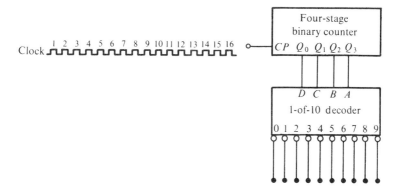

Figure 9-71.

CHAPTER 10

Sequential Logic: Registers

A category of sequential circuit known as the *register* is very important in applications involving the *storage* and *transfer* of data in a digital system. The basic difference between a register and a counter is that a register has no specified sequence of states except in certain very specialized applications. A register, in general, is used solely for the purpose of *storing* and *shifting* data (1s and 0s) entered into it from an external source and possesses no characteristic internal sequence of states.

10–1 The Storage Function of a Register

The storage capability of a register is one of its two basic functional characteristics and makes it an important type of *memory* device. Figure 10–1 illustrates the concept of "storing" a 1 or a 0 in a flip-flop. A 1 is applied to the input as shown, and a clock pulse is applied that stores the 1 by setting the flip-flop. When the 1 on the input is removed, the flip-flop remains in the SET state and the 1 is stored by it. The same procedure applies to the storage of a 0, as also illustrated in Figure 10–1.

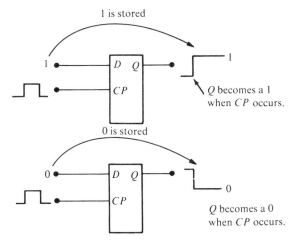

Figure 10–1. The Flip-Flop as a Storage Element.

315

The *storage capacity* of a register is the number of bits (1s and 0s) of digital data it can retain. Each stage of a shift register represents one bit of storage capacity, and therefore the number of stages in a register determines its total storage capacity.

Registers are commonly used for the *temporary* storage of data within a digital system (rather than more permanent storage usually accomplished with larger memories such as the magnetic core). Registers are normally implemented with a flip-flop for each stage, but other methods—such as utilizing the inherent capacitance in MOSFET devices and other types of capacitive and magnetic storage elements—can sometimes be employed.

10–2 The Shift Function of a Register

The *shift* capability of a register permits the movement of data stored from stage to stage within the register or into or out of the register. Figure 10–2 shows symbolically the types of data movement in shift register operations. The block represents any arbitrary register, and the arrow indicates direction and type of data movement.

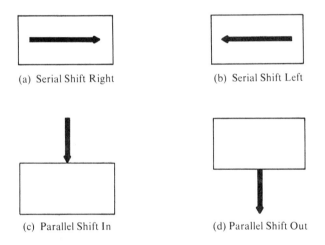

(a) Serial Shift Right

(b) Serial Shift Left

(c) Parallel Shift In

(d) Parallel Shift Out

Figure 10–2.   Illustration of Data Movement with Registers.

10–3 Serial In-Serial Out Shift Registers

This type of shift register accepts digital data serially—that is, one bit at the time. It produces the stored information on its output also in serial form. Let us first look at the serial entry of data into a typical shift register with the aid of Figure 10–3, which shows a four-bit device implemented with RS flip-flops.

With four stages, this register can store up to four bits of digital data; its *storage capacity* is four bits. We will illustrate the entry of the four-bit binary number 1010 into the register, beginning with the right-most bit. The 0 is put onto the

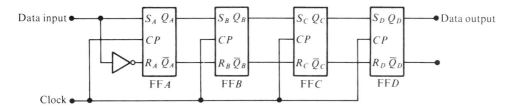

Figure 10-3.   Serial In–Serial Out Register.

data input line, making $S_A = 0$ and $R_A = 1$; when the first clock is applied, FFA is RESET, thus "storing" the 0 in FFA. Next the 1 is applied to the data input, making $S_A = 1$ and $R_A = 0$; $S_B = 0$ and $R_B = 1$, because they are connected to the Q_A and $\overline{Q}_A$ outputs, respectively. When the second clock pulse occurs, the 1 on the data input is "shifted" into FFA because FFA SETS, and the 0 that was in FFA is "shifted" into FFB. The next 0 in the binary number is now put onto the data input line and a clock pulse is applied. The 0 is entered into FFA, the 1 stored in FFA is shifted into FFB, and the 0 stored in FFB is shifted into FFC. Examination of the S and R inputs of each of the flip-flops will verify this operation. The

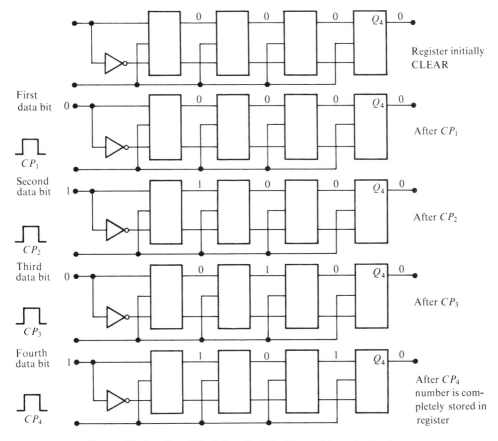

Figure 10-4.   Four Bits Being Serially Entered into the Register.

last bit in the binary number, a 1, is now applied to the data input, and a clock pulse is applied to the CP line. This time the 1 is entered into FFA, the 0 stored in FFA is shifted into FFB, the 1 stored in FFB is shifted into FFC, and the 0 stored in FFC is shifted into FFD. This completes the serial entry of the four-bit number into the shift register, where it can be retained for any length of time. Figure 10–4 illustrates each step in the shifting of the four bits into the register.

If we want to get the data out of the register, it must be shifted out serially and taken off the Q_D output. After CP_4 in the data entry operation described above, the right-most 0 in the number appears on the Q_D output. If clock pulse CP_5 is applied, the next bit appears on the Q_D output. CP_6 shifts the next bit to the output, and CP_7 shifts the last bit out, as illustrated in Figure 10–5. Notice that while the original four bits are being shifted out, a new four-bit number can be shifted in.

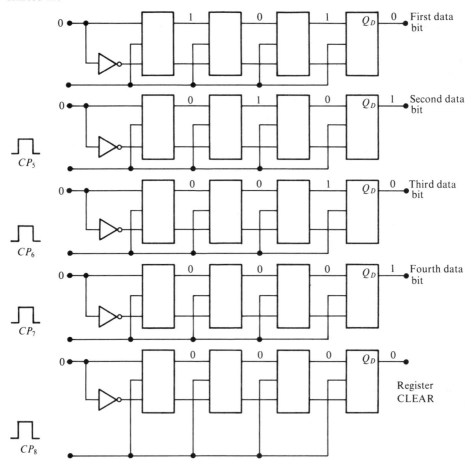

Figure 10–5.   Four Bits Being Serially Shifted Out of the Register.

Example 10–1   Show the states of the five-bit register in Figure 10–6(a) for the specified data input and clock waveforms. Assume the register is initially cleared (all 0s). D type flip-flops are used for this particular implementation.

Solutions:
See Figure 10–6(b).

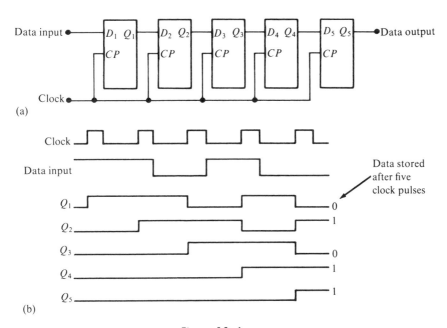

(a)

(b)

Figure 10–6.

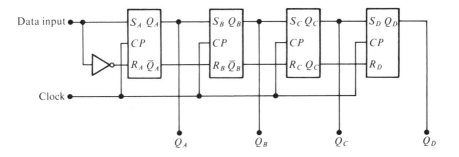

Figure 10–7.   Serial In-Parallel Out Register.

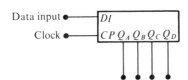

10-4   Serial In-Parallel Out Shift Register

Data are entered into this type of register in the same manner as discussed in the last section—serially. The difference is the way in which the data are taken out of the register; in the parallel output register, the output of each stage is available. Once the data are stored, each bit appears on its respective output line and all bits are available simultaneously, rather than on a bit-by-bit basis as with the serial output. Figure 10-7 shows a four-bit serial in-parallel out register and its equivalent logic block symbol.

Example 10-2   Show the states of the four-bit register for the data input and clock waveforms in Figure 10-8(a). The register initially contains all 1s.

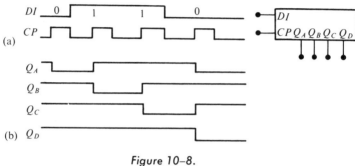

Figure 10-8.

Solutions:
The counter contains 0110 after four clock pulses. See Figure 10-8(b).

10-5   Parallel In-Serial Out Shift Register

For a register with parallel data inputs, the bits are entered simultaneously into their respective stages rather than on a bit-by-bit basis as with serial data inputs. The serial output is executed as described in Section 10-3 once the data are completely stored in the register. Figure 10-9 illustrates a four-bit parallel in-serial out register. Notice that there are four data input lines, D_A, D_B, D_C, and D_D, and a parallel enable line, PE, that allows four bits of data to be entered in parallel into the register. When the PE line is HIGH, gates G_1 through G_4 are enabled, allowing each data bit to be applied to the S input of its respective flip-flop and the data bit complement to be applied to the R input. When a clock pulse is applied, the flip-flops with a 1 data bit will SET and those with a 0 data bit will RESET, thereby storing the data word on one clock pulse.

When the PE line is LOW, parallel data input gates G_1 through G_4 are disabled and shift enable gates G_5 through G_7 are enabled, allowing the data bits to shift from one stage to the next. The OR gates allow either the normal shifting operation to be carried out or the parallel data entry operation to be accomplished, depending on which AND gates are enabled by the level on the PE line.

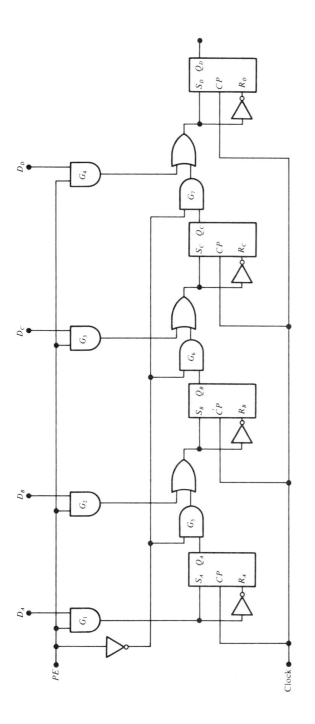

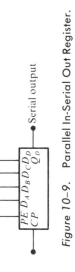

Figure 10–9. Parallel In-Serial Out Register.

321

Example 10-3 Show the Q_D waveform for a four-bit register with the input data, *PE*, and clock waveform as given in Figure 10–10(a). Refer to Figure 10-9 for the logic diagram. The register is initially cleared.

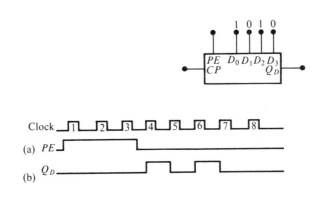

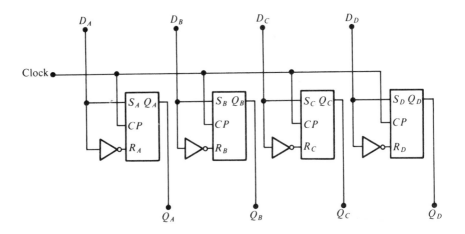

Figure 10–10.

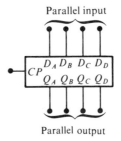

Figure 10–11.   Parallel In-Parallel Out Register.

Solution:

On clock pulses 1, 2, and 3, the same parallel data are respectively loaded into the register, keeping Q_D a 0. On clock pulse 4, the 1 is shifted onto Q_D; on clock pulse 5, the 0 is shifted onto Q_D; on clock pulse 6, the next 1 is shifted onto Q_D; and on clock pulses 7 and 8, all data bits have been shifted out and only 0s remain in register because no new data have been entered. See Figure 10-10(b).

10-6   Parallel In-Parallel Out Shift Register

Parallel entry of data was described in Section 10-5, and parallel output of data was also previously discussed. The parallel in-parallel out register employs both methods—immediately following the simultaneous entry of all data bits, they appear on the parallel outputs. This type of register is shown in Figure 10-11.

Example 10-4   Illustrate the transfer of four bits of binary data into and out of the register in Figure 10-12(a).

Solution:
See Figure 10-12(b).

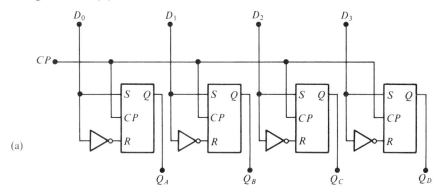

(a)

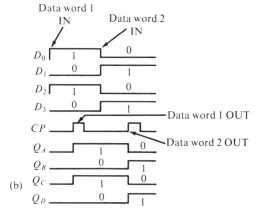

(b)

Figure 10-12.

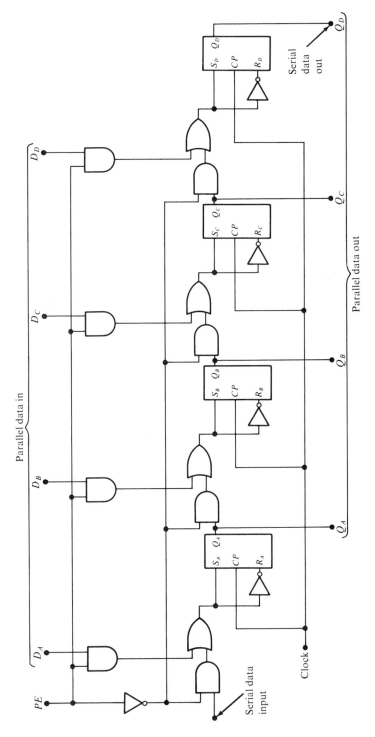

Figure 10–13.   Register with Serial and Parallel Input and Output Capacity.

10-7 Combination Registers

In the previous sections we discussed various shift registers and the means by which data are transferred into and out of them. It is, of course, possible to incorporate each method of data input and output in a single register; this is shown in the logic diagram of Figure 10-13 for a four-bit shift register. We are limiting our discussions to registers with only a few stages, but the same methods apply regardless of the number of stages.

Serial data can be entered when the *PE* line is LOW. A HIGH on the *PE* line allows parallel entry of four bits. The *Q* output of each stage is brought out, so a parallel output is always available. A serial output can be accomplished by taking the state of the Q_D output on a bit-by-bit basis as clock pulses are applied.

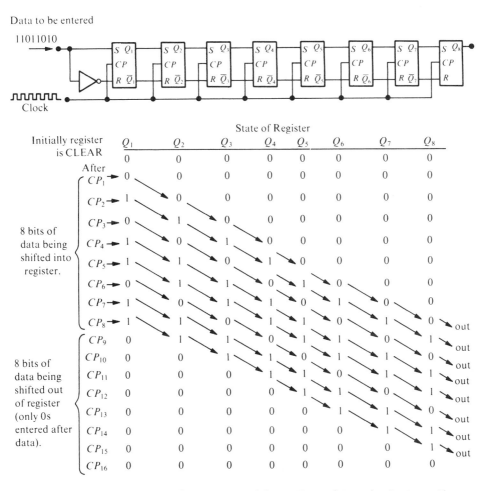

Figure 10-14. Illustration of a Data Word Being Entered into the Register, Then Shifted Right and Out.

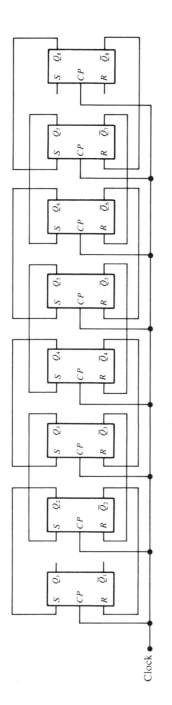

Clock

State of register

	Q_1	Q_2	Q_3	Q_4	Q_5	Q_6	Q_7	Q_8
Original data in register	1	1	0	1	0	1	0	1
After CP_1	1	0	1	0	1	0	1	
CP_2	0	1	0	1	0	1		
CP_3	1	0	1	0	1			
CP_4	0	1	0	1				
CP_5	1	0	1					
CP_6	0	1						
CP_7	1							
CP_8								

All data bits have been shifted out to the left.

Figure 10–15. Illustration of a Data Word Being Shifted Left in the Register. Method of Entry Is Not Shown.

326

10–8   The Shift-Right Function

In most of the registers discussed, a data bit can be shifted from one flip-flop to the next flip-flop to the right. This means that the data stored in a register can be moved from left to right through the register. This operation is illustrated in Figure 10–14 with an eight-bit serial shift register and its sequence of states for a specific data word. The eight-bit word is loaded from some unspecified source (perhaps another register) into the eight-bit shift register and shifted out to an unspecified destination (perhaps another register) for purposes of illustration.

10–9   The Shift-Left Function

Sometimes it is desirable to have the capability to move digital data from right to left within a shift register. This is done by implementing a register with a *shift-left* function, as shown in Figure 10–15. This operation is accomplished in a manner similar to the shift-right function, except that the register is essentially "turned around" by connecting the outputs of a flip-flop to the inputs of the preceding stage rather than to the following stage. Figure 10–15 shows the data in the register shifted out to the left.

10–10   The Bidirectional Shift Register

A *bidirectional* shift register is one in which the data can be shifted either left or right. This can be implemented by using gating logic that enables the transfer of a data bit from one stage to the next stage to the right or to the stage preceding it, depending on the level of a control line. A four-stage implementation is shown in Figure 10–16 for illustration purposes. A HIGH on the right/left control input allows data to be shifted to the right and a LOW enables a left shift of data. An examination of the gating logic should make the operation apparent. When the right/left control is HIGH, gates G_1 through G_4 are enabled and the state of the Q output of each flip-flop is passed through to the D input of the *following* flip-flop. When a clock pulse occurs, the data are then effectively shifted one place to the *right*. When the right/left control is LOW, gates G_5 through G_8 are enabled and the Q output of each flip-flop is passed through to the D input of the *preceding* flip-flop. When a clock pulse occurs, the data are then effectively shifted one place to the *left*.

Example 10–5   Determine the state of the shift register of Figure 10–16 after each clock pulse for the given right/left control input waveform in Figure 10–17(a). Assume $Q_A = 1$, $Q_B = 1$, $Q_C = 0$, $Q_D = 1$, and serial data in line is LOW.

Solution:
See Figure 10–17(b).

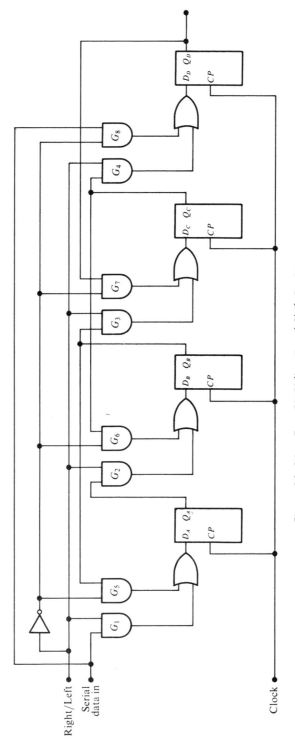

Right/Left

Serial
data in

Clock

Figure 10–16.   Four-Bit Bidirectional Shift Register.

328

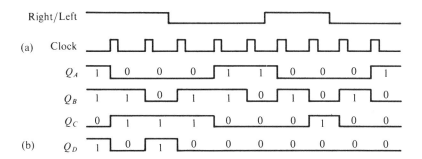

Figure 10–17.

10–11   Recirculating Shift Registers

In the serial output shift registers discussed in the previous sections, the data stored in the register are simply "lost" as they are shifted out unless they are transferred into another storage device. The purpose of a *recirculating* shift register is to restore the data that are being shifted out, thus retaining a data word instead of losing it. Recirculation of the contents of a register is accomplished by a feedback connection from the output to the input. As each bit is shifted out of the register, it is also shifted back into the input. When all of the bits in the register that make up the given data word have been shifted out, that data word has been reentered into the register. Figure 10–18 shows a four-bit shift register capable of recirculating its contents. The AND/OR logic on the input to the first stage permits serial entry of new data or recirculation of data already stored in the register, depending on whether the data control line is HIGH or LOW.

10–12   An Example of Data Transfer and Storage in a System of Registers

To illustrate how the various data transfer and storage capabilities of registers might be used to accept and store digital information, we will use a specific example. Suppose we have a digital system which must accept serial data in the form of BCD numbers, and temporarily store those data until they are needed for processing or display. This system must be able to accept and store a four-digit BCD number. Figure 10–19 shows one method of utilizing registers to perform this function.

In general, the operation of the system of Figure 10–19 is as follows: Four bits of data (one BCD digit) are serially clocked into register 1 by four clock pulses on clock line A. Next, a pulse on clock line B parallel shifts the four bits out of register 1 and into register 2. The next BCD digit (four bits) is then serially shifted into register 1 by four more pulses on clock line A; this is followed by a pulse on clock line B that parallel shifts the first BCD digit from register 2 to register 3 and the second BCD digit from register 1 into register 2. Next, the third BCD digit

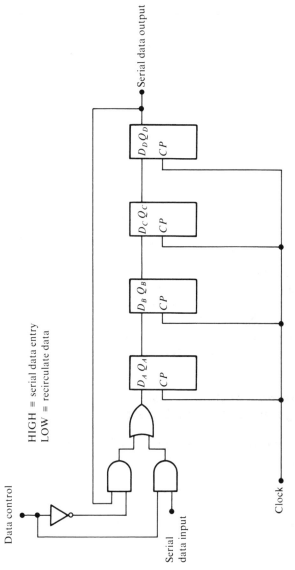

Data control

HIGH ≡ serial data entry
LOW ≡ recirculate data

Serial data input

Serial data output

Clock

Figure 10–18. **Four-Bit Shift Register with Data-Recirculating Capability.**

330

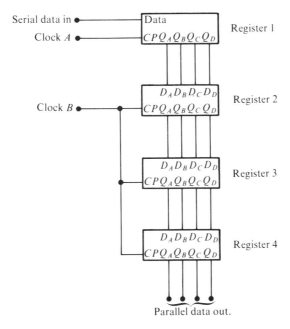

Figure 10–19. Register Configuration for Storing a Four-Digit Decimal Number in BCD Form.

is shifted into register 1 by four more clock pulses on clock line A. A pulse on clock line B now shifts the first BCD digit into register 4, the second BCD digit into register 3, and the third BCD digit into register 2. The fourth BCD digit is now shifted into register 1 by four more pulses on clock line A. At this time the four BCD digits representing a decimal number are stored in the four registers. This entire operation is illustrated in Figure 10–20.

When the data are required to be removed from the storage registers, each BCD digit (four bits) held in each register is sequentially shifted in parallel into register 4, and as each of the four bit groups enters this register, it can be taken out as parallel data or serially shifted out before the next digit is shifted into the register. This process is illustrated in Figure 10–21.

10–13   Dynamic Shift Registers

In integrated circuit form, shift registers are commonly implemented using MOSFET methods or bipolar transistor techniques such as TTL. The term *dynamic* usually refers to a type of MOS register that uses the inherent gate *capacitance* of a MOSFET as the basic storage element and takes advantage of the very high impedance of the MOSFET. The capacitance cannot retain its charge indefinitely and must be replenished periodically so that stored information will not be lost; this restriction requires that a shift register implemented with dynamic storage devices must be clocked at a specified minimum rate or higher to keep the data in the register from being lost as the charges on the capacitors leak off.

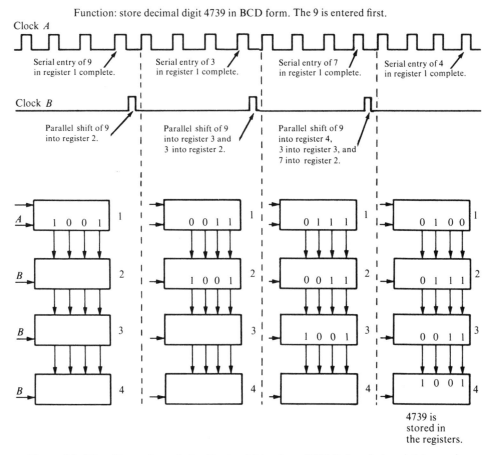

Figure 10–20. Illustration of the Decimal Number **4739** Being Entered into and Stored in the Storage Registers in BCD Form.

A single stage circuit implementation for an MOS dynamic shift register is shown in Figure 10–22. This type of shift register requires a *two-phase* clocking system; that is, two clock signals with pulses occurring at different times are used, as shown in Figure 10–22.

We will look at the basic operation of a single stage so that you will understand how a bit of data is entered into one stage and shifted out to another in a dynamic shift register. The operation is basically the same regardless of the number of stages, and therefore the following discussion applies to all stages in a register of any *length* (number of stages).

When a logic 1 (near 0 V) is applied to the stage input (the gate of Q_1) and a *phase-one* ($\phi 1$) clock pulse occurs, Q_1 is OFF because of the logic 1, and Q_2 and Q_3 are turned ON by the clock. The gate capacitance of Q_4 (C_1) charges near $-V_{DD}$ through Q_2 and Q_3, which represents a logic 1 being stored. When the *phase-two* clock pulse occurs, Q_4 is ON because of the negative charge on its gate, and both Q_5 and Q_6 are turned ON by the clock pulse. The gate capacitance of the input to the next stage discharges near 0 V through Q_4 and Q_6, thereby shifting the logic 1

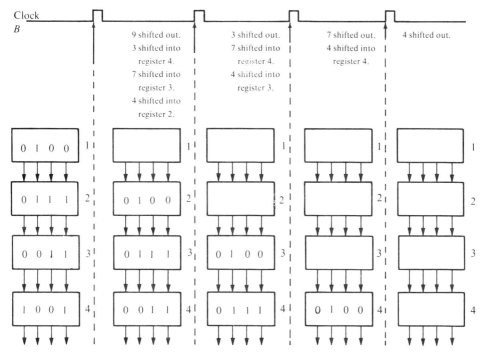

Figure 10–21. Illustration of Decimal Number 4739, Stored in BCD form, Being Shifted Out of the Storage Registers.

that was stored in this stage onto the input of the next stage. Now let us look at a condition for a logic 0 (negative voltage) on the gate of Q_1. When a *phase-one* clock pulse occurs, both Q_2 and Q_3 turn on. Q_1 is ON because of the logic 0 on its gate. The gate capacitance of Q_4 (C_1) then discharges to near ground potential through Q_1 and Q_3, which represents the storing of the logic 0. When a *phase-two* clock pulse occurs, both Q_5 and Q_6 are turned ON. The logic 0 on the gate of Q_4 keeps it OFF, and allows the gate capacitance of the input to the next stage to charge to near $-V_{DD}$. At this time the logic 0 that was stored in this stage appears on the input to the next stage.

Figure 10–23 is a logic block symbol for an *n*-bit dynamic MOS shift register. Notice that it has one data input, *two* clock inputs, and a data output. Remember, the important thing in the operation of a *dynamic* shift register is that *it cannot be operated below a specified clock frequency or the data will be lost* due to the capacitive storage.

10–14   Static Shift Registers

A second type of MOS register is known as a *static* shift register, which uses MOSFET flip-flops as storage elements so that data can be stored indefinitely. The static register requires no minimum clock frequency to maintain the stored data, as does the dynamic type.

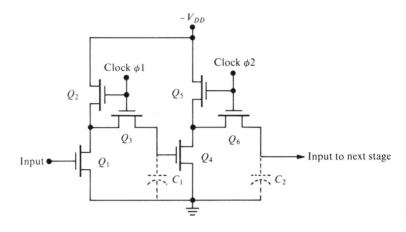

(a) Typical MOS Dynamic Shift Register Stage
(P-Channel Enhancement Mode FET)

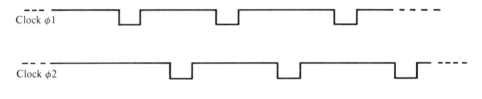

(b) Two-Phase Clock Waveforms

Figure 10–22.

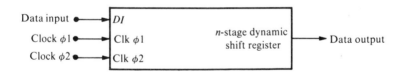

Figure 10–23.   Logic Block Diagram of an *n*-Stage Dynamic Shift Register.

10–15   The Shift Register as a Time-Delay Device

The serial in-serial out operation of a shift register is sometimes used to delay digital data for a fixed time. A bit of data entered into a shift register appears on the output of the last stage *n* clock periods later; this time delay between when the data bit goes into the register and when it comes out is illustrated in Figure 10–24. In this case, because we are using an eight-bit shift register, a 1 is clocked into the register on CP_1 and appears on the output of the register on CP_8.

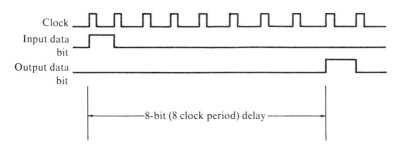

Figure 10–24. Operation of a Shift Register as a Time Delay Device.

PROBLEMS

10-1 Explain how a flip-flop can store a binary bit.

10-2 For the data input and clock timing diagram in Figure 10–25, determine the state of each flip-flop in the shift register of Figure 10–3. Assume the register contains all 1s initially.

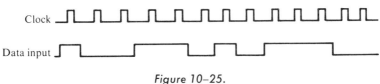

Figure 10–25.

10-3 Repeat Problem 10–2 for the waveforms in Figure 10–26.

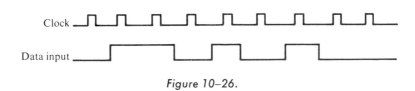

Figure 10–26.

10-4 What is the state of the register in Figure 10–27 after each clock pulse if it starts in the 101001111000 state?

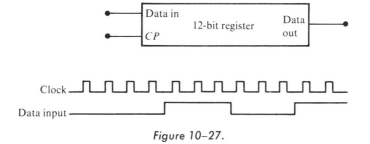

Figure 10–27.

10-5 For the serial in-serial out shift register, determine the data output waveform for the data input and clock waveforms in Figure 10–28. Assume the register is initially cleared and clocks on the positive edge.

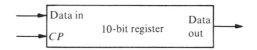

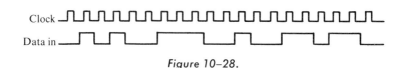

Figure 10–28.

10-6 Repeat Problem 10–5 for the waveforms in Figure 10–29.

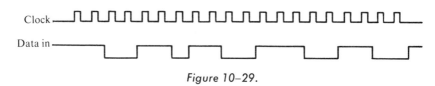

Figure 10–29.

10-7 Design an eight-stage serial in-serial out shift register using JK flip-flops.

10-8 The data output waveform in Figure 10–30 is related to the clock as indicated. What binary number was stored in the register if the first bit out is the LSB?

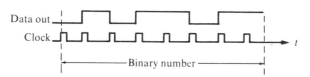

Figure 10–30.

10-9 For the eight-stage bidirectional register in Figure 10–31, determine the state of the register after each clock pulse for the right/left control waveform given. A 1 on this input enables a shift to the right and a 0 enables a shift to the left. Assume the register is initially storing a binary "76," with the right-most position being the LSB.

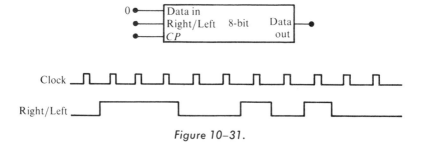

Figure 10–31.

10–10   Repeat Problem 10–9 for the waveforms in Figure 10–32.

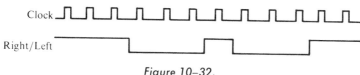

Figure 10–32.

10–11   For the recirculating shift register shown in Fig. 10–18, determine the serial data output waveform for 16 clock pulses if the register starts in the 1010 state and continuously recirculates the data.

10–12   For the register in Figure 10–18 and the waveforms in Figure 10–33, determine the serial data output waveform.

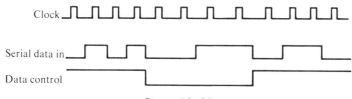

Figure 10–33.

10–13   For the system of registers shown in Figure 10–19, determine the decimal number stored at the completion of the clock pulse sequence given and for the data input waveform in Figure 10–34. Assume data are BCD, and the LSB is first.

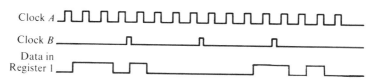

Figure 10–34.

10–14   Repeat Problem 10–13 for the waveform sequence in Figure 10–35.

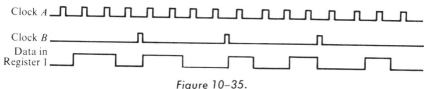

Figure 10–35.

10–15   Devise a system of registers to meet the following specifications:
   (a)   Store two four-digit decimal numbers.
   (b)   Each number is stored separately.
   (c)   The numbers are to be entered one digit at a time serially on the same data line.
   (d)   The numbers are to be shifted out one digit at a time serially on the same line.
   (e)   The numbers are coded in BCD.
   (f)   Show the timing required.

CHAPTER 11

Memories

The memory is a vital part of many digital systems. It is an arrangement of storage elements that is used to retain digital data. The individual element or cell in a memory system is capable of storing only a single bit of digital data. The total capacity of a memory is therefore dependent on the number of storage cells in the memory.

There are many types of memory elements in use today, and these elements are arranged in many ways to form complete memory systems. In this chapter we will examine some of the most important types of memory elements and systems.

11–1   The Magnetic Core

The *magnetic core* is one of several magnetic devices used as storage elements and is perhaps one of the most widely applied in memory systems, especially in large computers.

The typical magnetic core is toroidal (doughnut-shaped) and is composed of a ferromagnetic composition material. Memory cores are typically very small, with diameters ranging down to a few tenths of mils (millinches). Figure 11–1 is a pictorial representation of a magnetic core.

The magnetic properties of this type of core exhibit a two-state characteristic; that is, a core can be magnetized in two directions, making it ideal for the storage of binary information. This magnetic property can be described graphically by what is called a *hysteresis curve,* which is a plot of the *magnetic flux density B* versus the *magnetic force H,* as shown in Figure 11–2.

If sufficient magnetizing force is applied in a given direction, the magnetic flux density increases along the curve *abc.* The core is in a *saturated* condition from point *b* to point *c,* meaning that any further increase in the magnetizing force will produce very little change in the flux density. When the magnetizing force is removed, the flux density follows the *cbd* portion of the curve and stays at point *d,*

Figure 11–1.   Toroidal Magnetic Core.

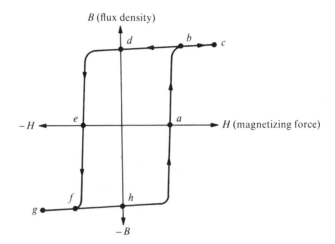

Figure 11–2.   Hysteresis Curve for a Square Loop Magnetic Core.

which is called the positive *remanent* flux state. Once the core is saturated in the positive direction, it will remain magnetized in that direction when the magnetizing force is reduced to zero. Thus, a core is capable of storing energy in the magnetic field without additional external energy. Now, if a sufficient magnetizing force is applied in the opposite direction (represented to the left of the origin of the graph), the flux density will change along the curve *defg* and will saturate in the opposite direction. When this magnetizing force is removed, the flux returns to point *h*, which is the negative remanent flux state. Thus, if the core is driven into one of its two magnetic states, it can remain there indefinitely without consuming any energy. In a digital memory, one of the core states represents a 1 and the other a 0.

How can a magnetic core be driven into either of its two states? Simply run a wire through the core and pass a sufficient amount of current through the wire to create a magnetizing force by virtue of the magnetic field around the current conductor. Figure 11–3 shows the direction of the magnetic field within the core for each direction of current through the wire. This direction can be remembered by using the "right-hand rule," which says that if the thumb of the right hand points in the direction of the current, the fingers point in the direction of the magnetic field.

The value of current required to switch a core into either of its two states is known as the *critical value* or the *full-select value* of magnetizing current, which we will designate I_m. As you can see, if the full-select current is made to flow mo-

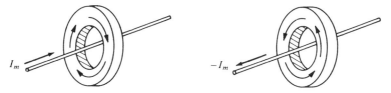

Figure 11-3.   Direction of the Magnetizing Current Determines the Direction of the Magnetic Field within the Core.

mentarily in one direction, the core will switch to the corresponding magnetic state. If a full-select value of current momentarily flows in the opposite direction, the core will go to its other magnetic state. By this basic method a 1 or a 0 can be stored, depending on the direction in which the current flows through the wire; entering or storing a 1 or a 0 in a memory core is termed *writing* into the memory.

Once a bit of information is written into a core, how do we detect what is stored? Notice that on the hysteresis curve, a large transition of magnetic flux density occurs when the core is driven from one of its remanent flux states to the other. This change in the magnetic field direction will induce a voltage in a wire passing through the core. Therefore, when the core is driven from one state to the other, the transition can be detected by measuring this induced voltage. If the positive remanent state is selected as the 1 state, the negative remanent state is then the 0 state. If a current is made to flow in a direction through the wire such that the core will always be driven to the 0 state, an induced voltage will occur on another wire if the core is in the 1 state and makes the transition to the 0 state. If the core is already in the 0 state, a much smaller induced voltage will occur due to a small transition from the 0 remanent state to a saturated state and back. By this basic method we can detect whether a 1 or a 0 is stored in the core; this process is called *reading* information from the core. Basically, two wires through the core are required—one to carry the *read* or *write* currents and the other to detect or sense the induced voltage. The latter is called the *sense line.* This is illustrated in Figure 11-4.

Figure 11-5(a) shows a typical curve for the voltage on the sense line for both a 1 and a 0 being read from the core. The voltage induced by a stored 1 is con-

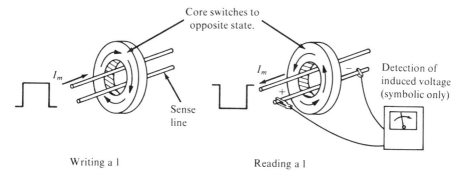

Writing a 1   Reading a 1

Figure 11-4.   Basic Illustration of Read and Write Operations with a Magnetic Core.

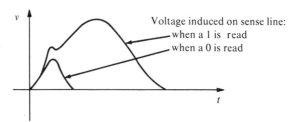

(a) Typical Curves of the Voltage Induced on the
Sense Line When the State of the Core Is Read

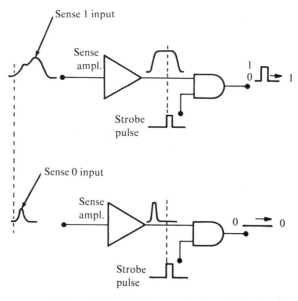

(b) One Method of Detecting the Sense Output
of a Magnetic Core

Figure 11–5.

siderably longer in duration and normally greater in amplitude than that for a
stored 0 and can be readily detected by a strobed amplifier connected to the sense
line. "Strobed" means that the sense amplifier circuit is enabled only at a specified
time to "look" for a voltage on the sense line. If the strobe pulse is delayed past
the duration of a 0 sense voltage, then a 1 sense voltage can be detected when it
occurs. Figure 11–5(b) illustrates one method of strobing a saturated sense amplifier to distinguish between a 1 and a 0 sense voltage.

The write and read functions can be performed using a single wire, as previously discussed, by properly timing each current pulse so that the time at which
a write pulse occurs is distinct from the time of a read pulse. This operation is
illustrated in Figure 11–6. As shown, a bit is stored in the core by a pulse of write

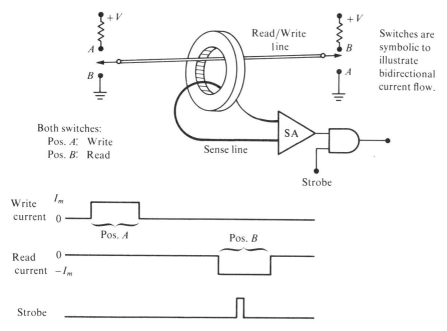

Figure 11-6.   Simplified Illustration of Read/Write Implementation and Timing.

current. Later a pulse of read current is applied in the opposite direction, and at an appropriate time the sense line is strobed.

11-2   Magnetic Core Memory Arrays

In most applications it is necessary to store large numbers of bits (1s and 0s), and therefore a magnetic core memory is made up of many individual cores. We will discuss the basic configuration of an array of cores in this section, using small arrays for clarity and purposes of illustration. The same principles apply to larger arrays.

Each core in the simplified array of Figure 11-7 (sometimes called a core plane) occupies a unique location that is identified by X and Y coordinates; the location is called the *address* of the core within the array. For instance, the co-ordinates $X_1 Y_1$ identify the core in the lower left corner of the array; the co-ordinates $X_4 Y_4$ identify the core in the upper right corner; and so on.

As previously discussed, a full-select current, I_m, is required to put the core into one of its magnetic states. If one-half of the current (the *half-select* current or $I_m/2$) is made to flow through one of the X lines, and the same value is made to flow through one of the Y lines in a corresponding direction, then the currents will intersect at a particular core dependent on the X and Y lines selected. When a half-select current passes through each of the select lines in a particular core, the two currents are additive and the core is switched to either the 1 or the 0 state, depending on the direction of the two currents. In this manner, information can be written into or read out of the core array at specified locations or addresses. The

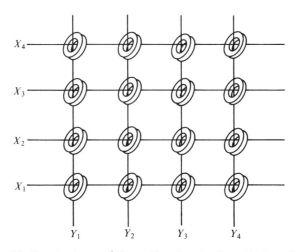

Figure 11–7.   An Array of Cores Showing the X and Y Coordinates.

basic operation is illustrated in Figure 11–8 where the $X_2 Y_3$ core is selected and the half-select currents through the X_2 line and the Y_3 line cause a 1 to be written into the core. Notice that a half-select current passes through each core "threaded" on the X_2 line and each core on the Y_3 line. Since these cores have only one-half of the critical value of current passing through each of them, they do not switch. Only the core through which the currents are *coincident* is affected; for this reason, a memory that uses this type of cell selection is called a *coincident-current* memory.

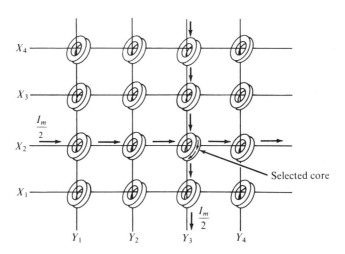

Figure 11–8.   A Core Array Showing the Selection of a Specific Core by Coincident Current on the X and Y Lines.

In addition to the X and Y select lines, a sense line is required to detect an induced voltage when the memory core array is being "read." A single wire is used for the sense line and threaded through *each* core as shown in Figure 11–9. Because only one core is selected at a time, only one sense line is required. Any induced voltage appearing on the sense line is a result of the particular core selected.

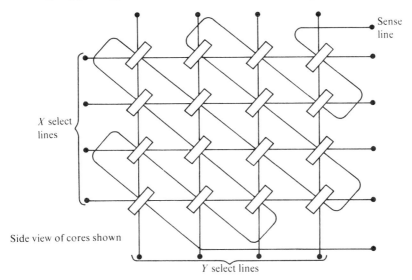

Figure 11–9.   Core Array Showing X and Y Select Lines and a Sense Line.

11–3   The Coincident-Current Core Memory

Magnetic core memories are normally composed of several core planes situated in a three-dimensional or "stack" arrangement. The reason for this is that binary words of several bits can be stored where each bit in a specific word occupies the same corresponding cell in each core plane, as illustrated symbolically in Figure

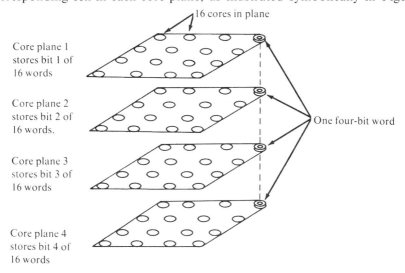

Figure 11–10.   Symbolic Representation of Four Core Arrays (Core Planes) Used to Form a Memory with a Capacity of 16 Four-Bit Words.

11-10. Each square represents a core plane, and a single bit of a given binary word is contained in each. Core plane 1 stores bit 1 of each of n words, core plane 2 stores bit 2 of each of n words, and so on. For example, if a core plane contains 16 cores, and there are four core planes in the memory, then the memory can store 16 four-bit binary words.

For a multiplane memory, each of the X and Y select lines "threads" through the corresponding row and column of *each* plane, as illustrated in Figure 11-11 with a single X select line and a single Y select line for three core planes. All other lines and cores are omitted for clarity. When a half-select write current is passed through the X line and a half-select write current is passed through the Y line, the corresponding cores of each plane are selected by the coincident write currents. In a practical memory, the corresponding cores in each plane are always driven toward the same state, say, a 1.

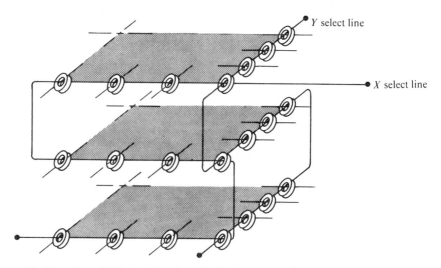

Figure 11-11.   X and Y Select Lines in a Multiplane Coincident-Current Memory.

To store a 0 in any of the cores, a half-select value of current called the inhibit current must be passed through the selected core in the opposite direction of the write currents; this cancels the effect of one of the half-select write currents so that the core will not be set to a 1. The inhibit current requires an extra line, called the *inhibit* line. A separate inhibit line for each plane in the memory is threaded through all of the cores in the plane so that a half-select value of the inhibit current can flow in the opposite direction of the write current. The inhibit is used only when a 0 is to be written into a given core; in other words, the inhibit current prevents a 1 from being stored. The operation is illustrated for a one-core plane in Figure 11-12.

Now we will examine the complete operation of a simplified coincident-current core memory in an effort to put together all of the topics discussed thus far and to introduce an additional concept. We will use a three-plane memory as an example, and restrict our discussion to writing and reading a binary word into a

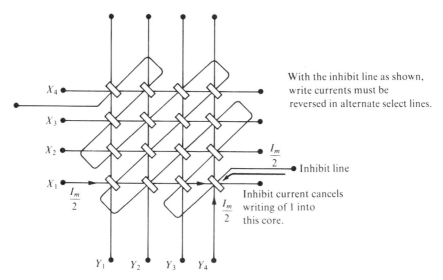

Figure 11–12. Core Array Showing How a 0 Is Written into a Selected Core by Current in the Inhibit Line.

single location (address) in the memory. The discussion, of course, will apply in general to any of the memory addresses.

A coincident-current core memory is shown in Figure 11–13 with only a single set of select lines pictured; the other select lines are omitted so the figure will not be overly complex. Each core plane has a single wire for its inhibit line that is connected to a driver circuit. The sense line for each core plane is connected to a sense amplifier that detects and amplifies the induced voltage from the selected core. As stated earlier, an induced voltage when a 1 is stored in the core is of longer duration than when a 0 is stored. In order to distinguish between a 1 and a 0, it is necessary to strobe the output of the sense amplifier after the 0 voltage has died but before the 1 voltage dies; this is done by pulsing the strobe gates at the appropriate time after the core is read. Flip-flops are used to temporarily store the bit that is read out of the core so that it can be written back into the core on the next write pulse (unless new information is to be entered). This method, *nondestructive readout,* prevents a binary word from being lost once it is read from the memory, and for this reason a read operation is always followed by a write operation.

We will now go through the basic operation of the memory of Figure 11–13 for a complete read/write cycle. The timing diagram for this memory cycle is shown in Figure 11–14.

We will begin by assuming that the binary word 101 is stored in the memory. A pulse of *read* current is applied to the X and Y select lines, *driving each of the three cores toward the 0 state.* The core in plane 1 will make a transition from the 1 state to the 0 state, inducing a voltage on the *bit 1 sense line.* This voltage is amplified by *sense amplifier 1* and sampled by the *strobe* pulse, which gates a HIGH level through gate SG_1 to SET *flip-flop 1.* Bit 1 has now been *read* out of the memory and temporarily stored in flip-flop 1. Simultaneously the 0 stored in core

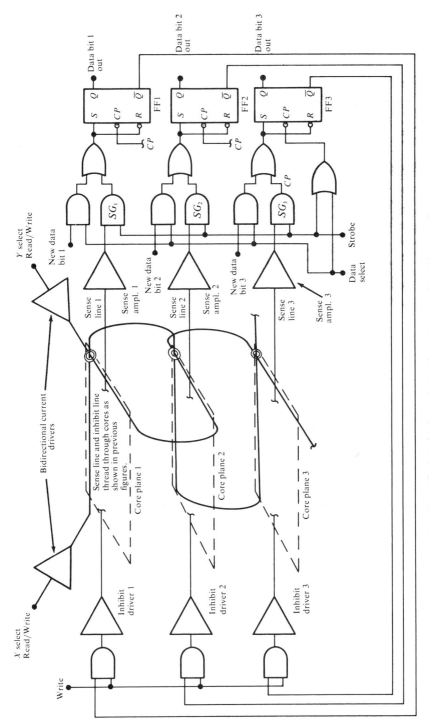

Figure 11–13. Simplified Multiplane Coincident-Current Core Memory System.

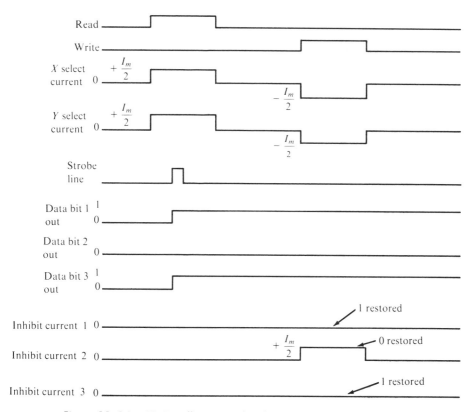

Figure 11–14.   Timing Illustration for the Memory of Figure 11–13.

plane 2 is *read out*. Since this is already in the 0 state, only a small induced voltage will occur on the *bit 2 sense line*. When the strobe pulse occurs, the output of *sense amplifier 2* is LOW, causing *flip-flop 2* to be RESET. Bit 2 has now been read out of the memory and temporarily stored in flip-flop 2, and the 1 stored in core plane 3 is *read out*. Since the core makes a transition from the 1 state to the 0 state, a large induced voltage will occur on *sense line 3*, causing a HIGH on the output of *sense amplifier 3* during the *strobe* pulse. As a result, *flip-flop 3* is SET and bit 3 is temporarily stored.

Next, a pulse of *write* current is applied to the X and Y select lines in the opposite direction of the read current, *driving each of the three cores toward the 1 state*. The bit 1 *inhibit-line driver* is disabled by the LOW on the $\overline{Q}$ output of flip-flop 1, and as a result, no *inhibit* current flows in *inhibit line 1*. The core is driven back to the 1 state by the *write* currents, and bit 1 is restored in the memory. The HIGH on the $\overline{Q}$ output of flip-flop 2 enables the bit 2 *inhibit-line driver* and causes an *inhibit* current to flow in a direction opposite to the *write* currents, preventing the core from switching back to the 1 state and resulting in the 0 being restored in the memory. The bit 3 *inhibit-line driver* is disabled by the LOW on the $\overline{Q}$ output of flip-flop 3, and as a result, no *inhibit* current flows in *inhibit line 3*. The core is driven back to the 1 state by the *write* currents, and bit 3 is restored in the memory. The binary word has been *read* from the memory for use elsewhere in the digital system and then restored in the memory by a *write* operation so it can be used again later. This read/write cycle is known as a *memory cycle,* and is a critical factor in memory design since it determines how quickly information can be retrieved from the memory. Notice that the gating logic on each flip-flop input in Figure 11–13 allows new data to be entered into the memory rather than restoring the information being read out. A HIGH on the data select line enables a new data bit into each flip-flop during the read operation. The write operation is identical to that previously described, except the flip-flops contain a new binary word instead of the binary word read from the memory.

We have used this illustration to give you a basic understanding of the operation of a coincident-current core memory; it is not intended as a lesson in core memory design, because there are many possible variations on our simplified implementation.

11–4   The Linear-Select Core Memory

This type of core memory is organized differently from the coincident-current memory. The *linear-select* memory is sometimes known as a *word-organized* memory. It requires a select line for *each* word of binary information in the memory,

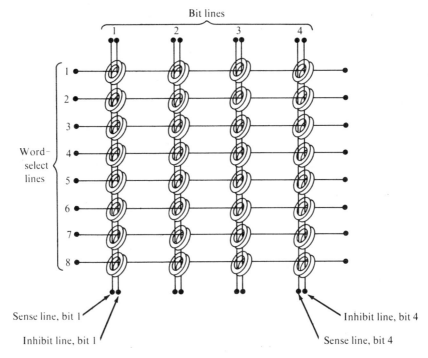

Figure 11–15.   Example of a Linear-Select Core Memory (Eight Words by Four Bits).

whereas the coincident-current type requires $x + y$ select lines for xy words. An eight-word linear-select memory array is illustrated in Figure 11–15. Each word contains four bits.

To *write* a four-bit word into the memory, a full-select current is passed through the specified select line, driving all four cores in that word location toward the 1 state. For any 0 bit, a current is passed through the appropriate inhibit line in opposition to the select current; this keeps the particular core in the 0 state (we are assuming the cores are all 0 from the previous read operation). To *read* the four-bit word from the memory, a full-select current is passed through the select line in the opposite direction of the write current; this *read* current drives each of the cores in the selected word toward the 0 state. If a core is in the 1 state, a large induced voltage occurs on the sense line (sometimes called bit line) when the core switches to the 0 state. If a core is already in the 0 state, a much smaller induced voltage occurs (as we have previously discussed). The basic sensing and non-destructive readout capability can be applied in a manner similar to that discussed for the coincident-current memory.

Example 11–1   Illustrate a linear-select memory with a 64-word, eight-bit capacity having separate inhibit and sense lines.

Solution:
See Figure 11–16.

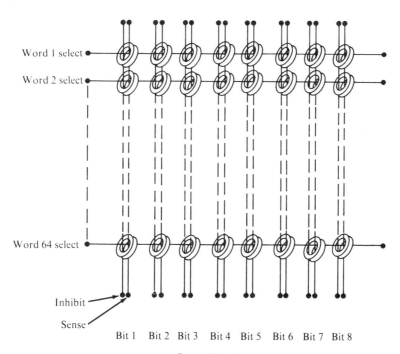

Figure 11–16.

11–5   Semiconductor Memories

Integrated circuit techniques have made possible a variety of semiconductor memories. Two basic technologies are presently used in memory fabrication—bipolar and MOS. The bipolar memories use bipolar transistor flip-flops as the basic storage cell, and are implemented with such techniques as TTL or ECL. The MOS memories use MOSFET circuits as the basic storage elements; these generally fall into two categories—static and dynamic. The static memories are implemented with MOS or bipolar flip-flops, and the dynamic memories utilize the inherent capacitance associated with MOSFET circuits as the basic storage element. In the following sections we will examine several important types of semiconductor memories.

11–6   Coincident-Select Flip-Flop Memories

This type of semiconductor memory is formed with an array of storage cells, where each cell is a flip-flop and each flip-flop addressed by an X select line and a Y select line. When a specified flip-flop is selected (addressed), a data bit can be either written into or read out of the cell. Figure 11–17 shows a 3-by-3 array to illustrate the basic concept of coincident selection. Here we are using a very straightforward logic implementation; later we will discuss another method more commonly used in integrated circuit bipolar memories.

Each X select line in Figure 11–17 is connected to all of the read gates and all of the write gates in a given row; each Y select line is connected to all of the read gates and all of the write gates in a given column. The read gates are designated RG and the write gates are WG.

The read input is connected to all of the read gates and the write input is connected to all of the write gates in the array; the *data-in* line also goes to each write gate. The output of each flip-flop is connected to its corresponding read gate, and the outputs of all of the read gates are ORed to produce the *data output*. We will describe the operation of this memory by first writing a 1 into the X_1Y_1 cell. HIGHs are applied to the X_1 line, to the Y_1 line, and to the write line; this allows the 1 on the data-in line to pass through gate WG_1 and SET the X_1Y_1 flip-flop.

To read the 1 for the cell, a HIGH is applied to the X_1 and Y_1 select lines as before, and a HIGH is applied to the read line with a LOW on the write line; this allows the 1 on the flip-flop output to pass through gate RG_1 and the OR gate to the data output. The read operation does not cause the data bit to be erased from the memory cell because it does not affect the state of the flip-flop. The memory is inherently nondestructive. The timing diagram for this read/write operation is shown in Figure 11–18.

It should be mentioned that this memory is a *static* memory because it will retain stored information indefinitely, as long as external power is supplied to the flip-flops, which can be either bipolar or MOS. The organization of this memory array is nine words with one bit per word. To increase the word length (number of bits per word), additional arrays can be stacked, as illustrated in Figure 11–19

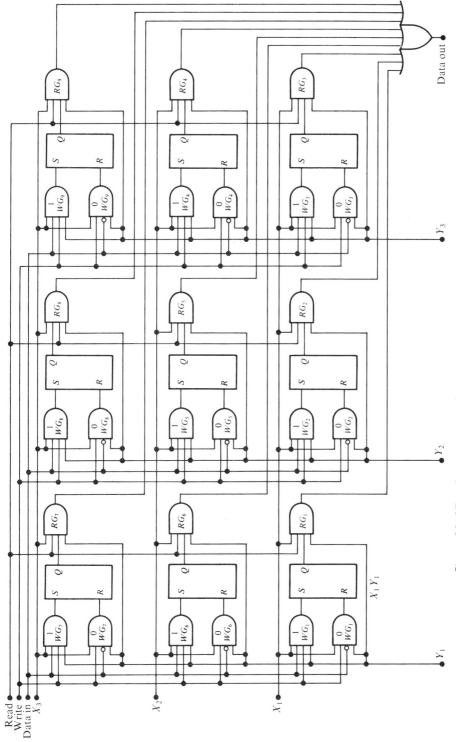

Figure 11–17. One Way to Implement a Coincident-Select Flip-Flop Memory.

353

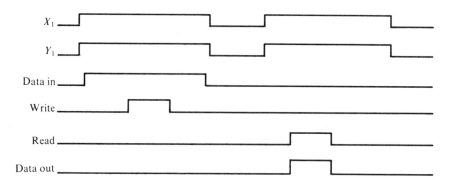

Figure 11–18.   Timing Diagram for the Memory of Figure 11–17.

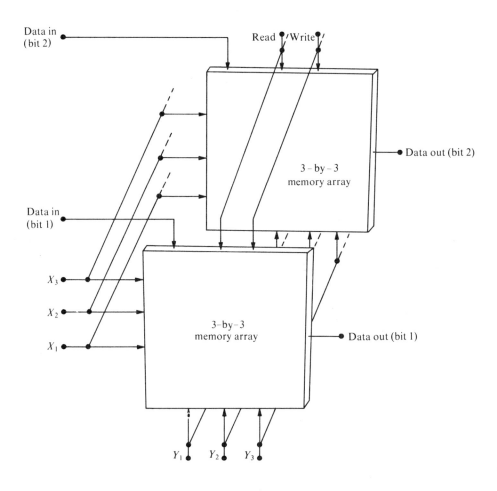

Figure 11–19.   Expansion of a Coincident-Select Memory to Accommodate More
                   Bits per Word.

where each square represents a single array. Notice that the organization of this coincident-select memory results in a "three-dimensional" configuration.

A method commonly found in integrated circuit technology for implementing a coincident-select bipolar memory is as follows. The basic flip-flop storage cell is pictured in Figure 11–20 and, as you can see, utilizes multiple-emitter bipolar transistors. Prior to addressing, the X and Y select lines are normally LOW, conducting current from the associated emitters of the ON transistor. To address or select the flip-flop, a HIGH is applied to both the X and the Y select lines, causing the emitter current of the ON transistor to be diverted to one of the sense lines (depending which transistor is ON). If Q_1 is ON, the 0 sense line conducts, activating a sense amplifier connected to that line; if Q_2 is ON, the 1 sense line conducts, activating the sense amplifier connected to that line. When one of the sense amplifiers is activated, its output level changes state. A level change on the output of the 0 sense amplifier indicates that a 0 bit has been read from the flip-flop storage cell. A level change on the output of the 1 sense amplifier indicates that a 1 bit has been read out. This type of readout is nondestructive since the state of the flip-flop remains unaffected by the operation.

In order to write a data bit into the memory cell, a HIGH must be applied to one of the write amplifiers; to store a 0, a HIGH is applied to the 0 write amplifier, and to store a 1, a HIGH is applied to the 1 write amplifier. Also, the flip-flop must be addressed by HIGHs on the X and Y select lines. When a 0 is being stored, the output of the 0 write amplifier goes LOW, causing transistor Q_1 to turn ON (thus storing a 0 in the flip-flop). When a 1 is being stored, the output of the 1 write amplifier goes LOW, causing transistor Q_2 to turn ON (thus storing a 1 in the flip-flop). The writing operation affects only the selected flip-flop, because at least one emitter on each flip-flop in each of the other cells in an array is already held LOW by the select lines. A 16-bit memory array is shown in Figure 11–21; each block is representative of the memory cell in Figure 11–20.

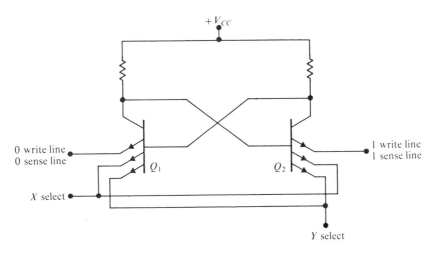

Figure 11–20.   A Typical T^2L Integrated Circuit Flip-Flop Memory Cell.

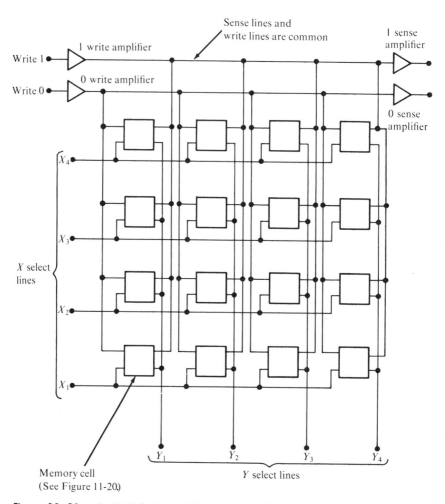

Figure 11–21. A 16-Cell Array Where Each Block Represents a Flip-Flop like That Shown in Figure 11–20.

Example 11–2 Illustrate a coincident-select memory that will store 16 four-bit words.

Solution:

Four memory planes like that in Figure 11–21 are connected as shown in Figure 11–22. Each bit plane contains a single bit for each of 16 words.

11–7   Linear-Select Flip-Flop Memory

Another type of organization found in semiconductor memories as in the core memory is the linear-select, in which each row of cells in the array is a data word and each column contains a bit in each word. An example array of three-bit words

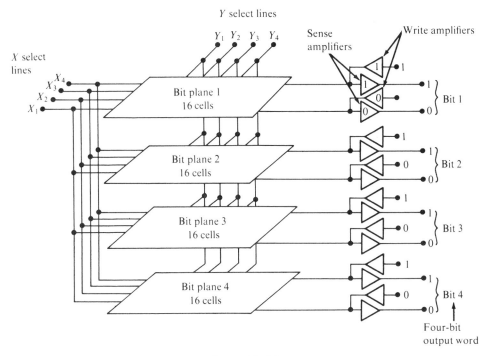

Figure 11–22.

is shown in Figure 11–23. Let us examine how a three-bit word, say, 101, can be written into address 1 (top row) and then read out. A HIGH on the word 1 select line and a HIGH on the write line enables write gates WG_1, WG_2, and WG_3. The 1 on the bit 1 line passes through gate WG_1 and SETS the flip-flop. The 0 on the bit 2 line results in the bit 2 flip-flop being RESET. The 1 on the bit 3 line passes through write gates WG_3 and SETS the bit 3 flip-flop. The data word is now stored in the memory.

To read the data word from the memory, a HIGH is applied to the read line and to the word 1 select line. The gates RG_1, RG_2, and RG_3 are enabled, and the bits of the stored data word appear on the data output bit lines.

Expansion of this type of memory to handle data words of more bits or additional words is shown in Figure 11–24 for six six-bit words. Each square represents a flip-flop storage cell and the associated logic gates. Notice that this type of organization results in a "two-dimensional" rather than a "three-dimensional" arrangement.

11–8 Dynamic Memories

Dynamic memories differ from the flip-flop or static memory in that the storage cells will not store a bit indefinitely and must be "refreshed" periodically to keep the stored data from disappearing. All dynamic memories utilize MOS circuits

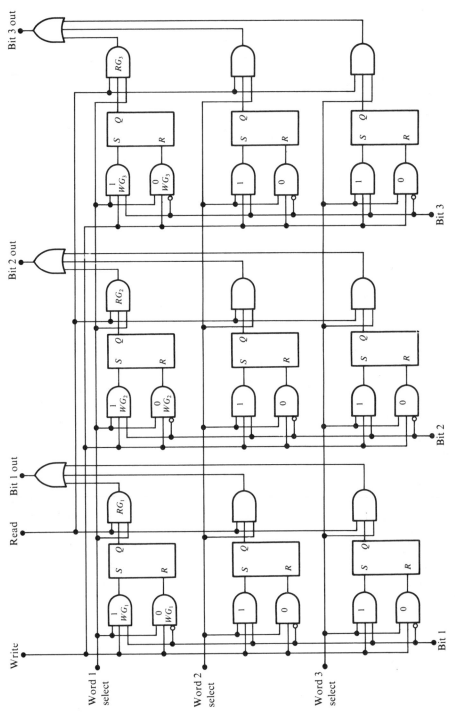

Figure 11-23. One Way to Implement a Linear-Select Flip-Flop Memory.

358

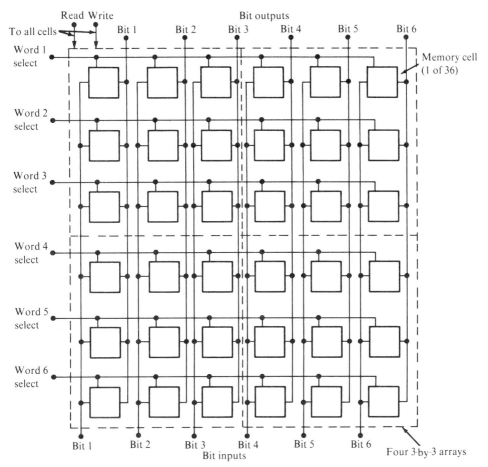

Figure 11–24.   Illustration of the Expansion of a Linear-Select Memory Using Four 3-by-3 Arrays to Form a Six-Word-by-Six-Bit Array.

because of the inherently high capacitance associated with MOSFETs. Bipolar transistors do not possess this high capacitance and are therefore not used for dynamic storage. The capacitance is the basic storage element in a dynamic memory cell, and because the stored charge will leak off over a period of time, it must be recharged (refreshed) at certain intervals in order to retain the stored information.

One of the advantages of MOS circuits over bipolar circuits is that a dynamic storage cell can be implemented with fewer integrated components and an MOS transistor requires a much smaller surface area than the bipolar transistor. Also, less power is consumed by MOS circuits than by an equivalent bipolar circuit. A disadvantage is that MOS circuits are normally slower (longer switching times) than bipolar, resulting in longer times required to put data into and take data out of a memory.

We will now examine a typical implementation of an individual MOS dynamic storage cell as shown in Figure 11–25. The basic operation is as follows.

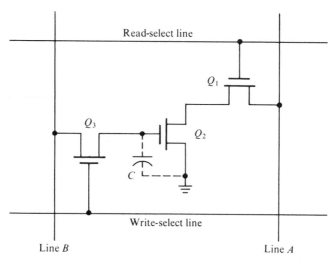

Figure 11–25. A Typical Dynamic MOS Storage Cell.

To read a data bit from the cell, line A is set to a negative potential, and the read-select line is enabled. If a 1 is stored by the gate capacitance of Q_2, the voltage on the gate turns Q_2 on and pulls line A to ground potential; this is detected by a sense amplifier connected to line A. If the capacitance is storing a 0, the voltage on the gate of Q_2 keeps Q_2 off, and line A remains at the negative potential. The sense amplifier detects this level and interprets it as a 0. To "refresh" the stored data bit after a read operation, the voltage on line A is inverted and applied to line B, and the write-select line is enabled, causing the capacitor to be recharged to the state detected on line A.

To write a new data bit into the cell, the state of line B is controlled independently of line A. The state of line B is stored by the capacitor when the write-select line is enabled. Figure 11–26 shows a 4-by-3 linear-select array, with each block representing an MOS dynamic storage cell.

11–9 The Random-Access Memory (RAM)

Random access refers to the way in which the cells in the memory are accessed or selected—it is a particular method of addressing a memory for a read/write operation. A random-access memory (RAM) is one in which any cell can be selected at random, independently of the previous cell addressed or the next cell to be addressed; in other words, no prescribed sequence of addressing is required. All RAMs have read/write capability, and can be either coincident-select or linear-select with either static or dynamic storage.

A simple random-access memory with a coincident-select configuration is shown in Figure 11–27. This is a 16-cell memory array, where each block represents a flip-flop storage cell like those in Figure 11–17. Keep in mind that we are using relatively small memory arrays for illustrative purposes. The same basic principles apply to memories with much larger capacities.

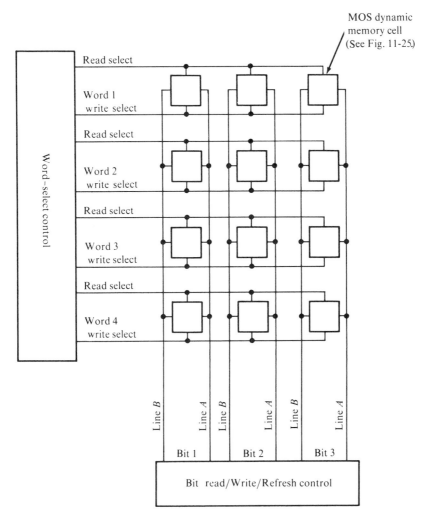

Figure 11–26.   Block Diagram of a Four-Word-by-Three-Bit Linear-Select Dynamic Memory.

As shown in Figure 11–27, the X decoding logic consists of four decoding gates with two address lines that are required to select one of the four X addresses. The Y decoding logic consists of an identical arrangement. This means that a total of four bits is required for the address code, which enables us to select any of 16 locations ($2^4 = 16$).

To illustrate how we can select a specific cell, let us store a 1 in cell 7. To address cell 7, a 0 is applied to address line A_3, a 1 to address line A_2, a 1 to address line A_1, and a 1 to address line A_0; this produces HIGHs on select line X and select line Y. Now, to enter the data 1 into the flip-flop, a HIGH is applied to the write line and the data 1 is applied to the data-in line. To read the data bit from this cell, a HIGH is applied to the read line with address 7 still selected, and the data 1 appear on the data-out line. The timing diagram for these operations is

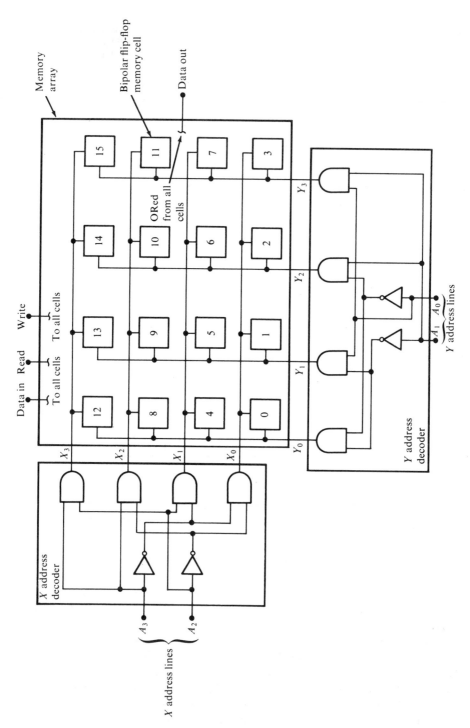

Figure 11-27. Coincident-Select RAM.

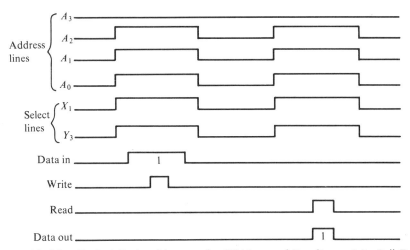

Figure 11-28. Typical Timing Diagram for Writing and Reading a 1 in Cell 7 of the RAM of Figure 11-27.

TABLE 11-1.   Address Code for the
Memory of Figure 11-25.

Binary Address Code				Selects Cell
A_3	A_2	A_1	A_0	No.
0	0	0	0	0
0	0	0	1	1
0	0	1	0	2
0	0	1	1	3
0	1	0	0	4
0	1	0	1	5
0	1	1	0	6
0	1	1	1	7
1	0	0	0	8
1	0	0	1	9
1	0	1	0	10
1	0	1	1	11
1	1	0	0	12
1	1	0	1	13
1	1	1	0	14
1	1	1	1	15

shown in Figure 11-28. Table 11-1 lists the address codes for the memory of Figure 11-27.

11-10   The Read-Only Memory (ROM)

A *read-only* memory is one that contains permanently stored data that can be read out as often as desired. It does not have the capability of accepting new data; that

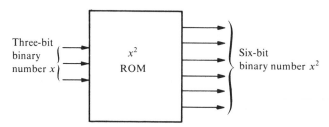

Figure 11–29. Functional Block Diagram for an x^2 "Look-Up Table" ROM.

is, there is no write operation. The ROM is preprogrammed to contain certain specified data required in a given application. Typical applications include code conversion, mathematical look-up tables, computer microinstructions, and generation of Boolean functions.

Several methods of implementing ROMs are commonly employed, including core, diode, bipolar transistor, and MOS transistor techniques. In order to gain a basic understanding of the ROM concept, we will first examine a diode matrix ROM. To start with an extremely simple example, let us use an ROM to store the squares of decimal numbers 0 through 7. If we apply a three-bit binary representation of any of these decimal numbers to the inputs of the ROM, we want a binary number that represents the square of the input to appear on the outputs. Since the highest value possible on the outputs is $7^2 = 49$, six bits are required (as shown in the block diagram of Figure 11–29).

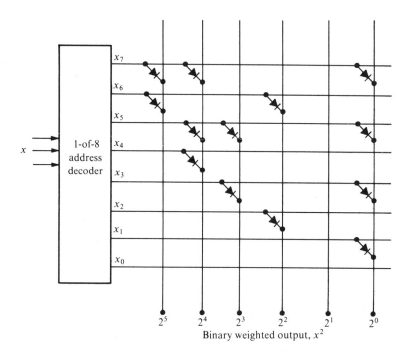

Figure 11–30. Diode Matrix Implementation of the x^2 ROM.

Basically, in a diode matrix, the presence of a diode represents a stored 1 at a given location and the absence of a diode represents a stored 0. When a certain input line is HIGH, all output lines that are connected to that input line through a diode will also be HIGH; the output lines that are not connected to that input line will remain LOW. An ROM implementation of our x^2 look-up table is shown in Figure 11-30. The decoder produces a HIGH on one of the x lines corresponding to the binary code on the inputs. Each of these x lines is connected through the diode matrix to the appropriate output lines. For example, when a binary 011 (decimal 3) is applied to the input lines, a HIGH occurs on the x_3 line. This HIGH is conducted through the diodes to the 2^3 and the 2^0 output lines, and the 2^4, 2^2, and 2^1 output lines remain LOW; this produces on the outputs a 01001 code, which represents decimal 9, the square of the input number (decimal 3). You can verify that for any digit 0 through 7 applied to the inputs, the binary code for the square of that number occurs on the outputs.

Another example of an ROM application is shown in Figure 11-31. This ROM is "programmed" as a four-bit binary-to-Gray-code converter. An inspection of this matrix will show that any four-bit binary code applied to the inputs will produce the corresponding four-bit Gray code on the outputs as shown in Figure 11-31.

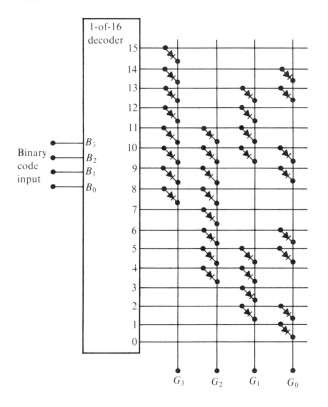

Binary				Gray			
B_3	B_2	B_1	B_0	G_3	G_2	G_1	G_0
0	0	0	0	0	0	0	0
0	0	0	1	0	0	0	1
0	0	1	0	0	0	1	1
0	0	1	1	0	0	1	0
0	1	0	0	0	1	1	0
0	1	0	1	0	1	1	1
0	1	1	0	0	1	0	1
0	1	1	1	0	1	0	0
1	0	0	0	1	1	0	0
1	0	0	1	1	1	0	1
1	0	1	0	1	1	1	1
1	0	1	1	1	1	1	0
1	1	0	0	1	0	1	0
1	1	0	1	1	0	1	1
1	1	1	0	1	0	0	1
1	1	1	1	1	0	0	0

Gray code output

Figure 11-31. A Diode Matrix ROM Programmed for Binary-Code-to-Gray-Code Conversion.

The integrated circuit ROMs currently available are either bipolar or MOS. In the bipolar ROMs, the base-emitter junctions of bipolar transistors are used as the diodes previously described. Two basic fabrication methods are normally used to program this type of ROM; the first method places transistors only at the appropriate locations in the matrix, utilizing certain integrated circuit fabrication techniques; the second method involves fabricating a transistor at all matrix locations, and then "burning" or "blowing" out the diode junctions at the locations where a 0 is required. MOS ROMs are either static or dynamic in operation, and are also programmed by the presence or absence of a transistor.

Example 11–3 Show how an ROM can be programmed to perform the following Boolean functions:

$$X = \overline{A}B\overline{C} + A\overline{B}C$$
$$Y = AB\overline{C} + \overline{A}BC + \overline{A}\,\overline{B}\overline{C}$$
$$Z = A\overline{B}C + \overline{A}\,\overline{B}\overline{C} + \overline{A}B\overline{C} + ABC$$

Solution:
See Figure 11–32.

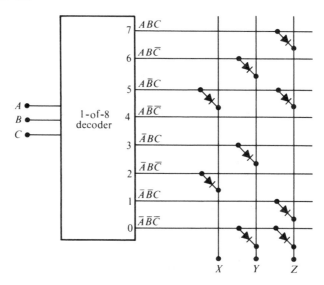

Figure 11–32.

11–11 Associative Memories

This type of memory basically allows external data to be compared with the data words stored in the memory; if a match occurs, an appropriate indication on the outputs is produced. We will examine the functional operation of one type of associative memory shown in the block diagram of Figure 11–33. The set of *enable* lines (E_0 through E_3) allow a comparison of any or all of the *data inputs* (D_0 through D_3) with the contents of any address selected by the address lines (A_0

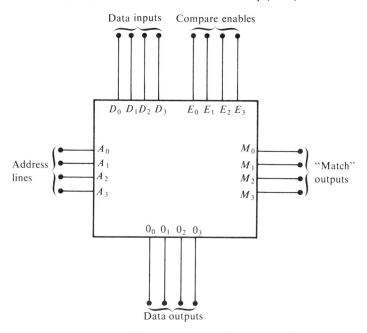

Figure 11–33.   Block Diagram of an Associative Memory.

through A_3) using a random-access method. If the associated stored data match the data bits on the inputs, the *match* outputs (M_0 through M_3) all go to a certain level, say, a HIGH. The process of addressing the memory to look for a match with the input data is sometimes called an *equality search*. In some memories of this type, more than one word in the memory can be addressed at a time and a simultaneous comparison made with the input data. The *data output* lines are available so that the data word being addressed can be read out.

11–12   The First-In–First-Out Memory (FIFO)

This type of memory is typically formed by an arrangement of shift registers. As we saw in the last chapter, the register is a form of storage or memory device and finds wide application in digital systems.

The term FIFO refers to the basic operation of this type of memory, where the first data bit to be written into the memory is the first to be read out. A simple analogy for this operation is a tube in which marbles are stored, as shown in Figure 11–34. The first marble to be put into the tube immediately rolls to the bottom end of the tube and is the first to come out when the lid is opened. As each marble is removed, room is created at the top end of the tube for a new marble.

In our analogy, the tube represents the storage device (FIFO register) and each marble represents a data bit. There is one important difference between a conventional shift register and a FIFO register: in a conventional shift register, a data bit moves through the register only as new data bits are entered; in a FIFO

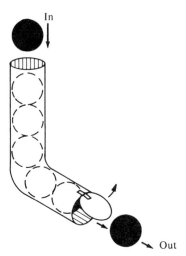

Figure 11–34. "Marble-in-Tube" Concept of the Operation of a FIFO Shift Register.

register, a data bit immediately goes through the register to the right-most empty stage available. This basic difference is illustrated in Figure 11–35.

Figure 11–36 is a block diagram of a 64-word-by-four-bit FIFO memory. This memory has four serial 64-bit data registers and a 64-bit control register (marker register). When data are entered by a shift-in pulse, it moves automatically under control of the marker register to the empty position closest to the output. Data cannot advance into occupied positions; however, when a data bit is shifted out by a shift-out pulse, the data bits remaining in the registers automatically move to the next position toward the output. Referring again to our marble analogy, when a marble is removed from the tube, the rest of the marbles move as close as possible to the output. The FIFO memory in Figure 11–36 is called an asynchronous type because data are shifted out independently of data entry—that is, two separate clocks are used.

Conventional shift register

Input	X	X	X	X	Output
0	0	X	X	X	⟶
1	1	0	X	X	⟶
1	1	1	0	X	⟶
0	0	1	1	0	⟶

X = unknown data bits

In a conventional shift register, data stay to left until "forced" through by additional data.

FIFO shift register

Input	—	—	—	—	Output
0	—	—	—	0	⟶
1	—	—	1	0	⟶
1	—	1	1	0	⟶
0	0	1	1	0	⟶

— = empty positions

In a FIFO shift register, data "fall" through—go right.

Figure 11–35. Input Data Storage in Conventional Shift Register Compared to That in a FIFO Shift Register. Courtesy Fairchild Semiconductor.

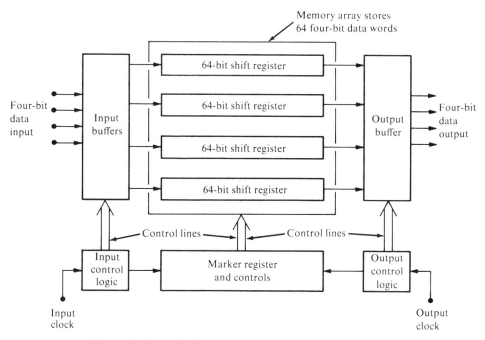

Figure 11–36. Block Diagram of a Serial FIFO Memory.

PROBLEMS

11-1 Assume a magnetic core is in its positive remanent flux state. If a full-select value of current is applied in a direction through the core winding to drive the core toward its negative state, indicate the general path on a hysteresis curve that the flux density would follow. Assume the current is removed after a period of time.

11-2 Assume 100 mA of current is required to switch a particular core. If a partial-select current of 75 mA is driven through one of the coincident-select lines as indicated in Figure 11–37, show the amount and direction of the current required through the other select line to switch the core to its opposite state.

75 mA

Figure 11–37.

11-3 Explain how a magnetic core is used to store binary information.

11-4 Explain the difference between the read and the write functions in a magnetic core, and show basically how each is accomplished.

11-5 Explain the purpose of the sense line in a magnetic core memory.

11-6 Explain the purpose of the inhibit line in a magnetic core memory.

11-7   A coincident-current core array has eight X select lines and eight Y select lines. What is the bit capacity of this array? How many sense lines are required?

11-8   A coincident-current core memory is composed of eight core planes, each consisting of 64 memory cells. How many binary words can be stored, and how many bits are in each word?

11-9   A storage capacity for 32 words of six bits each is required. Describe the arrangement of a coincident-current core memory to do this.

11-10   What is the difference between a coincident-current core memory and a linear-select core memory?

11-11   Show the arrangement of a linear-select core memory with a storage capacity of 12 words of six bits each.

11-12   Explain what is meant by a *memory cycle*.

11-13   Determine the number of *select* lines required for each of the following coincident-select memory arrays:
       (a)   16 cells   (b)   36 cells   (c)   64 cells   (d)   100 cells

11-14   Determine the number of *select* lines required for each of the following coincident-select memory arrays:
       (a)   8-words-by-4-bits   (b)   16-words-by-8-bits
       (c)   64-words-by-8-bits   (d)   128-words-by-16-bits

11-15   Explain the difference between a static storage cell and a dynamic storage cell.

11-16   Classify each of the following types of cells as static or dynamic:
       (a)   magnetic core   (b)   bipolar flip-flop
       (c)   capacitor   (d)   MOS flip-flop

11-17   Determine the number of binary *address* lines required for each of the following coincident-select type of memories. Assume a binary address code is used.
       (a)   16-words-by-4-bits   (b)   32-words-by-8-bits
       (c)   64-words-by-8-bits   (d)   128-words-by-8-bits
       (e)   256-words-by-10-bits   (f)   1024-words-by-16-bits

11-18   For the RAM of Figure 11-27, assume a 1 is stored in each of the odd-numbered cells and a 0 is stored in each of the even-numbered cells. For the address line sequence indicated by the waveforms in Figure 11-38, determine the data-out waveform during a read operation (read line HIGH). A_0 is the LSB.

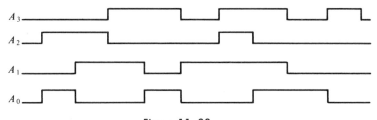

Figure 11-38.

11-19   What is meant by *nondestructive readout?*

11-20   Determine the number of binary *address* lines required for each of the following linear-select type of memories:
       (a)   16-words-by-4-bits   (b)   32-words-by-4-bits

(c)   64-words-by-8-bits   (d)   128-words-by-12-bits
(e)   512-words-by-16-bits   (f)   1024-words-by-32-bits

11-21 Place each of the following terms in the proper category(s) as listed below: RAM, coincident-select, associative, ROM, linear-select, static, core, dynamic, FIFO, flip-flop.

   (a)   memory organization   (b)   addressing method
   (c)   type of memory cell   (d)   read or write capability
   (e)   data input or output method

11-22 Sketch a block diagram for a coincident-select RAM with 64-word-by-four-bit capacity. Show all essential components and input and output lines.

11-23 Repeat Problem 11-22 for a linear-select RAM with the same capacity.

11-24 For the diode matrix ROM in Figure 11-39, determine the outputs for all possible input combinations.

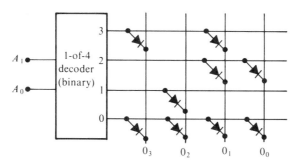

Figure 11-39.

11-25 Repeat Problem 11-24 for the diode matrix in Figure 11-40.

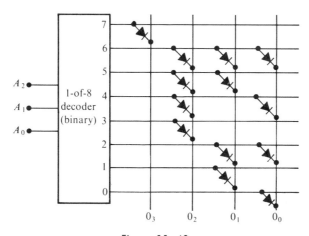

Figure 11-40.

11-26 Design a diode matrix ROM to convert BCD to Excess-3 code.

11-27 Assuming that the ROM is programmed by "blowing" or removing diodes, indicate which diodes you would blow in the matrix given in Figure 11-41 for an x^3 look-up table where x is represented by three binary bits.

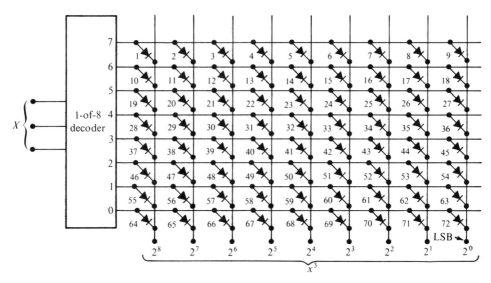

Figure 11-41.

11-28 Implement the following Boolean expressions with a diode matrix ROM:

$$x = AB + A(\bar{B} + C)$$
$$y = B(BC + CD)$$
$$z = ABC + \bar{A}B\bar{C}D + \bar{B}C + BC\bar{D}$$

11-29 Sketch a block diagram for a four-word-by-eight-bit FIFO memory.

11-30 Explain the basic difference between a FIFO shift register and a conventional shift register.

CHAPTER 12

Arithmetic Logic

In Chapter 7 we discussed some of the basic ways in which logic circuits can be used to perform arithmetic functions. The half-adder, full-adder, half-subtractor, full-subtractor, parallel adder, and a multiplier were examined. In this chapter we will look at some additional and more complete methods of performing addition, subtraction, multiplication, and division with digital logic circuits.

12–1 The Problem of Addition

When two numbers are added, four possible situations must be considered. These are

1. Addition of two *positive* numbers results in a positive sum of greater magnitude than either number. It is possible for the sum to be one digit larger than either of the numbers.

$$\begin{array}{r} 4 \\ +\ 3 \\ \hline 7 \end{array} \qquad \begin{array}{r} 9 \\ +\ 7 \\ \hline 16 \end{array}$$

2. Addition of two *negative* numbers results in a negative sum of greater magnitude than either number. It is possible for the sum to be one digit larger than either of the numbers.

$$\begin{array}{r} -6 \\ +\ -2 \\ \hline -8 \end{array} \qquad \begin{array}{r} -7 \\ +\ -4 \\ \hline -11 \end{array}$$

3. Addition of a *larger positive* number to a *smaller negative* number results in a positive sum of less magnitude than either of the numbers. The sum will always have the same number of digits as, or fewer digits than, either of the numbers.

$$
\begin{array}{r}
+5 \\
+ \quad -2 \\
\hline
+3
\end{array}
\qquad
\begin{array}{r}
+12 \\
+ \quad -\ 4 \\
\hline
+\ 8
\end{array}
$$

4. Addition of a *smaller positive* number to a *larger negative* number results in a negative sum of less magnitude than either of the numbers. The sum will always have the same number of digits as, or fewer digits than, either of the numbers.

$$
\begin{array}{r}
-6 \\
+ \quad +4 \\
\hline
-2
\end{array}
\qquad
\begin{array}{r}
-25 \\
+ \quad +20 \\
\hline
-\ 5
\end{array}
$$

In order for a logic circuit to handle addition completely, it is obvious that it must be capable of accommodating each of the four conditions. Of course, we must also be able to represent both positive and negative numbers in binary form.

12–2 Subtraction as a Special Case of Addition

It can be shown that subtraction is simply a form of addition, and that the subtraction of any two signed numbers can be accomplished by one of the four cases of addition discussed in the previous section. The following examples will illustrate this.

Subtraction	Addition

$$
\begin{array}{r}
+9 \\
- \quad -7 \\
\hline
+16
\end{array}
\qquad\qquad
\begin{array}{r}
+9 \\
+ \quad +7 \\
\hline
+16
\end{array}
$$

$$
\begin{array}{r}
-5 \\
- \quad +3 \\
\hline
-8
\end{array}
\qquad\qquad
\begin{array}{r}
-5 \\
+ \quad -3 \\
\hline
-8
\end{array}
$$

$$
\begin{array}{r}
+8 \\
- \quad +6 \\
\hline
+2
\end{array}
\qquad\qquad
\begin{array}{r}
+8 \\
+ \quad -6 \\
\hline
+2
\end{array}
$$

$$
\begin{array}{r}
-7 \\
- \quad -3 \\
\hline
-4
\end{array}
\qquad\qquad
\begin{array}{r}
-7 \\
+ \quad +3 \\
\hline
-4
\end{array}
$$

As you can see, each case of subtraction can be reduced to a case of addition and can be handled by the same logic circuit.

12–3 Representing Signed Numbers

To completely handle all arithmetic operations, we have to handle both positive and negative numbers—that is, we have to express the sign of the number in addition to its value (magnitude). A binary *sign-and-magnitude* notation can be used for this purpose. The general format of sign-and-magnitude notation is

$$A_s \cdot A_n A_{n-1} A_{n-2} \cdots A_1 A_0$$

The magnitude of the number is represented in fractional form by the bits to the right of the binary point, and the sign of the number is expressed by the single bit to the left of the binary point. Conventionally, a 0 sign bit represents a + (positive) and a 1 sign bit represents a − (negative) sign. Some examples will illustrate.

Example 12–1 Express the signed numbers 9, − 13, 25, and − 29 in binary sign-and-magnitude form.

Solutions:

$$+9 = 0.1001$$
$$-13 = 1.1101$$
$$+25 = 0.11001$$
$$-29 = 1.11101$$

12–4 Sign-and-Magnitude Addition

In the actual addition of numbers expressed in binary sign-and-magnitude form, the sign bit is handled as part of the addition process. *The magnitude of a negative number can be expressed in either 1's complement or 2's complement form before adding.* These two methods were covered in Chapter 2. We will go through several examples and then summarize the methods at the end of this section.

Example 12–2 Both numbers positive:

Decimal	Binary
+7	0.0111
+ +1	+ 0.0001
+8	0.1000

No negative numbers are involved; therefore no complements are required.

Example 12–3 Both numbers negative:

Decimal	1's complement	2's complement
−8	1.0111	1.1000
+ −5	+ 1.1010	+ 1.1011
−13	① 1.0001	✻ 1.0011
	+	
	→1	
	1.0010	1.1101
	1.1101	

In this case, both numbers are expressed in 1's complement form, an "end-around carry" is performed, and the sum has the proper sign but its magnitude is in 1's complement form.

For the 2's complement method, both numbers are expressed in 2's complement and added (ignoring the final carry). The sum has the proper sign but the magnitude is in 2's complement.

Example 12-4 Larger number positive and smaller number negative:

Decimal	1's complement	2's complement
+8	0.1000	0.1000
+ −5	+ 1.1010	+ 1.1011
+3	① 0.0010	∗ 0.0011
	+	
	→1	
	0.0011	

In this case, the negative numbers are expressed in complement form. The sum has the proper sign bit and the magnitude is in true form.

Example 12-5 Larger negative number and smaller positive number:

Decimal	1's complement	2's complement
−6	1.1001	1.1010
+ +4	+ 0.0100	+ 0.0100
−2	1.1101	1.1110
	1.0010	1.0010

In this case, no carries occur; the sum has the proper sign bit, but its magnitude is in complement form.

Notice that in each of the previous examples the sign bit was added as part of the number but was not included in determining complements.

When the number of bits in the sum exceeds the number of bits in each of the numbers added, it is called *overflow* and is illustrated by the following example:

Example 12-6

Decimal	Binary
+9	0.1001
+ +8	+ 0.1000
+17	1.0001
sign incorrect	magnitude incorrect

The overflow condition can occur *only* when both numbers are positive or both numbers are negative. It is indicated by an incorrect sign bit. The following is a summary of the four cases of sign-and-magnitude addition.

Both numbers positive

1. Add both numbers in *true* form, including the sign bit.
2. The sign bit of the sum will be 0 (+).

3. Overflow is possible. The sign bit and the magnitude of the sum will be incorrect.

Both numbers negative

1. Take the 1's complement or 2's complement of the magnitude of both numbers. Leave the sign bits as is.
2. Add the numbers in their complement form, including sign bits.
3. Add the end-around carry in the case of the 1's complement method; drop the carry in the case of the 2's complement method. The sign bit of the sum will be 1 (−), and the magnitude of the sum will be in complement form.
4. Overflow is possible. The sign bit and magnitude of sum will be incorrect.

Larger number positive, smaller number negative

1. Take the 1's complement or 2's complement of the magnitude of the negative number. Leave the sign bit as is. A positive number remains in true form.
2. Add the numbers, including the sign bits.
3. Add the end-around carry for the 1's complement method; drop the carry for the 2's complement method. The sign bit of the sum will be a 0 (+), and the magnitude will be in true form.
4. No overflow is possible.

Larger number negative, smaller number positive.

1. Take the 1's complement or 2's complement of the magnitude of the negative number. Leave the sign bit as is. A positive number remains in true form.
2. Add the numbers, including the sign bits.
3. No carries will occur. The sum will have the proper sign bit, and the magnitude will be in complement form.
4. No overflow is possible.

Now that we have examined the basic arithmetic processes by which two numbers can be added or subtracted, let us look at the addition of a string of numbers added two at a time. This can be accomplished by adding the first two numbers, then adding the third number to the sum of the first two, then adding the fourth number to this result, and so on. The addition of several numbers taken two at a time is illustrated in the following example, which is the basic way many adders operate in digital systems.

Example 12–7   Add the numbers 2, 4, 6, 5, 8, and 9.

Solution:

$$
\begin{array}{r}
2 \\
+ \quad 4 \\
\hline
6 \quad \leftarrow \quad \text{first sum} \\
+ \quad 6 \\
\hline
12 \quad \leftarrow \quad \text{second sum} \\
+ \quad 5 \\
\hline
17 \quad \leftarrow \quad \text{third sum} \\
+ \quad 8 \\
\hline
25 \quad \leftarrow \quad \text{fourth sum} \\
+ \quad 9 \\
\hline
34 \quad \leftarrow \quad \text{final sum}
\end{array}
$$

12–5   A Serial Binary Adder

The method of addition discussed here is termed *serial* because two numbers are added on a bit-by-bit basis, rather than simultaneously as in a parallel adder. Serial addition of two binary numbers is illustrated in Figure 12–1. The corresponding bits of each number are applied to the bit inputs of the full-adder, and the carry generated by the addition of the two previous bits is on the carry input. Every time a carry is generated, it must be delayed by one bit time so that it can be added to the next higher order bits when they come into the full-adder. During bit time 1, the A_0 and B_0 bits are added and a carry of 1 is generated. The carry bit is delayed or held until bit time 2 and added to bits A_1 and B_1. A carry of 1 results from the addition, and is held until bit time 3, when it is added to bits A_2 and B_2; this addition results in a 0 carry. During bit time 4, A_3 and B_3 are added, completing the addition of the two four-bit numbers. This operation is illustrated in Figure 12–1.

Additional functions must be implemented in order to have a complete adder that can handle each of the four cases of addition (including subtraction). Based on the fundamental processes for sign and magnitude addition, let us itemize all of the functions that an adder circuit must perform:

1. Acquire the two numbers.
2. Check the sign bit of each number. If it is a 0, the magnitude of the number is left in true form; if it is a 1, the magnitude of the number must be complemented by the 1's or 2's complement method.
3. Determine if addition or subtraction is to be performed. If subtraction is to be performed, the subtrahend must be complemented. If the subtrahend is negative, then the complementation done in Step 2 is canceled and the negative subtrahend is actually left in true form.
4. Add the LSBs of the two numbers. Continue to add a bit at a time until all bits (including the sign bit) have been added. This is the basic serial addition as illustrated in Figure 12–1.

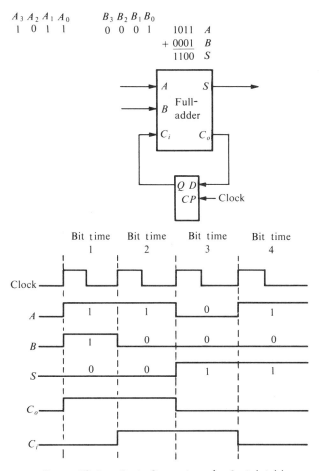

$A_3\ A_2\ A_1\ A_0$ $B_3\ B_2\ B_1\ B_0$
 1   0   1   1   0   0   0   1

$$\begin{array}{rl} 1011 & A \\ +\ 0001 & B \\ \hline 1100 & S \end{array}$$

Figure 12-1.   Basic Operation of a Serial Adder.

5. Add the end-around carry in the case of 1's complement, or drop the final carry in the case of 2's complement. The fact that the end-around carry addition is not required in the 2's complement method makes it the most widely used method in most systems.

6. If both numbers are negative or the larger number is negative, the sum is in complement form. This must be taken into account when the next number is added to this sum—that is, the complement operation must be omitted in this case.

7. Check for the correct sign bit. If incorrect, overflow has occurred.

With these required operations in mind, let us examine the serial adder of Figure 12-2. Initially, one of the numbers to be added is shifted into *register A* and the other number into *register B*. They are held in these registers until we are ready to start the addition process. Before the numbers are added, the *sign-bit comparison logic* determines if both of the numbers have the same sign. If they have

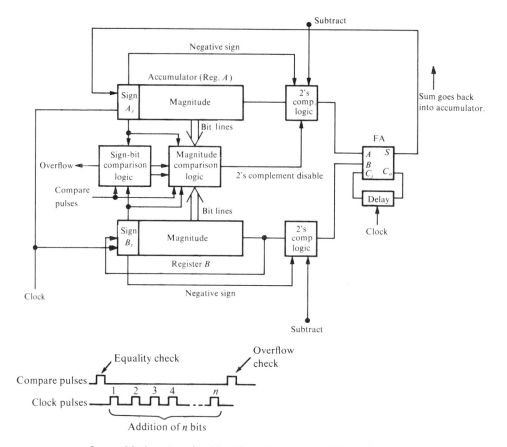

Figure 12-2. Serial Adder Block Diagram and Basic Timing.

the same, an overflow condition is possible, and the value (1 or 0) of the sign bit is stored so that it can be compared with the value of the sign bit of the sum after the numbers have been added. The logic also determines and remembers if both numbers are negative; also, if the numbers differ in sign, it activates the *magnitude comparison logic,* which determines if the larger number is negative.

A negative sign bit (1) on either of the numbers activates the respective *2's complement logic.* The addition process starts by shifting the LSBs of each number out of the shift registers, through the 2's complement logic, and onto the inputs of the full-adder. A sum bit is generated and fed back to the input of register *A* (sometimes called the *accumulator*). The next shift pulse puts the first sum bit into the accumulator and shifts the next bits of the two numbers into the full-adder. The process continues until all of the bits in the two numbers have been added and the sum has been stored in the accumulator. When the addition is complete, the sign-bit comparison logic checks the sign bit of the sum against the stored sign bit of the two numbers. If they are not the same, an overflow indication is produced.

If a string of numbers is to be added, the third number is loaded into register *B* and will be added to the sum of the previous addition, which is now stored in the *accumulator.* Before the previous addition, the magnitude comparison logic determined if the larger of two differently signed numbers was negative and the sign-bit comparison logic determined if both numbers were negative. In either case, the sum now stored in the accumulator is in complement form and negative. The 2's complement logic must therefore be disabled in this case, because the number is already complemented.

After the third number in the string has been loaded in register B, the entire operation is repeated and another sum is accumulated in register A. The next number in the string can now be loaded into Register B, the entire addition cycle repeated, and so on.

We have discussed the general operation of a 2's complement adder at the logic block diagram level to get a basic understanding of what takes place. A 1's complement adder can be implemented, but an additional complication arises because another add cycle is required to handle the end-around carry. Now that we generally understand the operation of a serial adder, let us look at ways to implement the logic required for some of the blocks in the adder of Figure 12–2.

The Serial Adder Registers

Shift registers have already been discussed in Chapter 10, so only a brief coverage of the requirements for this particular application is necessary here. As we have seen, two shift registers are required in the serial adder. The accumulator register must have a serial input to accept the sum bits as they are fed back from the full-adder output. Each register must have a serial output so the numbers stored in each can be shifted out one bit at a time. New numbers can be loaded into the

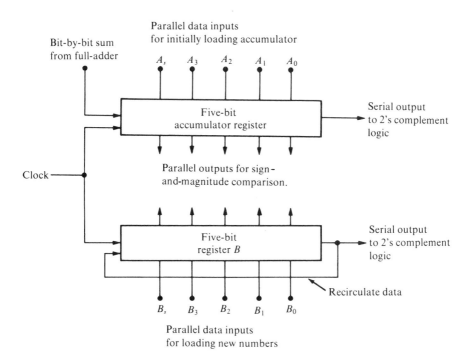

Figure 12–3.   Registers for a Five-Bit Serial Adder.

registers either in parallel or in series, depending on the particular system requirements. The capacity (number of stages) in each register must be equal to the number of bits in the magnitude of the numbers to be handled plus a stage to store the
sign bit. Figure 12–3 illustrates the operation of the two registers with a five-bit
capacity (four magnitude bits and one sign bit) and parallel loading. Register B
can recirculate the output to the input so the number can be saved as it is shifted
out.

Sign-Bit Comparison Logic

As we have seen, the purpose of this logic is to check the sign bits of the numbers
in the registers and, if they are equal, to store the value of the sign for a possible
overflow check at the end of the addition. If the sign of the sum is different from
the stored sign, an overflow indication is generated. Also, if the two numbers have
a negative sign, the 2's complement logic for the accumulator must be disabled
after the present addition has been completed and before the next addition starts.
The other function of the logic is to enable the magnitude comparison logic if the
two numbers differ in sign.

Figure 12–4 shows one way to implement this logic function. The operation is
as follows. After the two numbers are stored in the shift registers, the sign bits are
in the left-most position. The output of this stage of each register is connected to
the input of an exclusive-OR gate, which performs the basic comparison of the two
bits. The output of the exclusive-OR is inverted and is HIGH if the two sign bits
are equal. The output of this exclusive-OR gate is stored in the flip-flop by the
equality check pulse, and remains there until the end of the addition of the two
numbers. When the addition is complete, the sign bit now in register A is the sign
bit of the sum, and, since the content of register B was recirculated, the sign bit in
this register is the sign bit of the original number. The sign of the sum should be
the same as the sign of the number in register B if the flip-flop is SET; if they are
not the same, then all inputs to the overflow check gate are HIGH when the over-

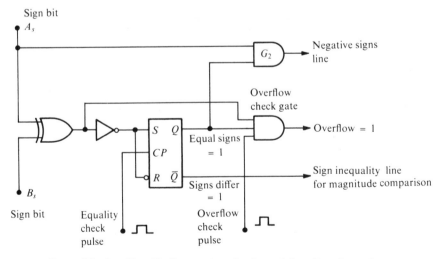

Figure 12–4.   Sign-Bit Comparison Logic and Overflow Detection.

flow check pulse occurs. If this condition exists, the output of the overflow check gate is HIGH, indicating an overflow condition—in which the sum is invalid. If the flip-flop was not originally SET or if the sign bit of the sum and the sign bit of the original numbers are the same, then no overflow indication occurs.

If the two sign bits are not equal, a HIGH appears on the exclusive-OR gate output, which is connected to the magnitude comparison logic via the sign inequality line. If the two sign bits are negative, a HIGH appears on the output of AND gate G_2, which is connected to the magnitude comparison logic via the negative signs line. The two signals are used by the magnitude comparison logic.

Magnitude Comparison Logic

The purpose of the magnitude comparison logic is to determine if the sum resulting from an addition is in its 2's complement form as it is stored in the accumulator. As we have seen, the condition can be predicted if the two numbers to be added are both negative or if the larger number is negative. The magnitude comparison logic records the existence of either of these two conditions prior to the addition of the two numbers. After the addition of the two numbers and before the addition of the next number to the resulting sum, the magnitude comparison logic

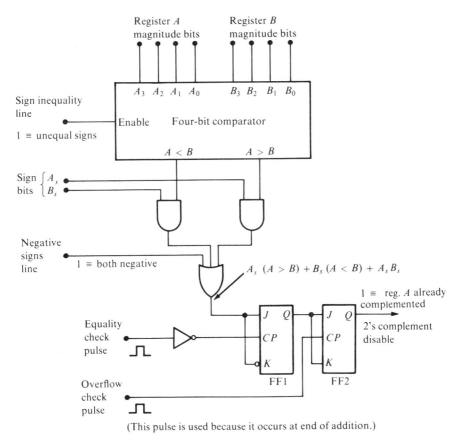

Figure 12–5.   Magnitude Comparison Logic.

disables the 2's complement logic for register A because A's content is negative and already in complement form. Figure 12–5 shows one way in which this function can be achieved. We are assuming for this particular case a four-bit magnitude, although the same principle can be extended to numbers of any size.

The four magnitude bits from each register are connected to the inputs of the magnitude comparator (comparators were discussed in Chapter 7). The sign inequality line is connected to the enable input of the comparator, and if it is HIGH, a comparison of the magnitudes of the two numbers is performed. If the number in register A is larger than the number in register B, the $A > B$ output goes HIGH. If the number in register B is larger, the $A < B$ output goes HIGH. The following expression gives the conditions for the case of a larger negative number, where A_s and B_s are the sign bits:

$$A_s (A > B) + B_s (A < B) \qquad (12\text{–}1)$$

This says that if $A_s = 1$ AND $(A > B) = 1$ OR if $B_s = 1$ AND $(A < B) = 1$, then the larger of the two numbers is negative. The function is implemented by the AND/OR logic as indicated in Figure 12–5. The negative signs line is also ORed with the above function, as shown. When the larger of the two numbers is negative or when both numbers are negative, the output of the OR gate is HIGH; this is stored in FF1 at the time of the equality check pulse prior to addition. The output of the flip-flop is transferred into FF2 at the end of the addition by the overflow check pulse. The output of FF2 is used to disable the 2's complement logic during the next addition.

2's Complement Logic

The 2's complement of a number can be achieved in a relatively easy fashion. Before we proceed with a description of the logic, let us look at the basic process involved. It is interesting to note that if we take a binary number and (going from right to left) copy each bit up to and including the first 1, and then write the complement of each bit thereafter, we get the 2's complement of the binary number. This is illustrated by the following examples.

Example 12–8 Determine the 2's complement of each of the following binary numbers: 1011, 101100, 11011000.

Solutions:

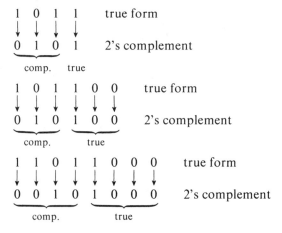

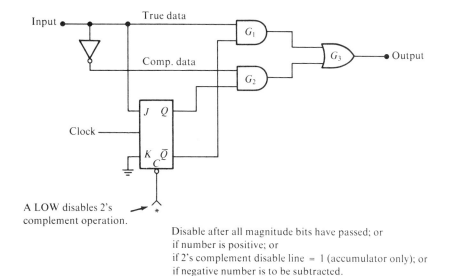

Disable after all magnitude bits have passed; or
if number is positive; or
if 2's complement disable line = 1 (accumulator only); or
if negative number is to be subtracted.

*The disable logic is left as an exercise at the end of the chapter.

Figure 12–6.   2's Complement Logic.

The logic required to perform this 2's complement operation is shown in Figure 12–6. Basically, the logic allows the bits to pass through to the output without being inverted until a 1 occurs; this first 1 is allowed to pass, but each bit thereafter is inverted.

The detailed operation for the particular implementation of Figure 12–6 is as follows. The data (numbers stored in the shift register) are shifted out of the register by the clock pulses as indicated; these serial data are applied to AND gate G_1 in true form and to AND gate G_2 in complement form. The flip-flop is initially RESET, thus enabling gate G_1 and disabling gate G_2. The bits pass through gate G_1 and OR gate G_3 uncomplemented until after the first 1 occurs; one bit time after the first 1, the flip-flop SETS because its J input is HIGH and K is LOW. Now gate

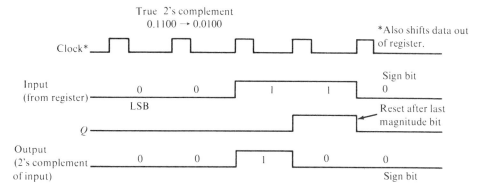

Figure 12–7.   Timing Diagram for the 2's Complement Circuit of Figure 12–6.

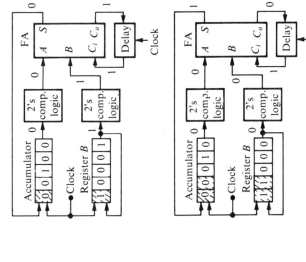

Initially

Numbers are in registers. Sign-and-magnitude comparisons are made (logic omitted here). 2's complement logic disabled because both numbers are positive. LSBs are added.

First clock pulse

Shifts both numbers to right one place. Least significant sum bit is now in first stage of accumulator (shaded cell). Least significant bit of number in register B is recirculated. Next LSBs are added.

Second clock pulse

Shifts both numbers to right one place. Next least significant sum bit is entered into accumulator (sum bits in shaded cells). Next bit in register B recirculated. Third bits added.

Third clock pulse

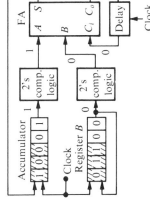

Shifts both numbers to right one place. Third sum bit is entered into accumulator (sum bits in shaded cells). Third bit in register B is recirculated. The MSBs added.

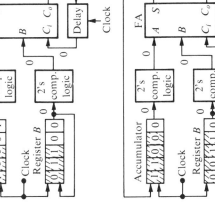

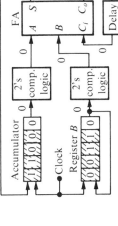

Fourth clock pulse

Shifts both numbers to right one place. The most significant sum bit is entered into the accumulator (sum bits in shaded cells). MSB in register B is recirculated. Sign bits are added.

Fifth clock pulse

Shifts last bits out of registers. Accumulator now contains the final sum. The original contents of register B have been restored. A sign check can now be made for overflow. The next number to be added can be loaded into register B.

Figure 12–8.

387

G_2 is enabled and gate G_1 disabled, allowing the complements of the remaining bits to pass through gates G_2 and G_3; this operation produces the 2's complement of the input number in serial form on the output line. The timing diagram in Figure 12–7 illustrates a particular four-bit number being converted to its 2's complement. The complementation process must be discontinued after all of the magnitude bits have passed through and before the sign bit comes in, because we do not want to complement the sign bit; this can be done by providing a reset pulse to the flip-flop immediately following the last magnitude bit, which is after the fourth bit in our example. When this is done, the sign bit will go through the 2's complement logic uncomplemented.

Let us look at the time required to complete an addition cycle with the serial adder. One bit time is required for the sign operation prior to adding the two numbers. Bit times equal to the total number of bits in each number (including the sign bit) are required to perform the actual addition. One bit time following the addition is required to check for an overflow condition. Therefore, a total of $n + 2$ bit times is used for a complete add cycle where n is the quantity of bits in the number.

Example 12–9   Show the adder states on a bit-by-bit basis to illustrate the serial addition of the following two numbers:

$$\begin{array}{r} 0.1001 \\ + \quad 0.0011 \\ \hline 0.1100 \end{array}$$

Solution:

For simplicity and clarity, the comparison logic is omitted. The fundamental operation in the addition process is shown in Figure 12–8.

12–6   Parallel Binary Ripple-Carry Adder

The parallel adder discussed in Chapter 7 is basically a ripple carry type of parallel adder. The carry out of each full-adder stage is connected to the carry input of the next higher order stage. The sum and carry outputs of any stage cannot be produced until the input carry occurs; this leads to a time delay in the addition process, as illustrated in Figure 12–9. The carry propagation delay for each full-adder is the time from the application of the input carry until the output carry occurs, assuming the A and B inputs are present.

Full-adder 1 cannot produce a potential carry out until a carry input is applied. Full-adder 2 cannot produce a potential carry out until full-adder 1 produces a carry out. Full-adder 3 cannot produce a potential carry out until a carry out is produced by full-adder 1 followed by a carry out from full-adder 2, and so on. As you can see, the input carry to the least significant stage (FA1) has to "ripple" through all of the adders before a final sum is produced. A delay through all of the adder stages actually is a "worst case" addition time. The total delay can vary depending on the carries produced by each stage. If two numbers are added such that no carries occur between stages, the add time is simply the propagation

Assume carry delay for each full-adder is 25 ns.

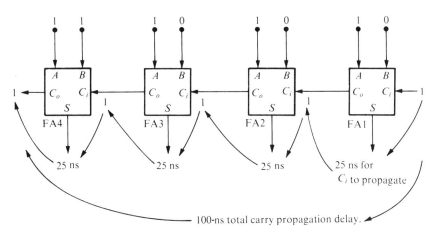

Figure 12–9. A Four-Bit Parallel Binary Ripple-Carry Adder Showing a Typical Carry Propagation Delay.

time through a single full-adder from the application of the data bits to the sum output.

A complete five-bit parallel binary adder is shown in Figure 12–10 in block diagram form. Negative numbers are handled easily in 1's complement form by inverting each bit of the negative number before adding, and connecting the carry output (C_o) of the last adder stage to the carry input (C_i) of the first stage to perform the end-around carry operation. This contrasts with the case of the serial adder where 2's complement operation is easier. Taking the 1's complement of a negative number can be done within the storage register, as we will see later. Sign-bit and magnitude comparison are also required, for the same purposes as in the serial adder.

The two numbers to be added are parallel-loaded into registers A and B. If the number has a negative sign (1), its magnitude is complemented within the register. A check is made for two negatively signed numbers and for a larger negative number. After a short time is allowed for these operations to take place and for the final sum to appear on the outputs, a pulse is applied to register A, causing the sum to be loaded back into the register so the next number can be added to it. Before the next addition, a check for an overflow condition is made. The next number is loaded into register B, and the process is repeated. The addition process in the parallel adder takes place very quickly compared to the serial adder—the major advantage is speed.

As previously mentioned, one way to accomplish the 1's complement of a negative number can be within the storage register by running the output of each stage to an exclusive-OR gate, as shown in Figure 12–11. When the complement control line is HIGH, the gates will complement each register output, and when the complement control line is LOW, the register outputs pass through the gates uncomplemented.

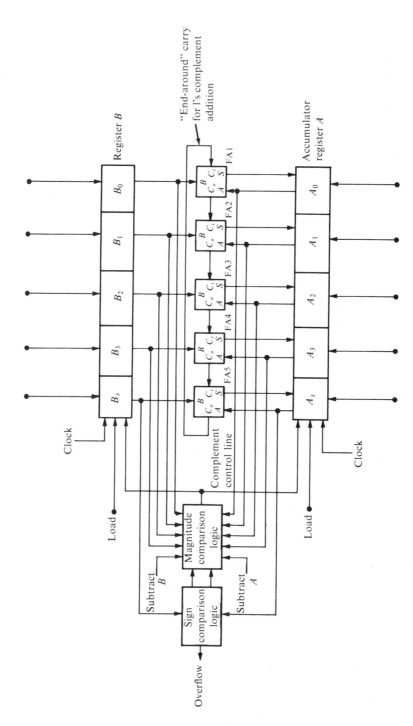

Figure 12–10. A Five-Bit Parallel Binary Adder.

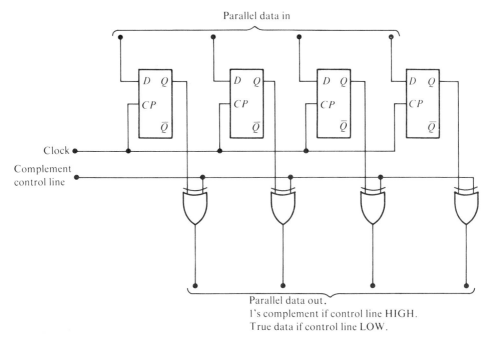

Parallel data in

Clock •

Complement
control line •

Parallel data out.
1's complement if control line HIGH.
True data if control line LOW.

Figure 12–11. One Method of Complementing the Contents of a Register.

Example 12–10 Show examples of the longest and shortest delay times through a five-stage parallel adder.

Solutions:

The longest delay possible occurs when a carry is propagated through each adder stage, as shown in Figure 12–12.

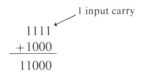

1 input carry

1111
+1000

11000

The total delay time is $5t_{dc}$, where t_{dc} is the input-carry-to-output-carry delay. The shortest possible delay occurs when there are no carries propagated, as shown in Figure 12–13.

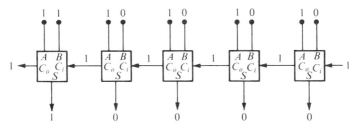

Figure 12–12.

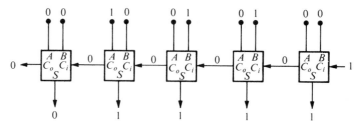

Figure 12-13.

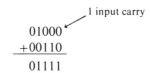

1 input carry

$$01000$$
$$+00110$$
$$01111$$

Note that t_{ds} is the total delay time between the application of the input bits and the occurrence of the sum on the output.

12-7 Carry-Look-Ahead Adder (CLA)

As you have seen in the discussion of the parallel adder, the speed with which an addition can be performed is limited by the time required for the carries to propagate or "ripple" through all of the stages of the adder. One method of speeding up this process by eliminating this carry "ripple" delay is called *carry-look-ahead* addition; this method is based on two functions of the full-adder called the *carry generate* and the *carry propagate* functions.

The *carry generate* function indicates when an output carry is produced (generated) by the full-adder. A carry is *generated* only when *both input bits are 1s*. This condition is expressed as

$$G = AB \qquad (12\text{-}2)$$

A carry input is *propagated* by the full-adder when either or both of the input bits are 1s. This condition is expressed as

$$P = A + B \qquad (12\text{-}3)$$

The *carry generate* and *carry propagate* conditions are illustrated in Figure 12-14.

How can the carry output of a full-adder be expressed in terms of the carry generate (G) and the carry propagate (P)? The output carry (C_o) is a 1 if the carry generate is a 1 OR if the carry propagate is a 1 AND the input carry (C_i) is a 1. In other words, we get an output carry of 1 if it is *generated* by the full-adder ($A = 1$ and $B = 1$) or if the adder can propagate the input carry ($A = 1$ OR $B = 1$) AND $C_i = 1$. This relationship is expressed as

$$C_o = G + PC_i \qquad (12\text{-}4)$$

Now we will see how this concept can be applied to a parallel adder whose individual stages are shown in Figure 12-15 for a four-bit example. For each full-adder, the output carry is dependent on its carry generate (G), its carry propagate

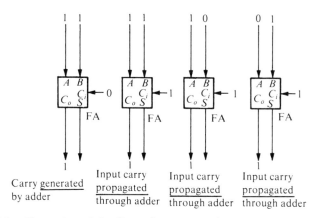

Figure 12–14. Illustration of the Carry Generate and Carry Propagate Conditions.

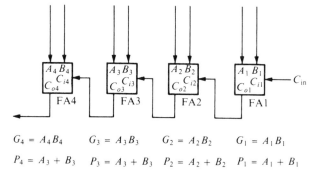

$$G_4 = A_4 B_4 \qquad G_3 = A_3 B_3 \qquad G_2 = A_2 B_2 \qquad G_1 = A_1 B_1$$

$$P_4 = A_3 + B_3 \qquad P_3 = A_3 + B_3 \qquad P_2 = A_2 + B_2 \qquad P_1 = A_1 + B_1$$

Figure 12–15. Carry Generate and Carry Propagate Functions in Terms of the Input Bits to a Four-Bit Adder.

(P), and its carry input (C_i). The G and P functions for each stage are immediately available as soon as the input bits A and B are applied because they are dependent only on these two bits. The carry input to each stage is the carry output of the previous stage. Based on this, we will now develop expressions for the carry out, C_o, of each full-adder stage for the four-bit example.

Carry out of FA1

$$C_{o1} = G_1 + P_1 C_{i1} \tag{12-5}$$
$$C_{i2} = C_{o1}$$

Carry out of FA2

$$
\begin{aligned}
C_{o2} &= G_2 + P_2 C_{i2} \\
&= G_2 + P_2 C_{o1} \\
&= G_2 + P_2(G_1 + P_1 C_{i1}) \\
&= G_2 + P_2 G_1 + P_2 P_1 C_{i1} \tag{12-6}
\end{aligned}
$$
$$C_{i3} = C_{o2}$$

Carry out of FA3

$$C_{o3} = G_3 + P_3 C_{i3}$$
$$= G_3 + P_3 C_{o2}$$
$$= G_3 + P_3(G_2 + P_2 G_1 + P_2 P_1 C_{i1})$$
$$= G_3 + P_3 G_2 + P_3 P_2 G_1 + P_3 P_2 P_1 C_{i1} \qquad (12\text{--}7)$$
$$C_{i4} = C_{o3}$$

Carry out of FA4

$$C_{o4} = G_4 + P_4 C_{i4}$$
$$= G_4 + P_4 C_{o3}$$
$$= G_4 + P_4(G_3 + P_3 G_2 + P_3 P_2 G_1 + P_3 P_2 P_1 C_{i1})$$
$$= G_4 + P_4 G_3 + P_4 P_3 G_2 + P_4 P_3 P_2 G_1 + P_4 P_3 P_2 P_1 C_{i1} \qquad (12\text{--}8)$$

Notice that in each of these expressions, the carry out for each full-adder stage is dependent only on the initial input carry (C_{i1}), its G and P functions, and the G and P functions of the preceding stages. Since each of the G and P functions can be expressed in terms of the A and B input to the full-adders, all of the output carries are immediately available (except for gate delays) and we do not have to wait for a carry to "ripple" through all of the stages before a final result is achieved. Thus the carry-look-ahead technique speeds up the addition process.

Equations (12–5) through (12–8) can be implemented with logic gates and connected to the full-adders to create a carry-look-ahead adder, as shown in Figure 12–16. To handle signed addition and subtraction, all of the auxilliary functions discussed in relation to the previous adders must be incorporated. They would be essentially the same as for the ripple-carry adder discussed in Section 12–6.

12–8   Carry-Save Adder (CSA)

Up to this point the methods discussed have involved the addition of two numbers at a time. As we have seen, a string of numbers can be added two at a time by accumulating the sum after each addition and adding the next number.

A method for adding three or more numbers at a time is called *carry-save* addition. This process is illustrated in Example 12–11, where the sum is generated and the carries are "saved" and added to the final sum.

Example 12–11   Add 0.0011, 0.0001, and 0.1001 using the carry-save method.

Solution:

$$
\begin{array}{ll}
\quad 0.0011 & A \\
\quad 0.0001 & B \\
+0.1001 & C \\
\hline
\quad 0.1011 & \text{sum, excluding carries} \\
+0.001 & \text{carries shifted left one place} \\
\hline
\quad 0.1101 & \text{final sum}
\end{array}
$$

A full-adder is used to add a bit from each of the three numbers, utilizing the carry input as one of the bit inputs. A sum from the addition of each of the three

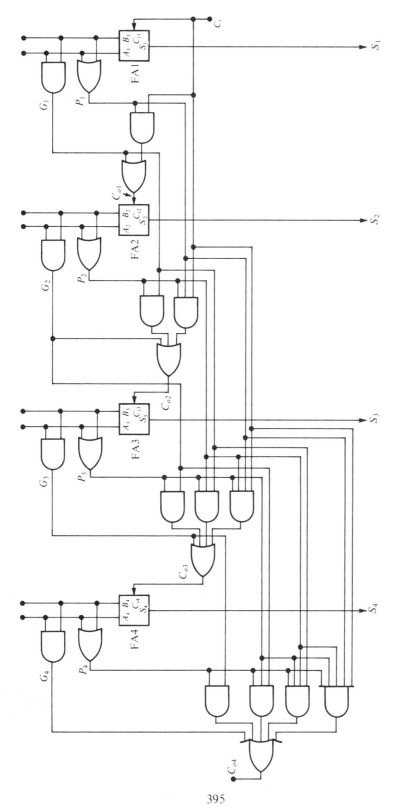

Figure 12–16. Logic Diagram for a Four-Stage Carry-Look-Ahead Adder.

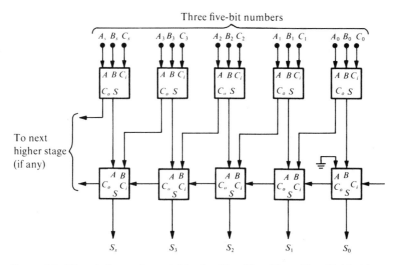

Figure 12–17.   A Carry-Save Adder for Handling Three Five-Bit Numbers.

bits is formed along with a carry. All of the carries are then added to the sum bits to form the final sum. Figure 12–17 shows a carry-save adder for the addition of three five-bit numbers. By increasing the number of carry-save levels, more than three numbers can be added at one time, as illustrated in Figure 12–18 for the addition of four numbers. As in the other types of adders, the auxilliary functions must be incorporated in order to handle all of the cases of addition of signed numbers. The carry-save method is a way to speed up addition at a cost of increased complexity.

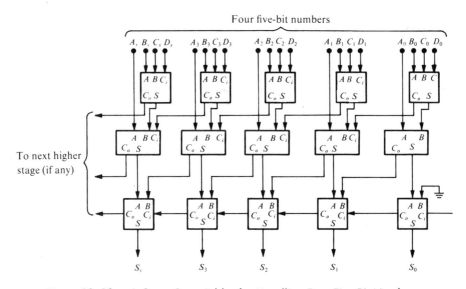

Figure 12–18.   A Carry-Save Adder for Handling Four Five-Bit Numbers.

12–9   A BCD Adder

In order to understand the requirements for a BCD adder, let us briefly review how BCD addition is accomplished. This topic was discussed in Chapter 3, and the rules are restated below for convenience.

1. Add the two numbers using the rules for binary addition.
2. If the four-bit sum is equal to or less than 9, it is *valid*.
3. If the four-bit sum is greater than 9 or if a carry is generated from the four-bit sum, it is *invalid*.
4. To correct the invalid sum, add 6 (0110) to the four-bit sum. If a carry results, add it to the next BCD digit.

Figure 12–19 shows a BCD adder. Functionally, the system operates as follows. Two four-bit BCD digits are entered into the adder along with any input

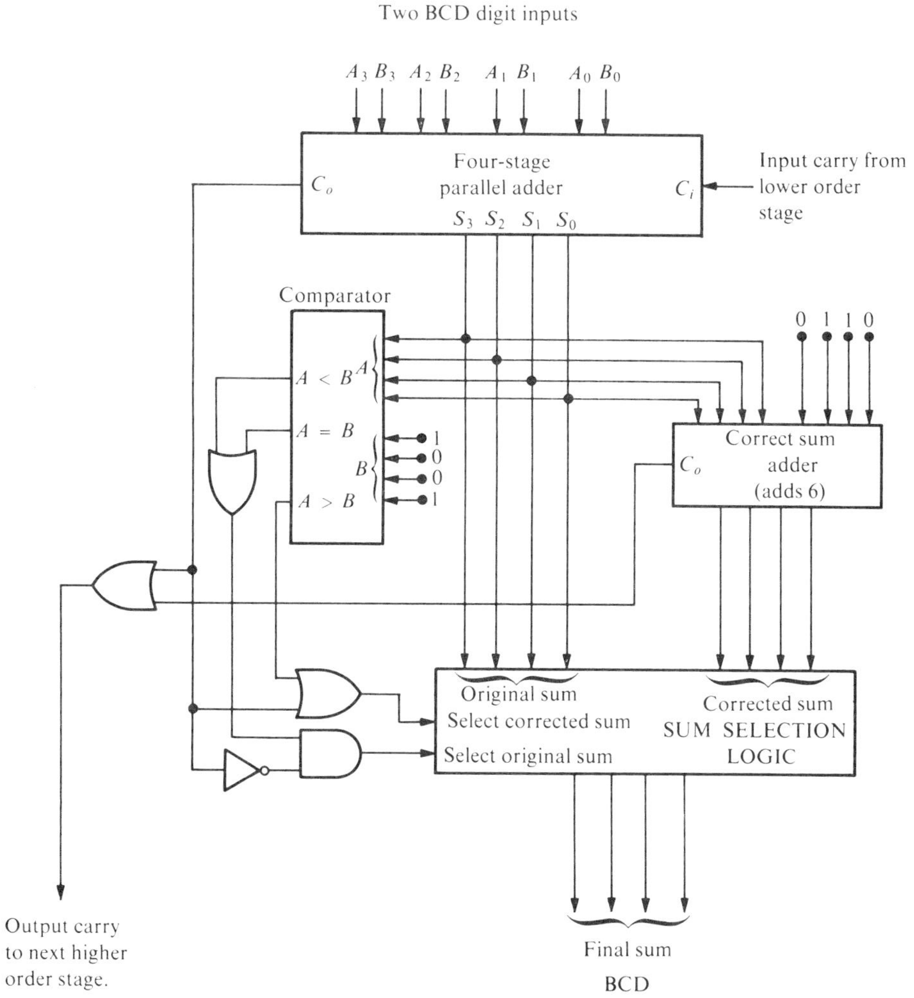

Figure 12–19.   Block Diagram for A BCD Adder.

carry. The resulting sum is compared to 9 (1001), and if it is equal to or less than 9 and no output carry occurs, it is gated through the sum selection logic to the final outputs. If the sum is greater than 9 or if an output carry occurs, it is invalid. The correct sum adder is enabled, adding 6 (0110) to the invalid sum; the corrected sum is then gated through the sum selection logic to the final outputs. The carry outputs of both adders are ORed and go to the next higher stage (if any).

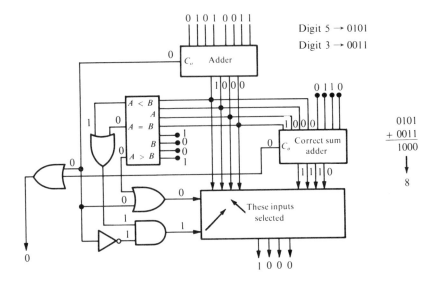

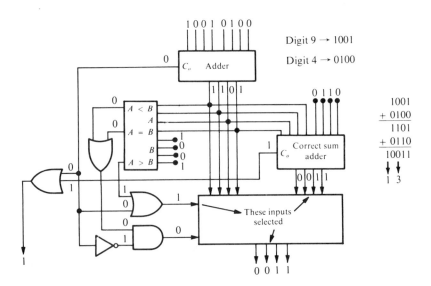

Figure 12-20.

Example 12–12   Show the states within a BCD adder for addition of the digits 5 and 3, and for 9 and 4. Refer to Figure 12–19 for a complete diagram.

Solutions:

See Figure 12–20.

12–10   Addition of Signed BCD Numbers

Addition of signed BCD numbers can be performed by using the 9's or 10's complement methods. A negative BCD number can be expressed by taking the 9's or 10's complement of each digit, as shown in Table 12–1. Our discussion of the BCD adder in the previous section assumed that negative numbers were already in complement form.

Conversion of a BCD digit to its 9's complement is not a straightforward process, as you can see by examination of the table. Notice, however, that a BCD digit expressed in Excess-3 form can easily be converted to its 9's complement by simply inverting each bit. Therefore, in order to avoid the difficulty of converting a BCD digit to its 9's complement, the Excess-3 code is often used. We will go through several examples to illustrate signed addition using Excess-3 code for each BCD digit.

Example 12–13

Decimal	BCD (9's complement)		Excess-3 (9's complement)	
5	0.0101		0.1000	
+ − 3	+1.0110	9's complement	+1.1001	9's complement
+2	1.1011		┌①0.0001	
	0110	add 6	└────→ +11	add 3
	┌①0.0001		0.0100	
	└───→ +1		+1	
	0.0010		0.0101	Excess-3 code for +2
−8	1.0001	9's complement	1.0100	
+6	+0.0110		+0.1001	
−2	1.0111	9's complement of result	1.1101	
	↓		−11	subtract 3
	1.0010		1.1010	9's complement of result
			↓	
			1.0101	Excess-3 code for −2
24	0.0010 0100		0.0101 0111	
−17	+1.1000 0010	9's complement	+1.1011 0101	9's complement
+7	1.1010 0110		┌①0.0000 1100	
	+0110	add 6	│ +11 −11	
	┌①0.0000 0110		│ 0.0011 1001	
	└────────→ +1		└────────→ +1	
	0.0000 0111		0.0011 1010	Excess-3 code for +7

TABLE 12–1.

Decimal	BCD	Excess-3	9's Complement BCD	9's Complement Excess-3	10's Complement
0	0000	0011	1001	1100	1010
1	0001	0100	1000	1011	1001
2	0010	0101	0111	1010	1000
3	0011	0110	0110	1001	0111
4	0100	0111	0101	1000	0110
5	0101	1000	0100	0111	0101
6	0110	1001	0011	0110	0100
7	0111	1010	0010	0101	0011
8	1000	1011	0001	0100	0010
9	1001	1100	0000	0011	0001

12–11 Multiplication

At this point you should review Section 7–11, where we discussed the multiplication process and presented a method of parallel multiplication. In this section we are going to look at multiplication in its simplest form by taking one multiplier bit at a time, just as we do in long hand multiplication.

As you know, multiplication can be performed by repeated additions of the partial products, with each successive partial product shifted to the left of the previous partial product. Let us take a look at an example of long hand multiplication and examine the steps involved.

$$
\begin{array}{ll}
1010 & \text{multiplicand} \\
\times 111 & \text{multiplier} \\
\hline
1010 & \text{step 1} \\
1010 & \text{step 2} \\
\hline
11110 & \text{step 3} \\
1010 & \text{step 4} \\
\hline
1000110 & \text{step 5}
\end{array}
$$

Step 1. The LSB of the multiplier is a 1, so we bring down the multiplicand. This is the first partial product.

Step 2. The next bit of the multiplier is a 1, so we bring down the multiplicand shifted one place to the left of the first partial product. This is the second partial product.

Step 3. We add the first and second partial products. We can also wait until all partial products are formed and then add, but logic can be more easily implemented by adding two numbers at a time.

Step 4. The MSB of the multiplier is a 1, so we bring down the multiplicand shifted one place to the left of the second partial product. This is the third partial product.

Step 5. We add the third partial product to the first two, and arrive at a final product.

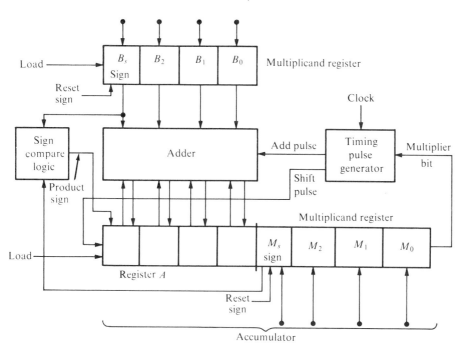

Figure 12–21. Block Diagram for a Three-Bit Multiplier.

This procedure can be implemented directly with logic elements to provide a straightforward binary multiplier circuit as shown in Figure 12–21.

This circuit operates as follows. The multiplicand is stored in the multiplicand register, and the multiplier is stored in the right-most positions of the accumulator. The accumulator register is used to store the partial products during the multiplication process and the final product when the operation is complete. When dealing with signed numbers, the sign-bit comparison logic checks the sign bits of the multiplier and the multiplicand, and stores the result. If the two signs are different, the product will have a negative sign, and if the two signs are the same, the product will have a positive sign. Once the sign determination is made, the sign bits of the multiplier and multiplicand stored in the registers are reset to 0 so they do not enter into the multiplication process.

The basic timing for the process is provided by the clock signal, and each step in the multiplication occurs at a clock pulse time. To begin, the LSB of the multiplier appears on the multiplier bit line; if it is a 1 when the clock occurs, an add pulse and a following shift pulse are generated. The add pulse enables the adder logic, the multiplicand is added to the contents of the accumulator (which is initially all 0s), and the result is put back into the accumulator. The shift pulse shifts the contents of the accumulator one place to the right; this is *effectively* the same as the left shift of the multiplicand in the long hand example. If the multiplier bit is a 0, no addition is performed but the accumulator contents are shifted. As the accumulator is shifted, the multiplier bits successively appear on the multiplier bit line, the add/shift operation is repeated for each multiplier 1, and a shift only is re-

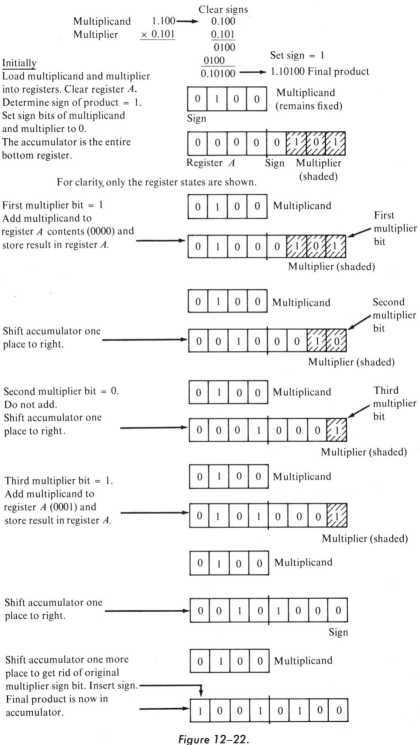

Multiplicand   1.100 ——→ 0.100   Clear signs
Multiplier   × 0.101   0.101
                           0100
                           0100
                         0.10100 ——→ 1.10100 Final product

Set sign = 1

Initially
Load multiplicand and multiplier into registers. Clear register *A*.
Determine sign of product = 1.
Set sign bits of multiplicand and multiplier to 0.
The accumulator is the entire bottom register.

| 0 | 1 | 0 | 0 | Multiplicand (remains fixed) |

Sign

| 0 | 0 | 0 | 0 | 0 | 1 | 0 | 1 |

Register *A*   Sign   Multiplier (shaded)

For clarity, only the register states are shown.

First multiplier bit = 1
Add multiplicand to register *A* contents (0000) and store result in register *A*.

| 0 | 1 | 0 | 0 | Multiplicand |

| 0 | 1 | 0 | 0 | 0 | 1 | 0 | 1 |

First multiplier bit

Multiplier (shaded)

| 0 | 1 | 0 | 0 | Multiplicand |

Second multiplier bit

Shift accumulator one place to right.

| 0 | 0 | 1 | 0 | 0 | 0 | 1 | 0 |

Multiplier (shaded)

Second multiplier bit = 0.
Do not add.
Shift accumulator one place to right.

| 0 | 1 | 0 | 0 | Multiplicand |

Third multiplier bit

| 0 | 0 | 0 | 1 | 0 | 0 | 0 | 1 |

Multiplier (shaded)

Third multiplier bit = 1.
Add multiplicand to register *A* (0001) and store result in register *A*.

| 0 | 1 | 0 | 0 | Multiplicand |

| 0 | 1 | 0 | 1 | 0 | 0 | 0 | 1 |

Multiplier (shaded)

| 0 | 1 | 0 | 0 | Multiplicand |

Shift accumulator one place to right.

| 0 | 0 | 1 | 0 | 1 | 0 | 0 | 0 |

Sign

Shift accumulator one more place to get rid of original multiplier sign bit. Insert sign.
Final product is now in accumulator.

| 0 | 1 | 0 | 0 | Multiplicand |

| 1 | 0 | 0 | 1 | 0 | 1 | 0 | 0 |

Figure 12–22.

402

peated for each multiplier 0. The operation continues until all multiplier bits have been shifted out and the accumulator contains the final product. Since the product can be twice the length of the multiplier, the accumulator must have a sufficient number of stages to accommodate it. We will now go through the multiplication of two four-bit words (including sign), and look at the state of the registers on a bit-by-bit basis in Example 12–14.

Example 12–14 Show each step in the multiplication of 1.100 and 0.101, using the basic multiplier of Figure 12–21.

Solution:
See Figure 12–22.

12–12 Division

In this section we will examine a basic method of division called *restoring division,* the basic method we all know because we use it in performing long hand decimal division. Other techniques are sometimes used in digital systems for binary division, but most of these are aimed at speeding up the process and are more complex than the restoring method. We will limit outselves to a discussion of the restoring method to gain a basic understanding of the process of binary division.

The basic procedure involves a trial subtraction of the divisor from the dividend or partial remainder. If the divisor can be subtracted from the partial remainder, the quotient bit is a 1; if not, then the quotient bit is a 0. Stated another way, if the divisor will go into the partial remainder, the quotient bit is a 1, and the if the divisor will not go into the partial remainder, the quotient bit is a 0. When the divisor is smaller than or equal to the partial remainder, subtraction takes place and the difference becomes the next partial remainder; when the divisor is greater than the partial dividend, the partial remainder is shifted one place to the left of the divisor.

The following example of long hand binary division will illustrate the procedure. Here we will divide 6 by 9, with both expressed in binary fractional notation. The result cannot exceed 0.1111 for a four-bit number because no magnitude bit can appear to the left of the binary point—only the sign bit:

$$0.0110 \div 0.1001$$

$$
\begin{array}{r}
0.1010 \leftarrow \text{quotient} \\
\text{divisor} \longrightarrow 1001 \overline{)\ 0110.0000} \leftarrow \text{dividend} \\
-1001 \\
\hline
1100 \\
-1001 \\
\hline
110 \quad \text{remainder}
\end{array}
$$

A step-by-step procedure of a logic divider circuit for the preceding division problem is as follows. Study this carefully so that you will understand exactly what the logic circuit must do.

			Register		
1.	Load the dividend into register A.	.0110	A		
	Load the divisor into register D.	.1001	D		
2.	Subtract.	$-.0011$	A	$A < D$; proceed	
3.	Add the divisor back to restore	$+.1001$	D		
	the dividend.	$+.0110$	A		
4.	Shift left.	$+.1100$	A		
5.	Subtract the divisor.	$-.1001$	D		
		$+.0011$	A	$A > D$	$Q_3 = 1$
6.	Shift left.	$+.0110$	A		
7.	Subtract the divisor.	$-.1001$	D		
		$-.0011$	A	$A < D$	$Q_2 = 0$
8.	Add the divisor back to restore	$+.1001$	D		
	the partial remainder.	$+.0110$	A		
9.	Shift left.	$+.1100$	A		
10.	Subtract the divisor.	$-.1001$	D		
		$+.0011$	A	$A > D$	$Q_1 = 1$
11.	Shift left.	$+.0110$	A		
12.	Subtract the divisor.	$-.1001$	D		
		$-.0011$	A	$A < D$	$Q_0 = 0$

At this point, we have generated four quotient bits, which is the limit for the four-bit system. The quotient is an approximation (due to the bit limitation), and would be more accurate if we could carry it out further. Also, in this example, the quotient has a positive sign because the signs of the divisor and the dividend are the same.

Figure 12-23 shows a block diagram of a restoring binary divider. The operation is as follows and results in the same procedure discussed in the preceding example.

First, the divisor is loaded into register D and the dividend is loaded into register A (accumulator). The Q register is cleared. A sign determination is made and stored so that the quotient can be given the proper sign at the end of the division process. If the signs of the divisor and dividend are the same, the quotient will have a positive sign; if they differ, the quotient will have a negative sign. Both the divisor and the dividend are applied to the adder, and the divisor is subtracted from the dividend. If the result is negative, the divisor is larger than the dividend and our quotient will be a fractional number—we may proceed with the division. If a smaller divisor is indicated, the procedure must be halted at this point and a correction made because we cannot have a quotient larger than 0.1111 for a four-bit fractional system. A 0 carry out of the adder indicates that the subtraction has resulted in a negatively signed number; this means that the divisor will not go into the dividend (or partial remainder), and a 0 quotient bit is entered into the Q register. The 0 on the output carry line also initiates a restore action by the control logic, causing the divisor to be added back to the contents of the accumulator. It should be pointed out that after each addition or subtraction, the result is put back into the accumulator. The contents of Register D remain fixed. The restore action is followed by a shift to the left of the contents of the accumulator and register Q.

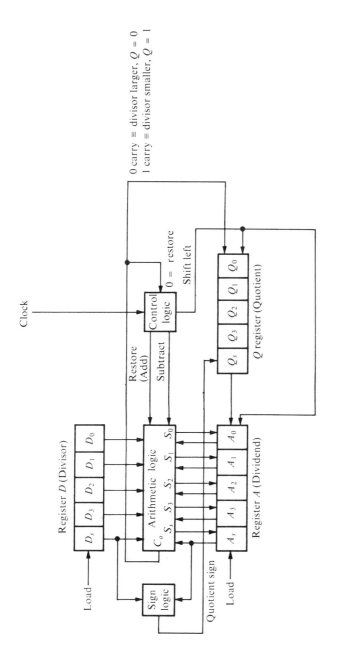

Figure 12–23. Block Diagram for a Four-Bit Divider.

405

A 1 output carry from the adder indicates that the subtraction has resulted in a positively signed number; this means that the divisor will go into the dividend or partial remainder and a quotient bit of 1 is entered into the Q register. The process continues until the Q register is filled. The proper sign bit is attached to the quotient to complete the operation.

PROBLEMS

12-1   Show how the following subtractions can actually be accomplished with addition only:
   (a) 4 – 2   (b) 8 – 3   (c) 15 – 6
   (d) –25 – 13   (e) 104 – 87   (f) –253 – 150

12-2   Determine the decimal numbers represented by the following sign-and-magnitude binary numbers:
   (a) 0.1101   (b) 0.1011   (c) 1.1110   (d) 1.10010
   (e) 0.00110   (f) 1.10101

12-3   Express the following signed decimal numbers in binary sign-and-magnitude notation.
   (a) +9   (b) +13   (c) –23   (d) –42   (e) +87   (f) –99

12-4   Perform the following binary additions:
   (a) 0.0010 + 0.0110   (b) 0.1000 + 0.1001   (c) 0.0111 + 0.0100

12-5   Perform the following additions using the 1's complement method:
   (a) 0.1001 + 1.0110   (b) 0.0110 + 1.0011   (c) 1.0100 + 1.0010
   (d) 1.1101 + 1.0001   (e) 1.1000 + 0.0111   (f) 1.0101 + 0.1010

12-6   Repeat Problem 12–5 using the 2's complement method.

12-7   In this chapter the serial adder using the 2's complement method was studied. Show a basic block diagram of a serial adder using the 1's complement method. Explain the advantages and/or disadvantages.

12-8   Show the adder states on a bit-by-bit basis to illustrate the serial addition of the following two numbers. Refer to Example 12–9.

$$0.0111 + 0.0100$$

12-9   Repeat Problem 12–8 for the following two numbers:

$$1.1000 + 0.0110$$

12-10   Explain how a string of several numbers is added with a serial adder.

12-11   Using the 2's complement logic of Figure 12–6, show the bit-by-bit conversion of the following binary numbers:
   (a) 10110100   (b) 11110000   (c) 10010011

12-12   Repeat Problem 12–11 for the following numbers:
   (a) 11000110   (b) 10000101   (c) 10000000

12-13   Explain the purpose of sign-and-magnitude comparison in an adder circuit.

12-14   Explain the purpose of overflow detection and the condition for the occurrence of overflow in an adder.

12–15 Illustrate an overflow condition and its detection by showing the states within the circuit of Figure 12–4.

12–16 Explain why the serial adder takes longer to complete an addition than a parallel adder.

12–17 Show a ripple-carry parallel adder for adding eight-bit numbers.

12–18 For the carry-look-ahead adder of Figure 12–16, verify proper operation for the addition of the following numbers by following the logic levels through the circuit:
(a) $1001 + 0011$ (b) $0100 + 0101$ (c) $1000 + 0111$

12–19 For the carry-save adder of Figure 12–17, verify that the addition of the following binary numbers results in the proper sum:
(a) $00001 + 01000 + 01110$ (b) $10000 + 10110 + 11100$
(c) $11001 + 11111 + 10001$

12–20 For the carry-save of Figure 12–18, verify proper operation for the following binary additions:
(a) $00010 + 00011 + 00100 + 00101$
(b) $01100 + 01001 + 01111 + 00110$
(c) $10101 + 11011 + 11101 + 11111$

12–21 Add the following BCD numbers:
(a) $0110 + 0010$ (b) $1000 + 0001$ (c) $0101 + 0111$
(d) $0111 + 0101$ (e) $1001 + 0100$ (f) $1001 + 1001$

12–22 Convert the following decimal numbers to BCD, and perform the indicated addition:
(a) $7 + 2$ (b) $12 + 9$ (c) $49 + 36$
(d) $9 + 5$ (e) $25 + 23$ (f) $192 + 148$

12–23 Show the stages within the BCD adder of Figure 12–19 for each of the following additions:
(a) $3 + 2$ (b) $7 + 6$ (c) $9 + 9$ (d) $5 + 4$
(e) $9 + 8$ (f) $5 + 5$

12–24 Repeat Problem 12–23 for the following additions:
(a) $1 + 4$ (b) $6 + 7$ (c) $5 + 1$ (d) $2 + 7$
(e) $3 + 9$ (f) $9 + 0$

12–25 Sketch a block diagram for a two-digit BCD adder.

12–26 Sketch a block diagram for a four-digit BCD adder.

12–27 Design the logic circuit for the sum selection logic block of Figure 12–19.

12–28 Show the states of the registers in a multiplication circuit for each step in the following multiplication:

$$\begin{array}{r} 1.110 \\ \times\,1.011 \\ \hline \end{array}$$

12–29 Repeat Problem 12–29 for the following multiplication:

$$\begin{array}{r} 0.111 \\ \times\,1.111 \\ \hline \end{array}$$

12-30 For a binary divider, show the states of each register for each step in the following division:

$$0.010 \div 0.100$$

12-31 For the 2's complement logic of Figure 12-6, design the logic required for disabling the circuit at the appropriate times.

CHAPTER 13

Miscellaneous Digital Circuits and Associated Topics

This is our "catchall" chapter where we will touch on several types of digital circuits that have not been discussed in the previous chapters. In general, the topics covered here are more specialized and are not as widely applied as those functions already covered, but they are nevertheless important in the study of digital systems.

The coverage includes the monostable multivibrator (one-shot), the astable multivibrator and other clock sources, the Schmitt trigger circuit, some aspects of digital data transmission including line drivers and line receivers, digital-to-analog and analog-to-digital conversion, and three-state logic.

You will note a shift back to detailed circuit descriptions in some sections presented to give you a basic feel for certain newly introduced circuits in contrast to the more functional approach we have used in the last few chapters.

13–1   The One-Shot

The one-shot is a type of multivibrator with a monostable characteristic; that is, it has only *one stable state.* In general operation this circuit stays in its stable state until triggered into its unstable state. The unstable state is *temporary,* and the circuit remains there for only a certain length of time (as determined by component values in the circuit). The circuit will always return to the stable state unless it is retriggered. Figure 13–1 shows the basic cross-coupled feedback configuration that is fundamental in one-shot operation. Only essential elements are shown for clarity. Although several variations of the implementation are possible, this cross-coupled feedback arrangement is essential to all.

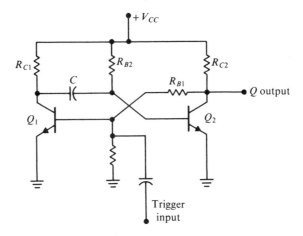

Figure 13–1. Basic One-Shot Circuit.

The basic operation is as follows. We begin by assuming the one-shot is in its stable state. In this state, the capacitor is fully charged to $V_{CC} - V_{BE2}$ and base current is supplied to Q_2 through R_{B2}, keeping the transistor saturated. The collector of Q_2, which is the true output of the one-shot, is LOW, keeping Q_1 OFF by reverse biasing its base-emitter junction. This stable condition is illustrated in Figure 13–2(a).

Now, if a short positive trigger pulse is applied to the trigger input, Q_1 is momentarily turned ON, causing its collector to go LOW. This HIGH-to-LOW transition on the Q_1 collector is instantaneously coupled through the capacitor to the base of Q_2, thereby turning Q_2 OFF, as shown in Figure 13–2(b). Now the capacitor begins discharging through R_{B2} and the saturated transistor Q_1, as indicated in Figure 13–2(c); this is the transitory condition that produces the unstable state of the one-shot. During the time that the capacitor is discharging, Q_2 is OFF and the Q (true) output of the circuit is HIGH. This unstable or transitory condition lasts until the voltage on the base of Q_2 reaches approximately +0.5 V. At this point, Q_2 turns ON, Q_1 turns OFF, and the one-shot returns to its stable state. The time the capacitor takes to return to the +0.5-V point on the base of Q_2 after the trigger pulse is applied (the length of time that Q_2 is OFF, dependent on the time constant $R_{B2}C$) determines the pulse width of the one-shot output. An approximate value for the pulse width is determined by the following analysis.

After the trigger is applied, the voltage on the base of Q_2 is an exponential curve that can be expressed by the following general formula where $V_0 = -(V_{CC} - 0.7)$ for a silicon transistor and $V_F = V_{CC}$:

$$v_{B2} = V_F + (V_0 - V_F)e^{-t/R_{B2}C} \qquad (13\text{--}1)$$

These values are shown on the curve in Figure 13–3.

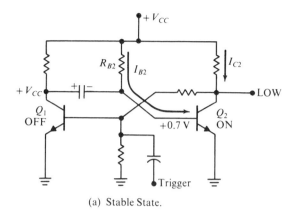

(a) Stable State.

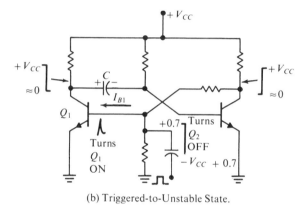

(b) Triggered-to-Unstable State.

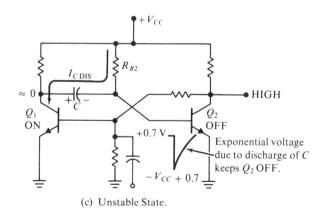

(c) Unstable State.

Figure 13–2.   One-Shot Operation.

411

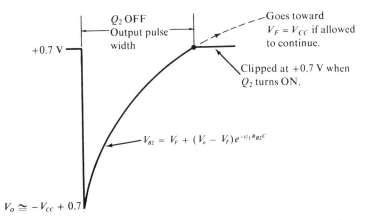

Figure 13–3. Voltage on the base of Q_2 in Figure 13–1 after the One-Shot Is Triggered into the Unstable State.

The next step in the procedure is to solve Equation (13–1) for the time required for the exponential voltage v_{B2} to reach +0.5 V for specified values of R_{B2} and C; this will be the pulse width. We proceed as follows:

$$v_{B2} = V_{CC} + (V_0 - V_{CC})e^{-t_{pw}/R_{B2}C}$$

$$e^{-t_{pw}/R_{B2}C} = \frac{v_{B2} - V_{CC}}{V_0 - V_{CC}}$$

$$-\frac{t_{pw}}{R_{B2}C} = \ln \frac{v_{B2} - V_{CC}}{V_0 - V_{CC}}$$

$$t_{pw} = -R_{B2}C \ln \frac{v_{B2} - V_{CC}}{V_0 - V_{CC}} \qquad (13\text{–}2)$$

Example 13–1 Calculate the pulse width t_{pw} of a one-shot having the circuit values $V_{CC} = 10\,\text{V}$, $R_{B2} = 5.6\,\text{K}$, and $C = 0.002\,\mu\text{F}$.

Solution:

$$t_{pw} = -R_{B2}C \ln \frac{v_{B2} - V_{CC}}{V_0 - V_{CC}}$$

$$V_0 = -(10 - 0.7) = -9.3$$

$$t_{pw} = -(5.6\,\text{K})(0.002\,\mu\text{F})\ln \frac{0.5 - 10}{-9.3 - 10}$$

$$= -(5.6\,\text{K})(0.002\,\mu\text{F})\ln 0.492$$

$$= (11.2 \times 10^{-6})(0.709)$$

$$= 7.941 \times 10^{-6}$$

$$= 7.941\,\mu\text{s}$$

It is interesting to note that the term in Equation (13–2)

$$\ln \frac{v_{B2} - V_{CC}}{V_0 - V_{CC}}$$

will always turn out to be approximately -0.7, independent of the value of V_{CC}. So Equation (13–2) can be restated in a simpler form:

$$t_{pw} \simeq 0.7 R_{B2} C \qquad (13\text{–}3)$$

Example 13–2   Given that a pulse width of 50 μs is required from a one-shot, determine the time constant necessary to produce it.

Solution:

$$t_{pw} \simeq 0.7 R_{B2} C$$

$$R_{B2} C \simeq \frac{t_{pw}}{0.7}$$

$$\simeq \frac{50\,\mu s}{0.7} = 71.4\,\mu s$$

From this value, we can select values for R_{B2} and C.

One remaining factor to consider when the one-shot returns to its stable state is the time required for the capacitor to recharge to its original condition through R_{C1} and the base-emitter junction of transistor Q_2. This is called the *recovery time,* and if it is short compared to the pulse width, it can normally be neglected. It does limit the rate at which the one-shot can be retriggered; if the capacitor is not allowed time to fully recharge to its original voltage after discharge, the one-shot will not produce a full-width output pulse when retriggered.

A typical logic symbol for a one-shot is shown in Figure 13–4. Both the Q and $\overline{Q}$ outputs are normally available, and there is a single trigger input. For one-shots in integrated circuit form, the entire circuit is contained on a single chip in an encapsulated package. It is apparent that for maximum flexibility the values of the resistor and capacitor that control the pulse width should be such that they can be changed easily; for this reason, integrated circuits normally have the appropriate points of the internal circuitry brought outside the package so an external resistor and capacitor can be connected for pulse width adjustment. This is indicated in Figure 13–4.

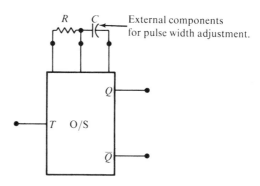

Figure 13–4.   Logic Symbol for One-Shot.

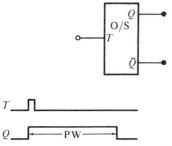

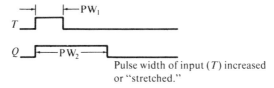

Time interval determined by pulse width.

(a) Time Interval Generation.

Pulse width of input (T) increased
or "stretched."

(b) Increasing Width of Input Pulse.

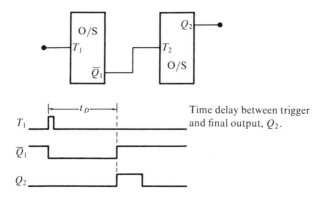

Time delay between trigger
and final output, Q_2.

(c) Time Delay Generation.

Figure 13–5. One-Shot Applications.

Some common applications of the one-shot in digital systems include generation of specified time intervals, pulse "stretching," and time delays; these are
illustrated in Figure 13–5.

13–2 The Astable Multivibrator

This multivibrator is sometimes called *free-running* and is a type of oscillator
circuit. It has a cross-coupled feedback arrangement, which differs from both the
bistable and monostable multivibrators already discussed. A basic circuit is shown
in Figure 13–6, which we will use to illustrate the fundamental operation of this
circuit.

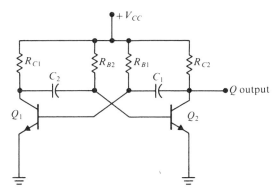

Figure 13–6. Basic Astable Multivibrator Circuit.

Notice that both cross-coupled feedback paths are capacitively coupled. The basic operation of this circuit can be understood by beginning with the assumption that Q_1 has just switched from OFF to ON (we have to start somewhere). We will use V_{CC} = 10 V in this discussion. As Q_1 switches from OFF to ON, the 10-V transition on the collector of Q_1 is coupled through C_2 to the base of Q_2, turning Q_2 OFF as indicated in Figure 13–7(a). Q_2 stays OFF until the voltage on its base reaches approximately +0.5 V. The OFF time of Q_2 is determined by the time constant of R_{B2} and C_2, and is calculated in the same manner as for the one-shot circuit.

$$t_{OFF2} \cong 0.7R_{B2}C_2$$

When the base of Q_2 reaches +0.5 V, Q_2 turns ON and the 10-V transition on its collector is coupled through C_1 to the base of Q_1, turning Q_1 OFF as shown in Figure 13–7(b). Q_1 stays OFF until the voltage on its base reaches approximately +0.5 V; this OFF time of Q_1 is determined by the time constant of R_{B1} and C_1 and is calculated as

$$t_{OFF1} \cong 0.7R_{B1}C_1$$

When Q_1 turns back ON, the cycle repeats itself and the circuit continues to oscillate back and forth with no external triggering required. Only a DC power source is necessary.

The frequency of oscillation is determined by the $R_{B1}C_1$ and the $R_{B2}C_2$ time constants. A period (T) of oscillation is equal to the OFF time plus the ON time of the transistor (Q_1 or Q_2). Taking the output off the Q_2 collector, we determine the frequency (f) as

$$T = t_{OFF2} + t_{ON2}$$
$$\cong 0.7R_{B2}C_2 + 0.7R_{B1}C_1$$

Then

$$f = \frac{1}{T}$$

If the two time constants are equal, a square wave is produced as the transistors switch ON and OFF. In this case the period is approximately $1.4R_{B2}C_2$.

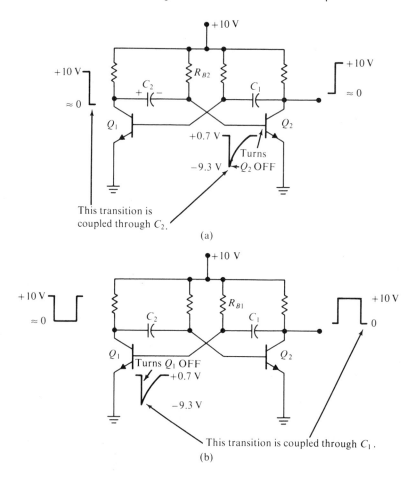

Figure 13–7.   Astable Multivibrator Operation.

The output from the collector of Q_1 is the complement of the output from the collector of Q_2 because when one transistor is ON, the other is OFF. The output waveforms for a square wave condition are shown in Figure 13–8. The rounding on the pulse edge is caused by the recovery time of the capacitors and depends on the $R_{C1}C_1$ and $R_{C2}C_2$ time constants (and has been neglected in the previous discussion for the sake of simplicity).

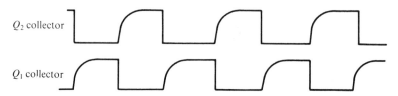

Figure 13–8.   Typical Astable Multivibrator Output Waveforms with 50 Percent Duty Cycle.

The astable multivibrator can be used as the main timing source or clock in a digital system. For a high degree of accuracy or frequency stability, other types of oscillator circuits (such as crystal-controlled oscillators, which we discuss in the next section) are sometimes used for timing sources.

Example 13-3   Determine the frequency and duty cycle of the output of the astable circuit in Figure 13–9.

OFF time of Q_2: $t_{OFF} \cong (0.7)(10\,K)(0.01\,\mu F) = 0.07 \times 10^{-3}\,s = 0.07$ ms
ON time of Q_2: $t_{ON} \cong (0.7)(20\,K)(0.01\,\mu F) = 0.14 \times 10^{-3}\,s = 0.14$ ms

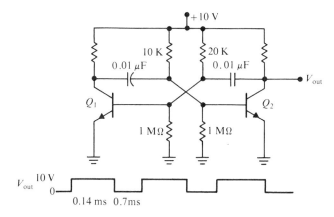

Figure 13–9.

Solution:

$$T = 0.14 \text{ ms} + 0.7 \text{ ms} = 0.21 \text{ ms} \qquad f = \frac{1}{T} = \frac{1}{0.21 \text{ ms}} = 4.76 \text{ kHz}$$

$$\text{Duty cycle} = \frac{t_{ON}}{T} \times 100 = \frac{0.14}{0.21} \times 100 = 66.7\%$$

13-3   A Crystal-Controlled Oscillator

A crystal-controlled (or simply crystal) oscillator utilizes a piezoelectric crystal as the resonant element. The crystal is normally made of quartz, exhibits a resonant characteristic, and is usually used with an external capacitance for best performance; crystals can be used in the place of tuned circuits in any LC oscillator and provide very good frequency stability. One type of crystal oscillator circuit is shown in Figure 13–10. The junction capacitances of the FET along with the crystal provide for resonant operation, and the frequency of oscillation can readily be changed by simply changing the crystal. Since the junction capacitances of the FET are relatively low, this circuit is most effective for high frequency applications. This is a linear type of circuit and therefore produces a sinusoidal output that can be adapted to digital applications by running the oscillator output through a "squaring" circuit or Schmitt trigger as indicated in Figure 13–11.

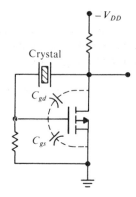

Figure 13–10.   A Type of Crystal-Controlled Oscillator.

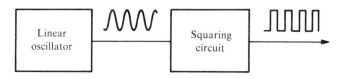

Figure 13–11.

As mentioned, a crystal is a device that acts as a tuned or resonant circuit. The advantage of using an oscillator circuit with crystal control of frequency rather than a discrete LC resonant circuit is the extremely good frequency stability characteristics of the crystal. That is, the frequency of oscillation will drift very little with time in a crystal oscillator (typically less than one part per million per day); this stability is very important in many digital systems where accurate timing is a necessity.

13–4   The Schmitt Trigger Circuit

The Schmitt trigger operates with two *threshold* or *trigger* points. When an increasing input signal reaches the *upper threshold point* (UTP), the circuit switches and the output goes to its HIGH level. When the input signal decreases to the *lower threshold point* (LTP), the circuit switches back and the output goes to its LOW level. This operation is illustrated in Figure 13–12(a).

As you can see, the Schmitt trigger circuit can be used to convert a sine wave into a pulse waveform. Also, it is useful in reshaping pulses that have been distorted in transmission, as illustrated in Figure 13–12(b).

Now let us look at the basic Schmitt trigger circuit to get an idea of how it works. The schematic is shown in Figure 13–13.

With no input signal, Q_1 is OFF, and the HIGH on its collector biases Q_2 ON; therefore, the output is normally at its lower level. The emitter current of Q_2 produces a voltage drop across R_E, which biases the emitter of Q_1 to voltage V_E; this voltage establishes the UTP. The input voltage must exceed V_E in order to turn Q_1 ON. When the input signal reaches the UTP, Q_1 turns ON, Q_2 turns

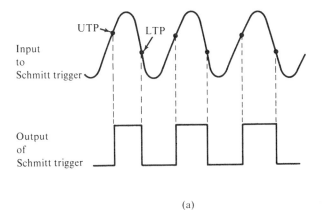

(a)

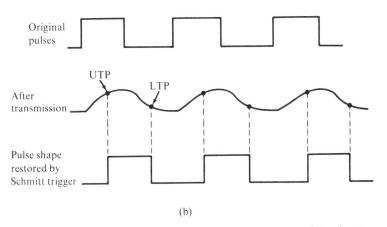

(b)

Figure 13–12.   Illustration of Schmitt Trigger Operation and Application.

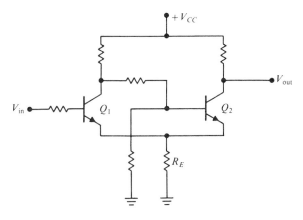

Figure 13–13.   A Basic Schmitt Trigger Circuit.

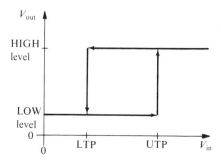

Figure 13–14.   Hysteresis Curve for a Schmitt Trigger.

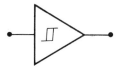

Figure 13–15.   Typical Logic Symbol for a Schmitt Trigger.

OFF, and the resulting output goes HIGH. Now the emitter bias voltage V_E is determined by the Q_1 emitter current since Q_1 is conducting; this voltage establishes the LTP. The input voltage must drop back below this LTP value for Q_1 to turn OFF and Q_2 to turn back ON.

The difference between the UTP and the LTP is called the hysteresis of the circuit, and a hysteresis curve is shown in Figure 13–14. It is necessary for the point at which the circuit output switches to a HIGH to be greater than the point at which it switches back LOW in order to prevent false triggering due to signal noise. A symbol sometimes used for the Schmitt trigger is shown in Figure 13–15.

Example 13–4   A Schmitt trigger has a UTP of 3 V and an LTP of 1.5 V. Determine the output for the input signal in Figure 13–16(a).

Solution:
See Figure 13–16(b).

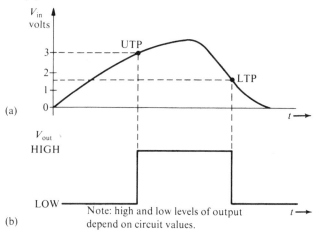

Figure 13–16.

13–5 Basic Digital Data Transmission Lines

In any digital system, data must be sent from one point to another; this transfer of digital information is called *data transmission.* The origin and destination of the digital data can be on the same assembly within the given system, or the origin can be on one assembly and the destination on another assembly in the same system, or the origin can be one digital system and the destination another system some distance from the first. There are two media commonly used for the transmission of data: electromagnetic radiation (radio frequency waves) and cable. In this section, we will be looking at the cable or direct connection between two points as a means of transmitting data. The RF link is normally used for longer distances. Data are transmitted from one point to another along a cable or wires called a *transmission line.* Depending on the particular application, the effects of a transmission line on the quality of the digital signal can become very complex, and a thorough analysis is beyond the intent of this book. Figure 13–17 shows a simple illustration of data transmission.

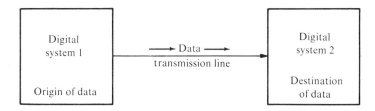

Figure 13–17. Data Transmission from One Point to Another.

The longer the transmission line, the greater effect it has on the shape of the pulsed data signal. Because of the distributed resistance, capacitance, and inductance of the line, the effects are greatly frequency-dependent. Since a digital signal is made up of many frequency components, the phase and attenuation of each component will be affected differently by the characteristics of the transmission line, and therefore the fast rise and fall times of a transmitted digital signal become progressively "rounded."

Let us take a look at a basic data transmission system as shown in Figure 13–18. A transmission line is connected between two points, and a signal is sent from the origin or transmitting system to the destination or receiving system over it. The transmission line itself can physically be one of several forms: a straight wire, an etched connection, a coaxial cable, a twisted pair, etc. The transmitting system can be a computer terminal or any number of other devices that produce a stream of bits at a fixed rate. The binary data stream controls the *line driver,* which converts the logic levels of the transmitting system (TTL, MOS, etc.) to the current and voltage levels required by the transmission line. Therefore, the line driver acts as an *interface* between the digital system and the transmission line. For short distances and in certain applications, the logic levels of the system can be sent directly over the transmission line without difficulty, and thus a logic gate can act as the line driver. For longer lines and in special applications, a special line driver circuit is required to change the logic levels, provide a greater amount of current, or match the transmission line impedance. The transmission line conveys the signals produced by the *line driver* to the *line receiver*. The line receiver makes a "decision" on the logic state (1 or 0) of the digital signal that it receives by comparing the signal to a predetermined threshold level, and it essentially restores the shape of the transmitted signal from the signal that has been distorted by the transmission line. A Schmitt trigger can be used as a line receiver in some applications.

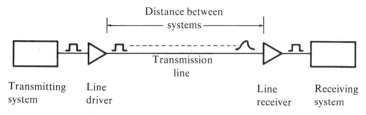

Figure 13–18.   Block Diagram of a Data Transmission System.

13–6   Forms of Data Transmission

There are two basic ways of transmitting a digital signal on a transmission line: *single-ended* and *differential*. The single-ended form uses a single conductor to carry the signal; the signal voltage is referenced to a common ground "return" conductor. This form, illustrated in Figure 13–19, is the simplest way to send digital data because it requires only one signal line per circuit. It does, however, have some disadvantages. In longer lines, induced voltages from external sources and noise on the ground return tend to add to the actual signal voltage, thereby producing a possibly erroneous signal at the receiving end. The form is often sufficient for short distances and for noise-free applications. There are, of course, several remedies for the noise problem; among these is the use of shielded cable to cut down on the induced noise voltage (this generally is more costly). Also, the output signal of the line driver can be increased in amplitude to overcome the noise (this requires more system power).

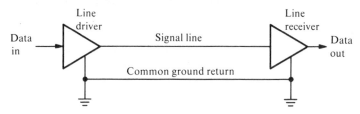

Figure 13–19.   Single-Ended Data Transmission.

To eliminate some of the noise disadvantages of the single-ended operation, the *differential* form of transmission can be used. As shown in Figure 13–20, a differential driver and a differential receiver are employed. The differential line driver is essentially two single-ended drivers, with one driver producing the complement of the output signal of the other. A transmission line pair is required in this case; it is composed of two signal conductors, one carrying the true signal and the other carrying the complement, and both referenced to a common ground return. Since the line receiver "sees" the difference between a signal and its complement, there is a much greater noise margin because any noise on the signal lines is common mode noise; that is, it appears on both lines with respect to the ground reference.

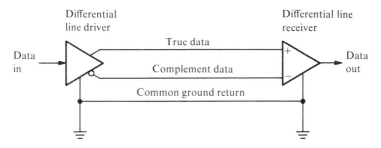

Figure 13–20. Differential Data Transmission.

13–7 Modes of Digital Transmission

There are several modes in which a data transmission system can operate. Each mode basically defines the direction or directions of data flow or movement from point to point along the transmission line.

The most fundamental mode of operation is called *simplex*. In this system, the data can move in only *one* direction, from the line driver to the line receiver, and the form of transmission can be either single-ended or differential. The simplex mode is illustrated in Figure 13–21 for both forms.

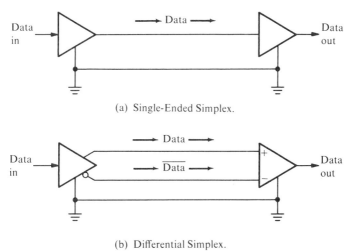

(a) Single-Ended Simplex.

(b) Differential Simplex.

Figure 13–21. Examples of Simplex Operation.

In some applications, more than one line receiver are used, as shown in Figure 13–22 for both single-ended and differential operation. This mode is called *simplex distribution bus.*

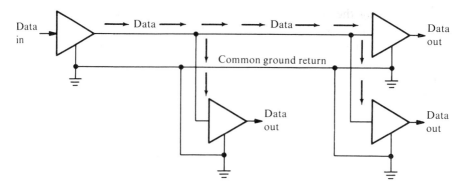

(a) Single-Ended Distribution Bus.

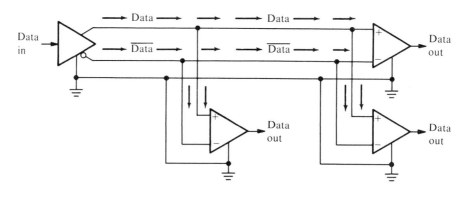

(b) Differential Distribution Bus.

Figure 13-22. Example of Simplex Distribution Bus Operation.

Another mode is called *half-duplex*. In this mode, data can move in either direction along the transmission line, but *cannot* move in both directions simultaneously. As shown in Figure 13-23, data can be sent from line driver *A* to line receiver *B*, or from line driver *B* to line receiver *A*. The line driver/receiver pair on each end of the line is called a *port*.

An extension of the half-duplex operation commonly employed is a *multiplex* operation, sometimes called *data bus* or *party line* operation. In this mode, nonsimultaneous two-way data transmission occurs among three or more ports (driver/receiver pairs). These ports are connected to the same transmission line, as illustrated in Figure 13-24. Data can be sent from any driver to any receiver along the transmission line. Control circuitry is required to select the driver and receiver for any given data transmission.

The most complex method we will mention is the *full-duplex* mode. This involves simultaneous, two-way data flow between two ports. Normally, in this mode, frequency separation of the simultaneous signals must be employed; this involves frequency division multiplexing, which is a process beyond the scope of this book.

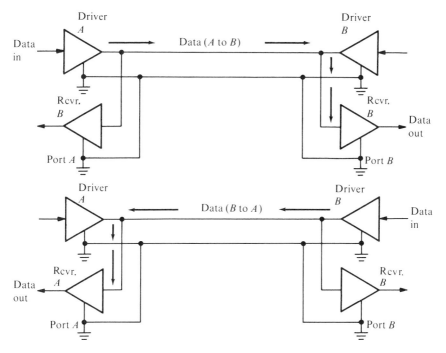

(a) Single-Ended Half-Duplex.

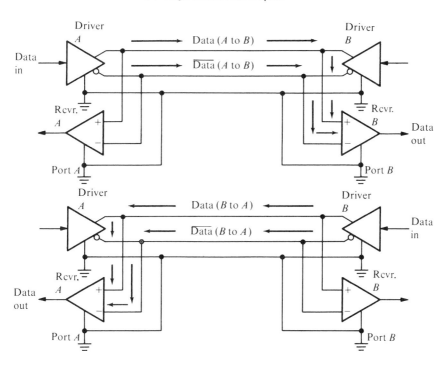

(b) Differential Half-Duplex.

Figure 13–23. Examples of Half-Duplex Operation.

425

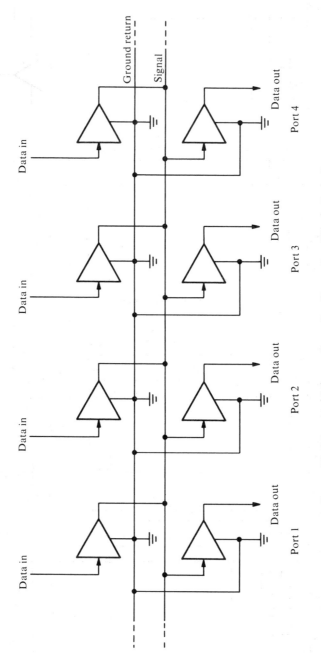

Figure 13–24. Multiplex or Data Bus Operation (Single-Ended).

13-8   Digital and Analog Representations

We all know how a quantity can be expressed in digital form, but what is an analog quantity? An analog function is one that has a continuous set of values, as contrasted with discrete sets of values for the digital case. To illustrate the difference between an analog and a digital representation, let us take the case of a voltage that varies over a range of 0 to +15 V. The analog representation of this voltage would take on all values between 0 and +15 V (an infinite number of values). In the case of a digital representation, if this same voltage range is represented by a four-bit binary code, only 16 values can be defined. We can represent more values between 0 and +15 V if the number of bits in the digital code is increased. So, as you can see, a continuous quantity (analog) can be represented to some degree of accuracy with a digital code. This concept is further illustrated in Figure 13-25, where the function shown is a smoothly changing curve (analog function) that takes on values between 0 and +15 V in the manner indicated on the graph. If we used a four-bit binary code to represent this curve, each binary number would represent a voltage level. To illustrate, the voltage on the analog curve is measured or "sampled" at each of 35 equal intervals. For simplicity, assume the voltage value at each interval can only be expressed as an integer value (no fractional values). At each interval, the voltage value can be represented by a four-bit binary code word, as indicated in Figure 13-25. What we have at this point is a series of binary numbers representing various voltage values along the analog curve, and we have essentially performed an analog-to-digital conversion.

An approximation of the analog function can be reconstructed from the sequence of digital numbers that have been generated. Obviously there is going to be some error in our reconstructed function, because only certain values are represented by the digital codes (35 in this example). If we now plot the values of each of the code words at each of the 35 intervals, a graph as shown in Figure 13-26 results. The graph only approximates the original curve because values between each interval are not known and must be represented by the 16 possible values obtainable with a four-bit code; this has illustrated the basic idea of a digital-to-analog conversion.

13-9   The Digital-to-Analog Converter (D/A)

Figure 13-27 shows a basic voltage divider arrangement that forms a simple four-bit digital-to-analog converter. Each of the inputs is one of the four digital bits. In a binary number, the 2^0 bit (LSB) has a weight that is one-eighth that of the 2^3 bit (MSB). The 2^1 bit has a weight one-fourth that of the MSB. The 2^2 bit has a weight one-half that of the MSB. In the resistive voltage divider circuit, the value of each input resistor has a value chosen so that its binary input has the proper weight in the output. This requires that the 2^0 input resistor have a value eight times that of the 2^3 input resistor, so the weight of the 2^0 bit is one-eighth. The 2^1 input resistor has a value four times that of the 2^3 input resistor, and the 2^2 resistor has a value twice that of the 2^3 input resistor. The value of the load resistor, R_L, is assumed to be much larger than any of the input resistors for simplicity of analysis.

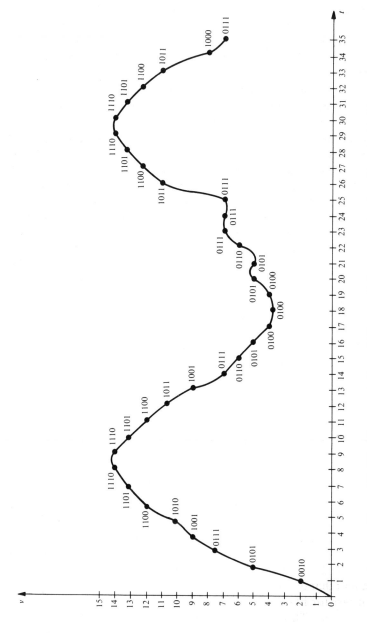

Figure 13-25. Representation of Discrete Points on an Analog Signal with Binary Numbers.

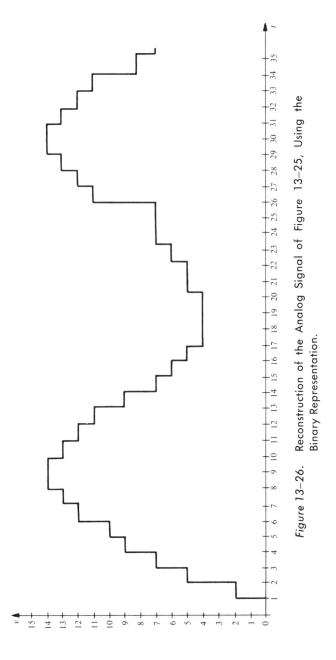

Figure 13-26. Reconstruction of the Analog Signal of Figure 13-25, Using the Binary Representation.

429

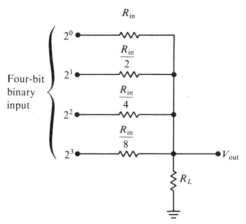

Figure 13–27.   A Simple Voltage Divider D/A Converter.

An output is produced with values proportional to those of the binary inputs. To illustrate the operation, let us assume that a binary 1 is represented on the input by a +15-V level, and a binary 0 is represented by a 0-volt level (ground). If all 1s (binary 15) are applied to the inputs, a +15-V level appears on the output ($R_L \gg R_{in}$), as illustrated by the equivalent voltage divider circuit in Figure 13–28(a) and shown by the following analysis:

$$V_{out} = \frac{R_L}{R_{eq} + R_L} V_{in} \qquad R_L \gg R_{eq}$$

$$\simeq \frac{R_L}{R_L} V_{in} = V_{in}$$

If a binary 7, for example, is applied to the inputs, the voltage divider appears as the equivalent circuit of Figure 13–28(b). The analysis for this situation is as follows:

$$V_{out} \simeq \left(\frac{R_{in}/8}{R_{in}/7 + R_{in}/8} \right) V_{in} \quad \text{neglecting } R_L, \text{ since } R_L \gg R_{in}$$

$$= \left(\frac{R_{in}/8}{\dfrac{8R_{in} + 7R_{in}}{56}} \right) V_{in}$$

$$= \left(\frac{R_{in}/8}{15R_{in}/56} \right) V_{in}$$

$$= \left(\frac{7R_{in}}{15R_{in}} \right) V_{in}$$

$$= \frac{7}{15} V_{in}$$

For $V_{in} = +15$ V,

$$V_{out} = \frac{7}{15}(+15) = +7 \text{ V}$$

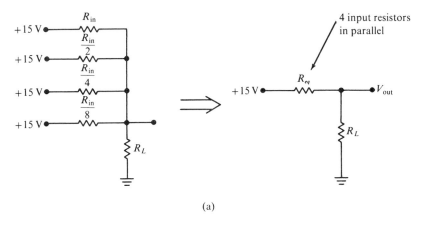

(a)

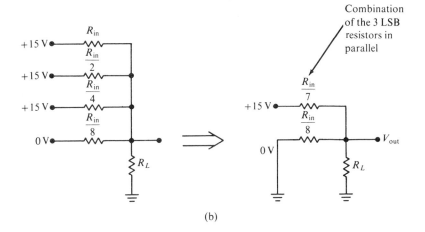

(b)

Figure 13–28.

Repeating this analysis for each of the remaining 16 possible combinations of four bits, the output voltage will have values for each case as listed in Table 13–1.

Example 13–5   The following sequence of four-bit binary numbers is applied to the resistive voltage divider circuit in Figure 13–29(a) at 1-microsecond intervals. Sketch the analog output. Assume for simplicity that a binary 1 = +15 V and a binary 0 = 0 V.

      0000, 0001, 0010, 0100, 0101, 0011, 0100, 0111, 0111, 1001, 1000,
      1010, 1100, 1011, 1011, 1110, 1111, 1101, 1100, 1011, 1000, 0110,
      1000, 1000, 0110, 0101, 0100, 0011, 0010, 0000

Solution:
See Figure 13–29(b).

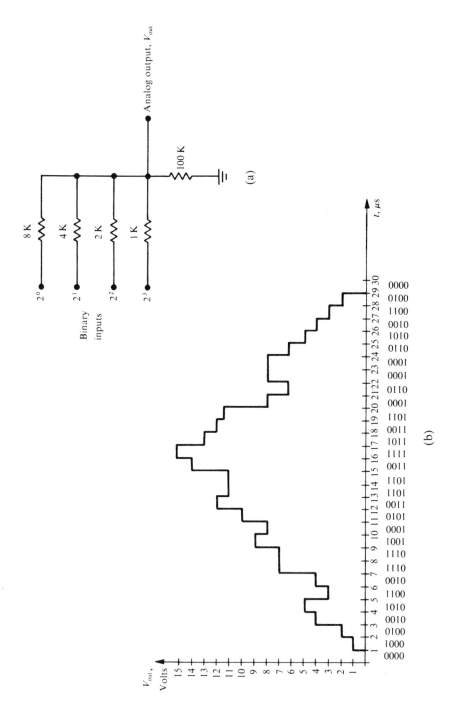

Analog output, V_{out}

100 K

(a)

Binary inputs

2^0 8 K
2^1 4 K
2^2 2 K
2^3 1 K

V_{out}, Volts

t, μs

(b)

Figure 13–29.

432

TABLE 13-1. Output Voltage
Values for the Example D/A
Converter Circuit of Figure 13-27.

Binary Input				Output Voltage, V
2^3	2^2	2^1	2^0	V_{out}
0	0	0	0	0
0	0	0	1	1
0	0	1	0	2
0	0	1	1	3
0	1	0	0	4
0	1	0	1	5
0	1	1	0	6
0	1	1	1	7
1	0	0	0	8
1	0	0	1	9
1	0	1	0	10
1	0	1	1	11
1	1	0	0	12
1	1	0	1	13
1	1	1	0	14
1	1	1	1	15

In our discussion we have used a four-bit example for illustration. If more bits are used in the binary number, the changes or "steps" in the output voltage will be smaller, resulting in a smoother curve and a closer approximation to the analog function that we are trying to represent. With the four bits, each step in the output is 1/15 the voltage level representing a binary 1, which in our example was +15 V. With five bits each step would be 1/31; with six bits each step would be 1/63; and so on.

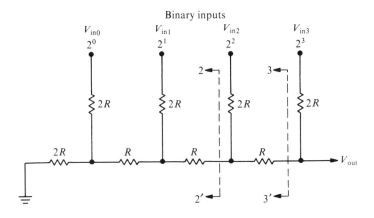

Figure 13-30.   Resistive Ladder Network, Four-Bits.

The voltage divider arrangement just discussed has certain disadvantages. Many values of resistors are required (depending on the number of bits), and the current requirement of each input differs because each resistor value differs. It is, however, the simplest arrangement and serves to illustrate the basic digital-to-analog conversion concept. We will now look at a resistive *ladder* network that has certain practical advantages over the voltage divider approach.

A four-bit resistive ladder network is shown in Figure 13–30. Let us see how each input produces a properly weighted output.

Starting with the most significant input, 2^3, the circuit to the left of the dashed line 3-3' looks like $2R$. Therefore, the equivalent circuit for the 2^3 input is as shown in Figure 13–31(a). V_{out} is then determined as follows using the voltage divider method:

$$V_{out} = \frac{2R}{2R + 2R} V_{in3}$$

$$= \frac{2R}{4R} V_{in3} = \frac{V_{in3}}{2}$$

Moving to the next input, 2^2, the circuit to the left of the dashed line 2-2' looks like $2R$. The resulting equivalent circuit looking from this input is shown in Figure 13–31(b) in several stages of simplicity. The voltage at point ② is calculated as follows, using the simplest form of the equivalent circuit:

$$V_2 = \frac{\dfrac{6R}{5}}{2R + \dfrac{6R}{5}} \cdot V_{in2} = \frac{\dfrac{6R}{5}}{\dfrac{10R + 6R}{5}} \cdot V_{in2} = \frac{6R}{16R} \cdot V_{in2} = \frac{6}{16} V_{in2}$$

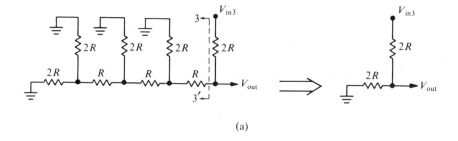

(a)

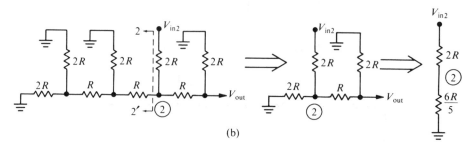

(b)

Figure 13–31.

Next, the output voltage is determined, using the middle form of the equivalent circuit as follows:

$$V_{out} = \left(\frac{2R}{R + 2R}\right)\left(\frac{6}{16}\right)V_{in2} = \left(\frac{2R}{3R}\right)\left(\frac{6}{16}\ V_{in2}\right) = \frac{V_{in2}}{4}$$

Continuing with the same approach, it can be shown that the output voltage due to the 2^1 input is $V_{in}/8$ and that the output voltage due to the 2^0 input is $V_{in}/16$. These analyses are left as exercises at the end of the chapter. As you can see, each input produces a properly weighted output.

To complete our discussion of the basics of digital-to-analog conversion, let us look at a complete D/A converter. The resistive ladder is the device that provides the actual conversion, but normally other auxiliary functions are required. Figure 13–32 shows a basic ten-bit D/A converter. The register is for holding the digital data, the gating logic is used to switch the data through to the buffer circuits that provide the voltage and current levels required by the resistive ladder network, and the output driver produces an appropriate level to any external circuitry.

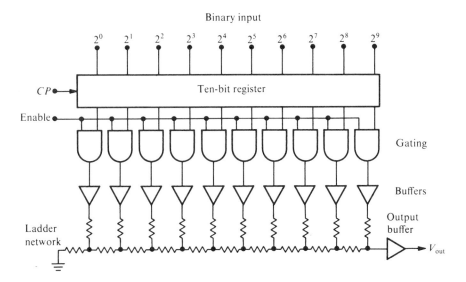

Figure 13–32. A Ten-Bit Digital-to-Analog Converter.

13-10 The Analog-to-Digital Converter (A/D)

In many applications, an analog function is converted to digital form so it can be processed by a digital system. There are several methods for analog-to-digital conversion, and all are more complex than the D/A conversion process. We will discuss two basic techniques in this section. First is a method of *simultaneous A/D conversion,* and a three-bit converter will serve to illustrate the principles involved; this three-bit A/D converter appears in Figure 13–33.

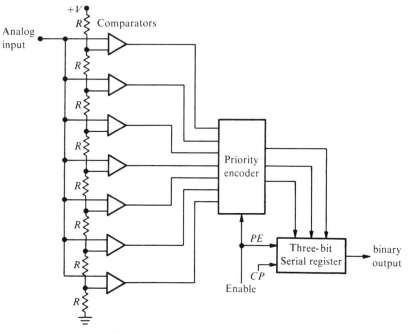

Figure 13–33.   A Simultaneous Analog-to-Digital Converter.

There are seven differential comparator circuits that compare a reference voltage against the analog input signal, and when the analog signal equals or exceeds the reference, a HIGH level output is produced. The reference voltage for each comparator is set by the resistive voltage divider network. The output of each comparator is connected to an input of the priority encoder (priority encoders were covered in Chapter 7). The encoder is "sampled" by a pulse on the enable input, and a three-bit binary code proportional to the value of the analog input appears on the encoder outputs. This binary code is determined by the highest order input having a HIGH level. The binary code is converted from parallel to serial form by the register as it is shifted out for processing. The encoder is sampled at a fixed rate, so that many points on the analog signal will be converted to binary code, as illustrated in Figure 13–34. In general, for this method, $2^n - 1$ comparators are required for conversion to an n-bit binary code. An example will illustrate this method of A/D conversion.

Example 13–6   Determine the encoded binary sequence for the analog signal in Figure 13–35, using the simultaneous three-bit conversion technique. Assume a sampling rate of 1 MHz.

Solution:

Sequence:

001, 010, 010, 011, 011, 010, 010, 010, 011, 101, 110,
111, 110, 100, 010, 000, 001, 010, 010, 010, 001, 000

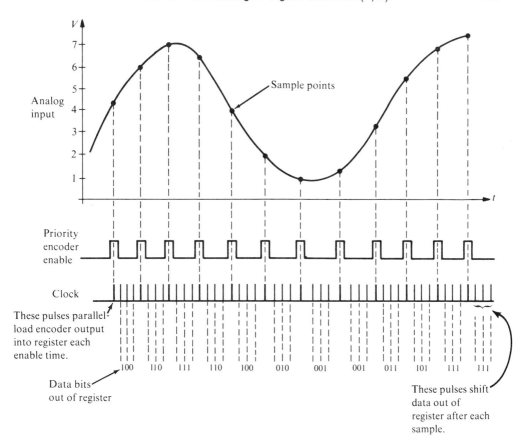

Figure 13–34.   Example of Conversion Timing for the A/D Converter of Figure 13–33.

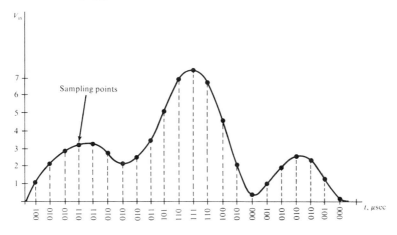

Figure 13–35.

Another method of A/D conversion, sometimes called the *counter method,* requires only one comparator circuit, and thereby avoids the use of a large number of comparators. In this technique, a binary counter and a resistive ladder network are used to generate a variable reference voltage, which is in the form of a "stair-step" and is applied to one input of the comparator. The analog signal to be converted is applied to the other comparator input, and the comparator provides an indication on its output when the reference equals or exceeds the analog voltage. When equality occurs, the binary state of the counter is proportional to the value of the analog signal. Figure 13–36 shows a four-bit A/D converter of this type.

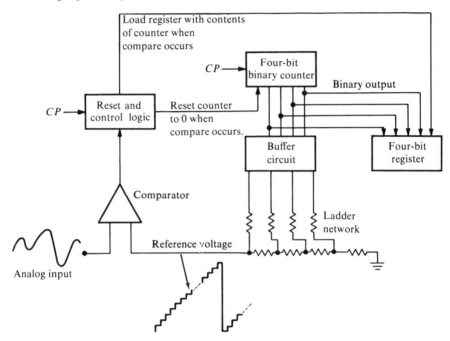

Figure 13–36. A Counter Analog-to-Digital Converter.

As the counter advances through its binary states, the "stair-step" reference voltage is generated by the ladder network. The counter will continue to advance from one binary state to the next, producing successively higher steps in the reference voltage until the instantaneous value of the analog signal is equaled or exceeded. At this point the comparator will cause the reset and control logic to load the binary number in the counter into a register for processing and to reset the counter, and another sampling sequence will start. This method is slower than the simultaneous method because, for each sampling of the analog signal, the counter must sequence through a possible 16 states (in this case) before a comparison occurs. If we use more bits in the conversion, the number of states in the counter sequence will increase and more time will be required. Figure 13–37 illustrates this conversion technique.

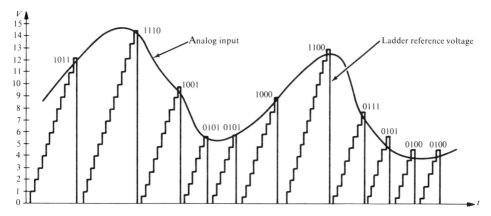

Figure 13-37.   Example of Conversion for the Circuit in Figure 13-36, Showing the Analog Signal and Ladder Reference Voltage Superimposed.

13-11   Three-State Logic

Three-state logic exhibits three possible output conditions: 1, 0, or open. When enabled, this type of logic circuit can provide a 1 or 0 as a two-state logic gate does. When disabled, its output is open or completely disconnected from its external load. The three-state gate can be considered a regular gate with a switch in its output line, as shown in Figure 13-38(a). A typical symbol is shown in Figure 13-38(b).

(a) Three-State Gate Analogy

(b) Three-State Gate Symbol

Figure 13-38.   A Three-State Logic Gate.

The circuit is useful in data bus multiplexing applications because any number of gates can be connected to a single line, with only one at a time electronically switched onto the line for data transfer while the rest are disconnected by virtue of the open output state; this allows use of a common bus to transfer data from several points in a system to one or more destinations, as illustrated in Figure 13-39.

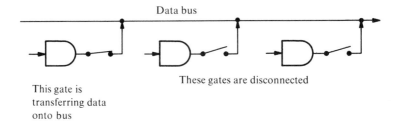

This gate is
transferring data
onto bus

These gates are disconnected

Figure 13–39.   Example of Three-State Application.

PROBLEMS

13–1   Determine the pulse width of the basic one-shot circuit in Figure 13–40.

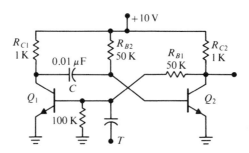

Figure 13–40.

13–2   What effects can the following changes have on the circuit of Problem 13–1?
     (a)   increase R_{B1} (b)   increase R_{B2} (c)   decrease R_{C1}
     (d)   decrease R_{C2} (e)   increase C

13–3   Show the output of the second cascaded one-shot in Figure 13–41. Each one-shot triggers on a positive-going edge. What potential application might this have?

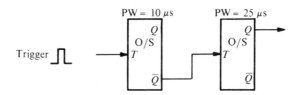

Figure 13–41.

13–4   Discuss the operation of the one-shot circuit connected as shown in Figure 13–42. Draw the output waveform after application of an initial trigger pulse. For simplicity, neglect the recovery time.

PW = 100 μs

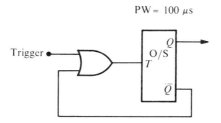

Figure 13–42.

13–5   Repeat Problem 13–4 for the circuit in Figure 13–43.

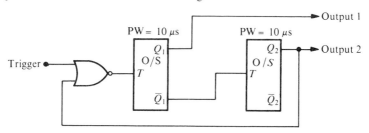

Figure 13–43.

13–6   Determine the Q_2 output waveform for the astable multivibrator of Figure 13–6 for each set of component values listed below.
  (a) $R_{B1} = 10$ K, $R_{B2} = 10$ K, $C_1 = 0.02\,\mu$F, $C_2 = 0.02\,\mu$F
  (b) $R_{B1} = 10$ K, $R_{B2} = 20$ K, $C_1 = 0.01\,\mu$F, $C_2 = 0.02\,\mu$F
  (c) $R_{B1} = 50$ K, $R_{B2} = 25$ K, $C_1 = 0.001\,\mu$F, $C_2 = 0.002\,\mu$F
  (d) $R_{B1} = 12$ K, $R_{B2} = 12$ K, $C_1 = 0.005\,\mu$F, $C_2 = 2000$ pF

13–7   For the astable multivibrator of Figure 13–6, what effects on the frequency, period, duty cycle, and general waveshape will the following changes have?
  (a)   increase R_{B1} (b)   decrease C_1 (c)   decrease R_{B2}
  (d)   increase C_2 (e)   increase R_{C1}

13–8   Determine the duty cycle of each of the astable multivibrator outputs in Figure 13–44.

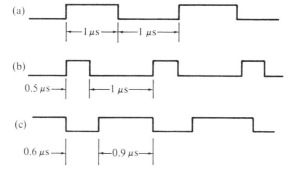

Figure 13–44.

13-9 For a Schmitt trigger circuit with a UTP of 1 V and an LTP of 0.5 V, sketch the out-output for the input in Figure 13–45.

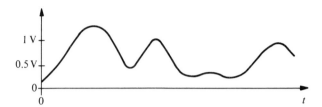

Figure 13–45.

13-10 For a Schmitt trigger circuit with a UTP of 2.5 V and an LTP of 2 V, sketch the output for the input in Figure 13–46.

Figure 13–46.

13-11 Suppose the input to the Schmitt trigger of Problem 13–9 has noise transitions as shown in Figure 13–47. Sketch the output. How can the effect of this noise on the output be eliminated?

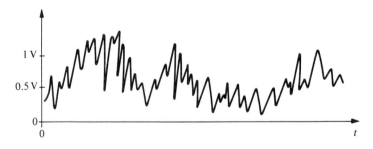

Figure 13–47.

13-12 Determine the UTP and the LTP of the Schmitt trigger circuit in Figure 13–48.

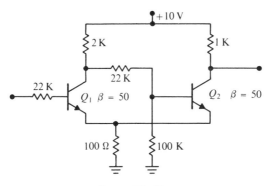

Figure 13-48.

13-13 What is a transmission line?

13-14 What is the main purpose of a line driver and line receiver in data transmission applications?

13-15 Explain the difference between single-ended and differential data transmission.

13-16 Classify the configurations in Figure 13-49 according to the mode of transmission.

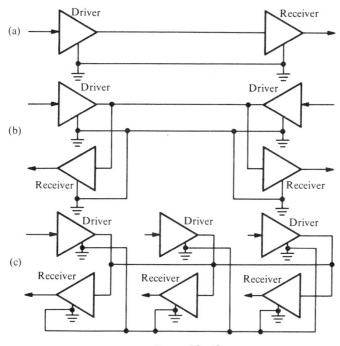

Figure 13-49.

13-17 For each sequence of three-bit binary numbers, determine the reconstructed analog output of a resistive voltage divider used as a D/A converter. Assume a binary 1 is +7 V and a binary 0 is 0 V. The binary numbers appear on the inputs at a 1-kHz rate.

    (a)   000, 010, 010, 001, 010, 011, 100, 011, 100, 101,
            110, 111, 111, 110, 100, 100, 100, 101, 011, 000.

      (b)   000, 001, 001, 110, 101, 110, 111, 110, 100, 001,
             010, 011, 111, 111, 101, 110, 000, 001, 100, 110.

13–18   Verify each value of V_{out} in Table 13–1 by the procedure illustrated in Section 13–9.

13–19   For each sequence of four-bit numbers, determine the reconstructed analog output of a resistive voltage divider used as a D/A converter. Assume a binary 1 is $+15$ V and a binary 0 is 0 V. The conversion rate is 10 kHz.
      (a)   0000, 0001, 0010, 0011, 0010, 0100, 0110, 1000,
             0101, 0101, 1000, 1001, 0001, 0001, 0011, 0111,
             1010, 1011, 1100, 1000, 1000, 1001, 1111, 1110,
             1101, 1000, 0111, 0101, 0101, 0111, 0100, 0010,
             0000.
      (b)   0000, 1000, 1111, 0001, 0001, 0100, 1110, 0101,
             0111, 0111, 0111, 0110, 0110, 0111, 1111, 1100,
             1101, 1011, 1001, 0000, 0000, 0010, 0000, 0000.

13–20   Show that $V_{in}/8$ is the weighted output of a four-bit ladder network for the 2^1 input.

13–21   Show that $V_{in}/16$ is the weighted output of a four-bit ladder network for the 2^0 input.

13–22   Explain the simultaneous method for A/D conversion.

13–23   Determine the encoded binary sequence for the analog signal in Figure 13–50, using the simultaneous conversion technique for three bits. Assume a sampling rate of 100 kHz, with the first sample pulse at 0.

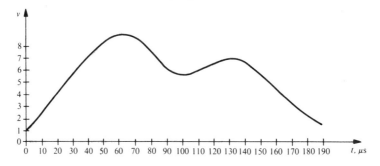

Figure 13–50.

13–24   Repeat Problem 13–23 for the analog waveform in Figure 13–51.

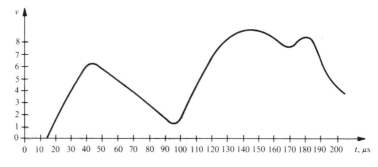

Figure 13–51.

13–25 Draw the logic diagram for a four-bit simultaneous A/D converter.

13–26 Explain the counter method of A/D conversion.

13–27 For a three-bit counter A/D converter, the ladder reference signal advances one step every microsecond. Determine the encoded binary sequence for the analog signal in Figure 13–52.

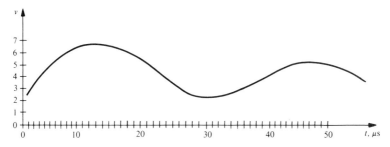

Figure 13–52.

13–28 For a four-bit counter A/D converter, the ladder reference signal advances one step every microsecond. Determine the encoded binary sequence for the analog signal in Figure 13–53.

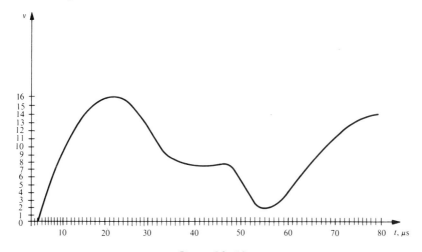

Figure 13–53.

CHAPTER 14

Digital Integrated Circuits

All of the fundamental logic circuits that we have discussed in this text are available in some type of integrated circuit (IC) form and in varying levels of complexity. Most modern digital systems utilize ICs to a large extent in their designs. In most cases, ICs have size, power, reliability, and cost advantages over discrete type circuitry (except in very specialized applications where a circuit must be "custom made" to meet unique requirements).

Because the integrated circuit has revolutionized the field of electronics, it is necessary that the technologist and engineer keep abreast of IC technology. In this chapter we will discuss some of the important aspects of ICs to give you a basic understanding of ICs and knowledge of available types and their characteristics.

14-1 IC Construction

A monolithic integrated circuit is an electronic circuit that is constructed entirely on a single small chip of semiconductor material. All of the components that make up the circuit—transistors, diodes, resistors, and capacitors—are an integral part of this single chip.

Typical chip sizes range from about 40 × 40 mils (a mil is 0.001 inch) to about 250 × 250 mils, depending on the complexity of the circuit. Anything from a few to several thousand components can be fabricated on a single chip. To get an idea of the size of a chip (or *die,* as it is called), let us look at Figure 14-1. Approximately 400 chips of 50 square mils each can be cut from a silicon wafer 1 inch in diameter. Circuitry is then formed on each of the chips, and each chip is encapsulated in one of several possible package configurations, as illustrated.

The integrated circuit is formed on the chip, which is a piece of P- or N-type silicon material cut from the original wafer. This chip is called the *substrate,* and is the "foundation" for building the integrated circuit. The transistors, diodes, resistors, and capacitors that make up a given circuit are built up on the substrate by

447

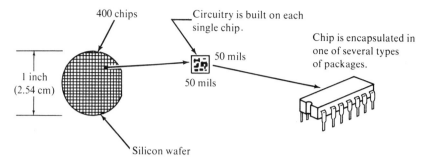

Figure 14-1.   Stages of IC Fabrication.

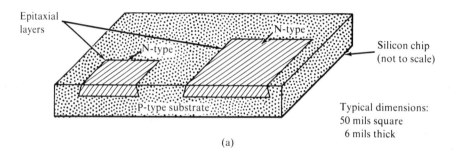

(a)

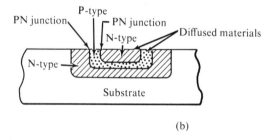

(b)

Figure 14-2.   Formation of PN Junctions with Epitaxy and Diffusion.

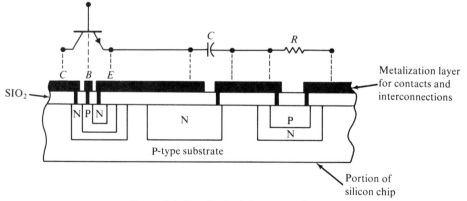

Figure 14-3.   Typical Structure of an IC.

epitaxial growth and diffusion of doped semiconductor materials. The *epitaxial* process consists of the addition of a layer of silicon deposited on the surface of the substrate. The desired resistivity is achieved by controlling the amount of impurity in the epitaxial layer. See Figure 14–2(a). Next, the desired components are formed by *diffusion* of additional layers of materials. Diffusion is the process of imparting N- or P-type atoms into the crystal structure of a material. It is used to form PN junctions in the epitaxial layers. The location of these PN junctions is controlled by a complex process called *photomask*. Figure 14–2(b) illustrates diffusion of PN junctions.

Next, a layer of silicon dioxide (SiO_2) is deposited, and a metalization layer is formed with contacts to the diffused P and N areas, as shown in Figure 14–3. The metalization provides for interconnection of the various components in the circuit and for external connections. A resistor can be formed by making two ohmic contacts to the P-type region, as indicated in Figure 14–3. A capacitor may be formed by using the SiO_2 layer as a dielectric and the metalization contacts as plates, as also shown in Figure 14–3.

This is a very basic and brief explanation of a very complex and involved process, but the intent is to give you some "feel" for what an IC is.

14–2 Small-Scale Integration (SSI)

The least complex ICs are placed in the SSI category. These are circuits with up to 10 or 12 equivalent gate circuits on a single chip, and include such basic functions as NAND, NOR, NOT, AND, OR, Exclusive-OR, and the flip-flops and one-shots. Each SSI function is packaged in one of two main configurations, the dual-in-line package (DIP) or the flat pack. Figure 14–4 illustrates these packages in both the 14-pin and the 16-pin versions.

Figure 14-5 shows a cutaway view of a DIP with the IC chip within the package. Leads from the chip are connected to the package pins to allow inputs and outputs to the logic. Commonly available SSI logic functions are shown in Figure 14–6. The diagrams indicate how the logic circuits are connected to the external pins. For simplicity, no connections are shown from the power pin (V_{CC}) or from the ground pin. Power and ground are, of course, connected to each circuit on the chip. Many of the functions shown are available in several IC implementations such as DTL, TTL, ECL, and CMOS.

14–3 Medium-Scale Integration (MSI)

The next classification according to complexity is called medium-scale integration (MSI). These are circuits with complexities ranging from 10 or 12 up to 100 equivalent gates on a chip. MSI circuits include the more complex logic functions such as encoders, decoders, counters, registers, multiplexers, arithmetic circuits, small memories, and others. MSI functions are available in 16-pin or 24-pin DIPS and some in flat packs. A 24-pin DIP is shown in Figure 14–7, and Figure 14–8 shows some currently available MSI functions.

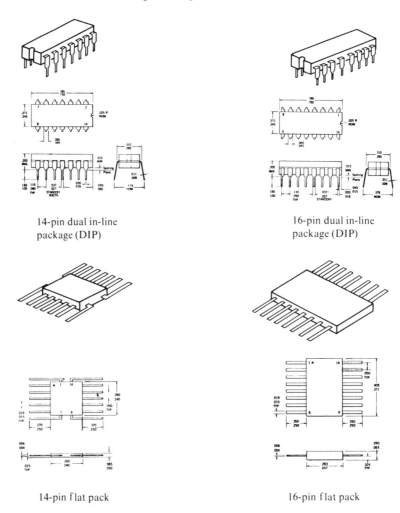

14-pin dual in-line
package (DIP)

16-pin dual in-line
package (DIP)

14-pin flat pack

16-pin flat pack

Figure 14-4.   Common Integrated Circuit Packages.

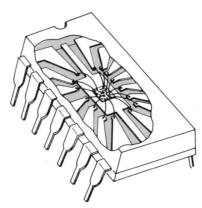

Figure 14-5.   Cutaway View of a Dual-In-Line Package Showing the IC Chip
Mounted Inside with Leads to Input and Output Pins.

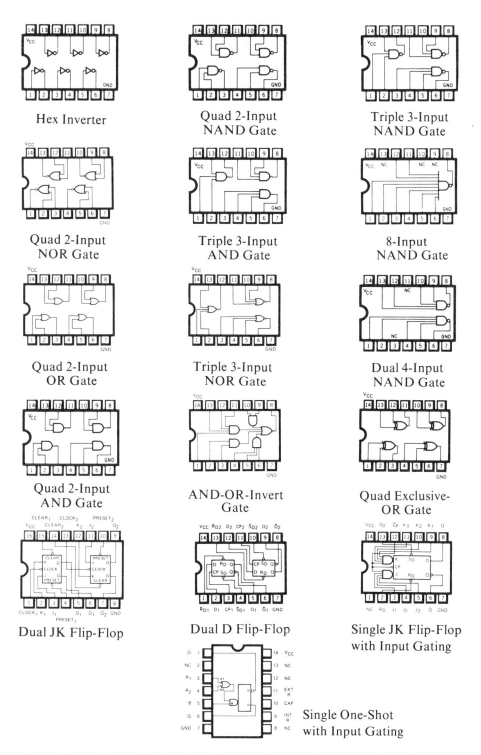

Hex Inverter

Quad 2-Input
NAND Gate

Triple 3-Input
NAND Gate

Quad 2-Input
NOR Gate

Triple 3-Input
AND Gate

8-Input
NAND Gate

Quad 2-Input
OR Gate

Triple 3-Input
NOR Gate

Dual 4-Input
NAND Gate

Quad 2-Input
AND Gate

AND-OR-Invert
Gate

Quad Exclusive-
OR Gate

Dual JK Flip-Flop

Dual D Flip-Flop

Single JK Flip-Flop
with Input Gating

Single One-Shot
with Input Gating

Figure 14–6. Some SSI Functions Showing Package Interconnections. Courtesy Fairchild Semiconductor.

451

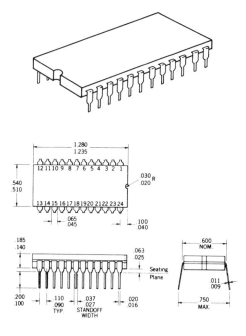

Figure 14–7.   24-Pin Dual-In-Line Package.

14–4   Large-Scale Integration (LSI)

Circuits with complexities of greater than about 100 equivalent gates per chip, including large memories and processors, generally fall into the LSI category. Some typical LSI functions are shown in Figure 14–9.

Now that we have examined how digital integrated circuits are classified according to their complexity, let us turn our attention to several IC technologies and see how various circuits are used to implement logic gates. Some of the circuits have been discussed in Chapter 4, but you will learn more about them in the following sections.

14–5   Bipolar ICs

Bipolar is one of two broad types of integrated circuit technology; the other is MOS. As the name implies, bipolar ICs utilize the bipolar transistor (the emitter, base, collector type), and are implemented using the bipolar technologies RTL, DTL, TTL, ECL, and I^2L. We will discuss each of these bipolar implementations in the following sections.

14–6   RTL

Resistor-transistor logic was one of the original circuits to be used in integrated circuit form. Today RTL is rarely used and is available only in limited varieties of logic functions. A basic RTL NOR gate is shown in Figure 14–10, and its opera-

LOGIC DIAGRAM

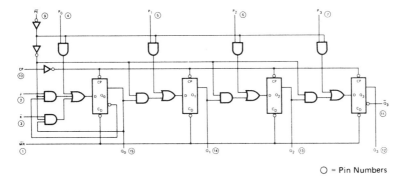

○ = Pin Numbers

(a)   Universal 4-Bit Shift Register

LOGIC DIAGRAM

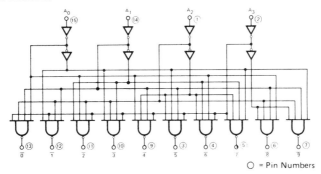

○ = Pin Numbers

(b)   One-of-Ten Decoder

LOGIC DIAGRAM

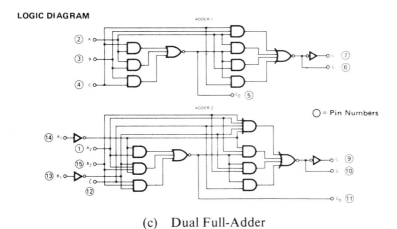

○ = Pin Numbers

(c)   Dual Full-Adder

Figure 14–8.   Examples of MSI Functions. Courtesy Fairchild Semiconductor.

453

LOGIC DIAGRAM

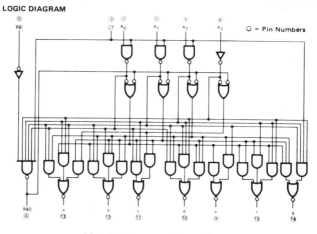

(d) 7-Segment Decoder

LOGIC DIAGRAM

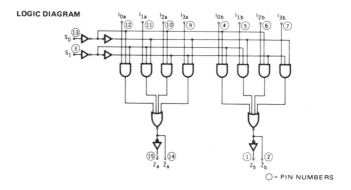

(e) Dual 4-Input Multiplexer

LOGIC DIAGRAM

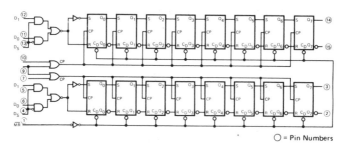

(f) Dual 8-Bit Shift Register

Figure 14–8, Cont.

454

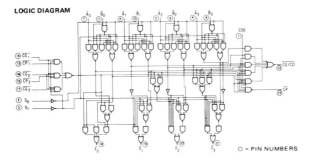

(g) 4-Bit Arithmetic Logic Unit with Carry Lookahead

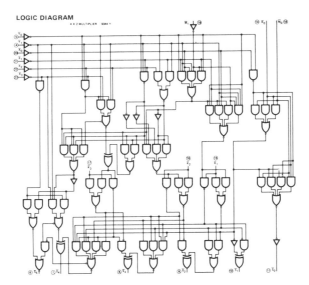

(h) Binary (4-Bit by 2-Bit) Full Multiplier

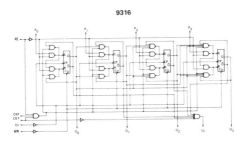

(i) 4-Bit Binary Counter

Figure 14–8, Cont.

455

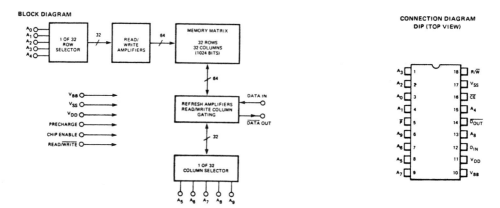

(a) 1024 by 1 Dynamic Random Access Memory

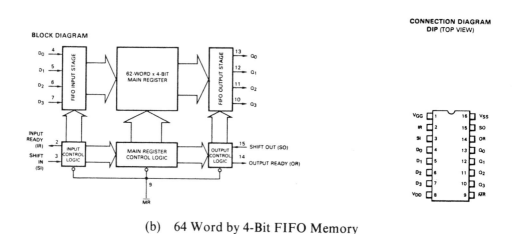

(b) 64 Word by 4-Bit FIFO Memory

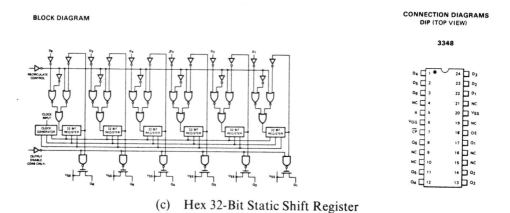

(c) Hex 32-Bit Static Shift Register

Figure 14-9. Examples of Some LSI Functions. Courtesy Fairchild Semiconductor.

LOGIC BLOCK DIAGRAM

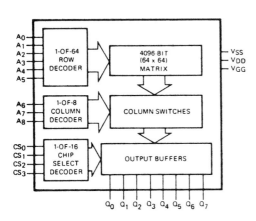

(d) 4096-Bit Read Only Memory

BLOCK DIAGRAM

(e) 4096-Bit Random Access Memory

Figure 14–9, Cont.

457

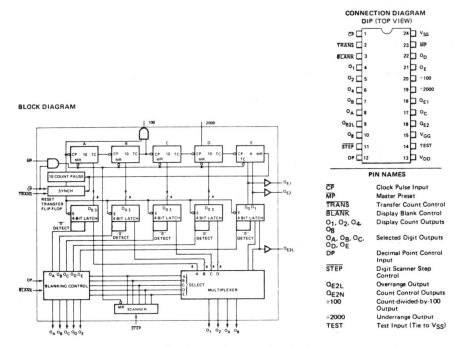

CONNECTION DIAGRAM
DIP (TOP VIEW)

CP	1	24	V$_{SS}$
TRANS	2	23	MP
BLANK	3	22	O$_D$
O$_1$	4	21	O$_E$
O$_2$	5	20	÷100
O$_4$	6	19	÷2000
O$_B$	7	18	Q$_{E1}$
O$_A$	8	17	O$_C$
Q$_{E2L}$	9	16	Q$_{E2}$
O$_8$	10	15	V$_{GG}$
STEP	11	14	TEST
DP	12	13	V$_{DD}$

BLOCK DIAGRAM

PIN NAMES

CP	Clock Pulse Input
MP	Master Preset
TRANS	Transfer Count Control
BLANK	Display Blank Control
O$_1$, O$_2$, O$_4$, O$_8$	Display Count Outputs
O$_A$, O$_B$, O$_C$, O$_D$, O$_E$	Selected Digit Outputs
DP	Decimal Point Control Input
STEP	Digit Scanner Step Control
Q$_{E2L}$	Overrange Output
Q$_{E2N}$	Count Control Outputs
÷100	Count-divided-by-100 Output
÷2000	Underrange Output
TEST	Test Input (Tie to V$_{SS}$)

(f)   Digital Voltmeter Logic Array

BLOCK DIAGRAM

CONNECTION DIAGRAM
DIP (TOP VIEW)

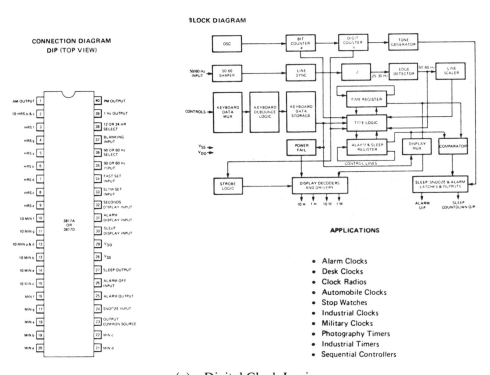

APPLICATIONS

- Alarm Clocks
- Desk Clocks
- Clock Radios
- Automobile Clocks
- Stop Watches
- Industrial Clocks
- Military Clocks
- Photography Timers
- Industrial Timers
- Sequential Controllers

(g)   Digital Clock Logic

Figure 14–9, Cont.

tion is as follows. A HIGH on input A will turn on Q_1, making the output LOW. A HIGH on input B will turn on Q_2, making the output LOW. A HIGH on both inputs will also make the output LOW. A LOW on both inputs keeps both transistors Q_1 and Q_2 OFF, thereby making the output HIGH. The use of the base resistors causes the switching speed of this gate to be relatively slow—on the average, about 45 nanoseconds. An increase in the number of inputs per gate is accomplished by adding more transistors, as shown in Figure 14–11.

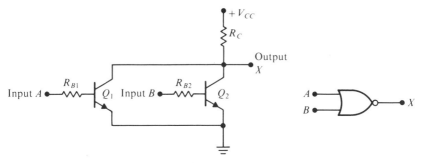

Figure 14–10.   RTL NOR Gate.

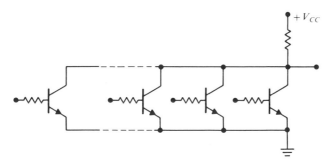

Figure 14–11.   Additional Inputs on RTL NOR Gate.

14-7   DTL

Diode-transistor logic was the next family of ICs to evolve. DTL exhibits improved switching speed and better noise margins than the RTL, and is available in a larger variety of functions. A basic DTL two-input NAND gate is shown in Figure 14–12.

The DTL NAND gate operates as follows. A LOW either on input A or input B or on both will forward bias the input diode(s) and keep transistor Q_1 OFF. Q_1 being OFF keeps Q_2 OFF, and the resulting output is HIGH; this state is illustrated in Figure 14–13(a). A HIGH on inputs A and B reverse biases both input diodes, causing current to flow through R_1 to the base of Q_1, keeping Q_1 ON. Emitter current from Q_1 flows into the base of Q_2, keeping it ON and making the output LOW; this state is shown in Figure 14–13(b).

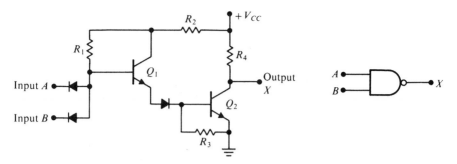

Figure 14–12.   DTL NAND Gate.

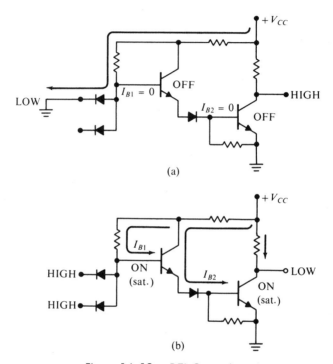

Figure 14–13.   DTL Operation.

The number of inputs to a gate is expanded by additional diodes, as shown in Figure 14–14. As with RTL, DTL outputs can be connected to create a wire-OR or a wire-AND as illustrated in Figure 14–15. The feature is convenient because additional gates are not required to create ORed or ANDed terms.

The typical DTL gate exhibits a propagation delay of about 30 nanoseconds and a power dissipation of 8 to 9 milliwatts.

14–8 TTL

Transistor-transistor logic, designated TTL or sometimes T^2L, is one of the most widely used bipolar integrated circuit forms. TTL exhibits a faster switching speed

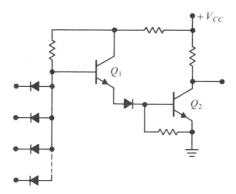

Figure 14–14.   Additional Inputs on DTL Gate.

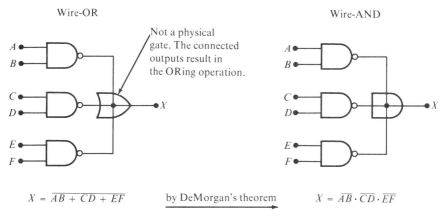

Wire-OR   Wire-AND

Not a physical gate. The connected outputs result in the ORing operation.

$X = \overline{AB + CD + EF}$ by DeMorgan's theorem $X = \overline{AB} \cdot \overline{CD} \cdot \overline{EF}$

Figure 14–15.   Wire-OR and Wire-AND Functions. They Are Physically Identical; the Difference Is the Boolean Interpretation.

(less propagation delay) than RTL and DTL, and is available in a greater variety of logic functions than any other type of bipolar IC. A basic two-input TTL NAND gate schematic is shown in Figure 14–16. All TTL integrated circuits are derived from this type of gate.

The operation of the TTL gate of Figure 14–16 is as follows. A LOW on input A or on input B will cause current to flow through R_1 and out of the forward biased base-emitter junction of the multiple-emitter input transistor Q_1; this deprives Q_2 of base current, thereby keeping it OFF. When Q_2 is OFF, base current is provided through R_2, keeping Q_3 ON and (since no base current is provided to Q_4) Q_4 OFF; in this state, the output level is HIGH and current is provided to a load, as indicated in Figure 14–17(a). The gate is said to be "sourcing" current.

If both inputs A and B are HIGH, the base-emitter junction of transistor Q_1 is reverse biased; in this state, base current is provided to Q_2 through R_1 keeping Q_2 ON. When Q_2 is ON, Q_3 is OFF and Q_4 is ON. The saturated Q_4 keeps the output LOW, and current flows into its collector from the load (usually other TTL inputs), as shown in Figure 14–17(b). The gate is said to be "sinking" current.

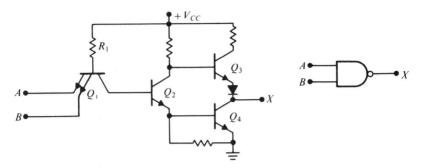

Figure 14–16.   TTL NAND Gate.

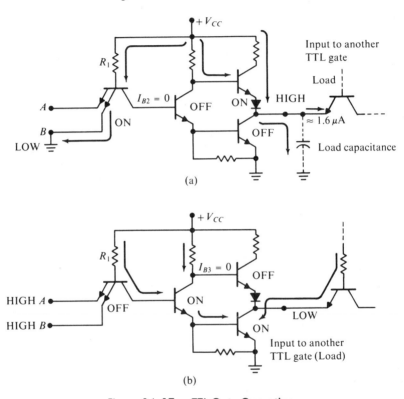

Figure 14–17.   TTL Gate Operation.

An increase in the number of inputs to the TTL gate is achieved by fabricating more emitters on the input transistor, as shown in Figure 14–18.

There are several variations on the basic TTL gate just discussed. First, most TTL circuits have input-clamping diodes on their inputs that considerably reduce the ringing resulted from long lines and impedance mismatches (negative voltage excursions are clipped off), as shown in Figure 14–19.

A second variation of the TTL circuit is designed to consume less power by increasing the resistor values. The decrease in power consumption is paid for by an

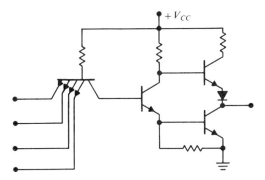

Figure 14–18.   Additional Inputs on TTL Gate.

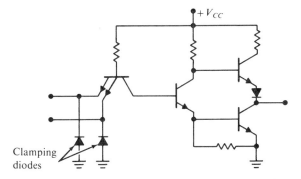

Figure 14–19.   TTL Gate with Input-Clamping Diodes.

increase in switching time because the higher resistor values along with the transistor-junction capacitances cause longer switching time constants.

Another variation of the basic TTL gate is the *open-collector* version shown in Figure 14–20; this circuit is commonly used to drive higher voltage level loads or loads requiring relatively high currents, such as lamps or relays.

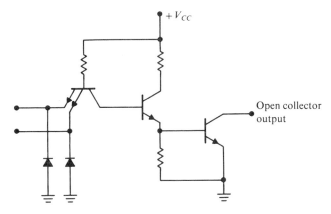

Figure 14–20.   TTL NAND Gate with Open-Collector Output.

The fastest switching TTL is called the *Schottky* TTL, which overcomes the limitation on switching speed imposed by a saturated transistor. In order to turn a saturated transistor from ON to OFF, the excess base charge must be removed; this takes time and results in a considerable delay when the transistor is switching. Schottky transistors use a surface barrier diode (Schottky diode) as a bypass between the base and collector, as shown in Figure 14–21(a). A symbolic representation of this arrangement is shown in Figure 14–21(b).

Figure 14–21. Schottky Transistor Symbols.

When the transistor is about to become saturated, the excess input charge is routed through the Schottky diode to the collector (rather than going into the base), which prevents the transistor from becoming fully saturated and thereby increases its switching speed. A basic Schottky TTL gate is shown in Figure 14–22. Table 14–1 compares the propagation delays and the power dissipation of three forms of TTL gate.

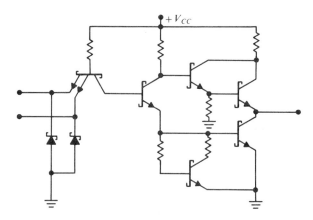

Figure 14–22. Schottky TTL NAND Gate.

TABLE 14–1. Comparison of Three Types of TTL Gates.

Type of TTL Gate	Typical Propagation Delay per Gate	Typical Power Dissipation per Gate
Standard TTL	8 to 10 ns	10 mW
Low Power TTL	20 ns	2 mW
Schottky TTL	3 ns	22 mW

14-9 ECL

Emitter-coupled logic (ECL) is a nonsaturating circuitry. As mentioned previously, saturation causes a buildup of charge in the base region of the transistor
that results in an excess switching delay. ECL allows for very fast switching speeds
by keeping the transistors out of saturation, and is the fastest logic circuit available. A basic ECL inverter is shown in Figure 14-23.

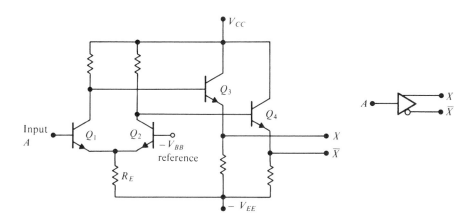

Figure 14-23.   ECL Inverter Circuit.

ECL operation is based on the differential transistor pair Q_1 and Q_2, and is
described as follows. For purposes of explanation, we will assume $V_{CC} = 0$ V,
$V_{EE} = -5$ V, and $V_{BB} = -1$ V. With input A at a LOW level, (we will define LOW
level shortly), transistor Q_1 is OFF. Q_2 is held ON but not saturated by the fixed
bias voltage V_{BB}, and its emitter is at approximately -1.7 V due to the V_{BE} drop.
Since Q_1 is OFF, its collector is at ground (0 V) potential (keeping Q_3 ON but not
saturated), and the X output is therefore at -0.7 V (one V_{BE} drop below 0 volts).
This is the HIGH logic level for this circuit. Now let us assume the collector of Q_2
is at -0.8 V; the $\overline{X}$ output is then at -1.5 V because of the V_{BE} drop of Q_4. This
is the LOW logic level for this circuit. When a HIGH (-0.7 V) is applied to in-
put A, its emitter is at -1.4 V due to the V_{BE} drop. The emitter voltage biases Q_2
OFF, and the $\overline{X}$ output goes HIGH (-0.7 V). Now if we assume the collector of
Q_1 is at -0.8 V, the X output is at -1.5 V, which is the LOW level. Notice that
the more negative of the two voltage levels is our LOW level, and the other is our
HIGH (although it is still a negative level). Figure 14-24 illustrates the operational
states of the basic ECL inverter circuit. To achieve multi-input gates, more transistors are connected in parallel with Q_1 as illustrated in Figure 14-25. A typical ECL
gate propagation delay is approximately 2 nanoseconds, which is faster than the
Schottky TTL. Notice also that both the true and complement outputs are available from ECL gates.

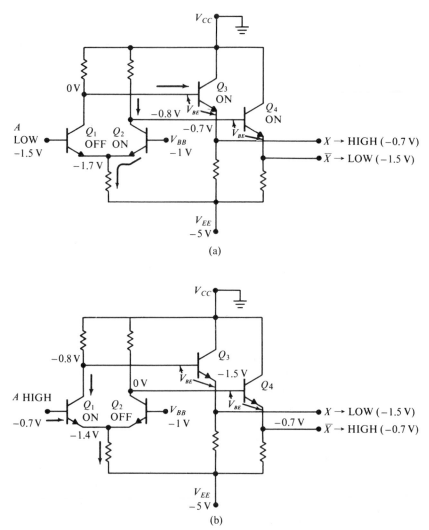

Figure 14–24. Basic ECL Operation.

14–10 I²L

Integrated injection logic (I²L) is a bipolar technology that allows extremely high component densities on a chip (up to ten times that of TTL). I²L is being used for complex LSI functions such as microprocessors and is simpler to fabricate than either TTL or MOS. It also has a low power requirement and reasonably good switching speeds that are improving all the time.

The basic I²L gate is extremely simple, as indicated in Figure 14–26. Transistor Q_1 acts as a current source and active pull-up, and the multiple-collector transistor Q_2 operates as an inverter.

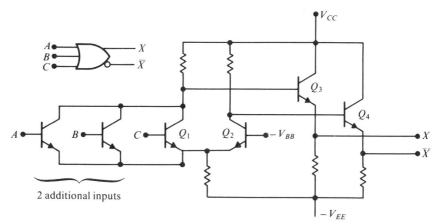

Figure 14–25.   An ECL NOR Gate (OR Also).

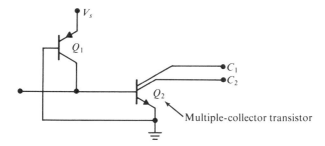

Figure 14–26.   I²L Gate.

Because the base of Q_1 and the emitter of Q_2 are common and the collector of Q_1 and the base of Q_2 are common, the entire I²L gate (constructed on a silicon chip) takes only the space of a single TTL multiple-emitter transistor.

Transistor Q_1 is called a current-injector transistor because, when its emitter is connected to an external power source, it can supply current into the base of Q_2. Switching action of Q_2 is accomplished by steering the injector current as follows. A LOW on the base of Q_2 will pull the injector current away from the base of Q_2 and through the low impedance path(s) provided by the driving gate(s), thus turning Q_2 OFF. When the output transistor of the driving gates are OFF (open), it corresponds to a HIGH input; this causes the injector current to be steered into the base of Q_2, which turns it ON. This action is illustrated in Figure 14–27. Figure 14–28 shows an example of an I²L implementation of a logic function.

14–11   MOS ICs

In the previous sections we looked at several types of bipolar integrated circuits. In this section we will examine various types of ICs that fall into the second broad category known as MOS (metal oxide semiconductor), in which the FET (field-

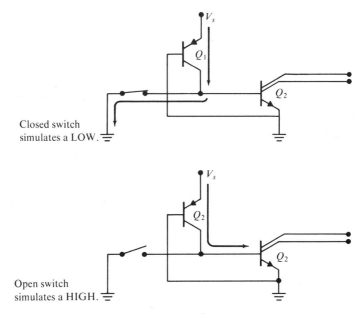

Closed switch
simulates a LOW.

Open switch
simulates a HIGH.

Figure 14-27. Basic I²L Operation.

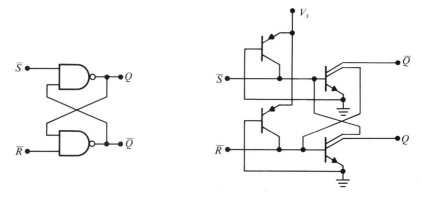

Figure 14-28. An I²L Flip-Flop.

effect transistor) is the main element. MOS devices exhibit extremely low power requirements.

Ten nanowatts per gate is typical for CMOS and 6 milliwatts per gate is typical for PMOS and NMOS circuits. Also, large numbers of devices can be fabricated in an extremely small chip area, making MOS technology widely used for complex logic functions that fall into the LSI and MSI categories (although SSI functions are also available). Reduced power requirements are paid for by sacrificing switching speed. MOS logic is considerably slower than most bipolar logic, but consumes much less power than equivalent bipolar circuits. The exception is the bipolar I²L logic, which is comparable in power consumption to and faster in speed than MOS.

14–12   PMOS and NMOS

Many types of logic functions are available in PMOS and NMOS circuitry. PMOS circuitry utilizes P-channel MOSFETs and NMOS circuitry uses N-channel MOSFETs. A basic three-input PMOS NAND gate circuit is shown in Figure 14–29(a). The funny looking symbol for the resistor is a MOSFET permanently biased on creating a resistive element. In this type of circuit the HIGH level is

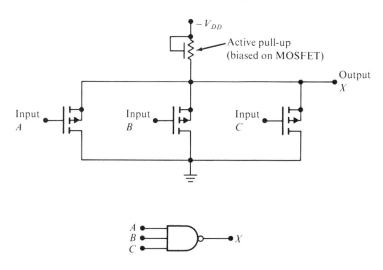

(a) A Three-Input PMOS NAND Gate

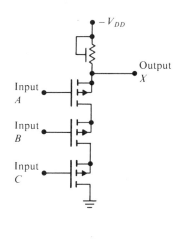

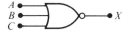

(b) A Three-Input PMOS NOR Gate

Figure 14–29.

close to ground, but usually some small negative value, and the LOW level is the more negative of the two voltages. The operation of this NAND circuit is as follows. A LOW level applied to any of the inputs biases the corresponding MOSFET(s) ON, and the output goes to the HIGH (toward ground) level through the ON resistance of the transistor(s). A HIGH level on all inputs turns all MOSFETs OFF, and the output goes LOW (toward $-V_{DD}$).

Figure 14–29(b) shows a three-input PMOS NOR gate, which operates as follows. LOW levels on all inputs bias each of the MOSFETs ON, making the output go HIGH (toward ground). A HIGH on any of the inputs will turn the corresponding transistor(s) OFF, and the output, which no longer has a path to ground, will go LOW (toward $-V_{DD}$).

A similar operational description applies to NMOS gates, except that voltage polarities are opposite those of the PMOS logic.

14–13   CMOS

Complementary metal oxide semiconductor (CMOS) logic utilizes a complementary arrangement of PMOS and NMOS FETs as shown in the basic inverter circuit of Figure 14–30.

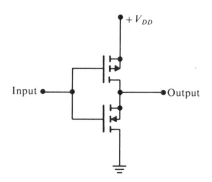

Figure 14–30.   Basic CMOS Inverter.

The operation is as follows. A LOW on the input turns transistor Q_1 (P-channel MOSFET) ON and transistor Q_2 (N-channel MOSFET) OFF; this connects the output to V_{DD} through the ON resistance of Q_1, thus producing a HIGH level output. A HIGH on the input turns Q_1 OFF and Q_2 ON, thereby connecting the output to ground through the ON resistance of transistor Q_2.

The schematic for a two-input CMOS NAND gate is shown in Figure 14–31(a). The logical operation for each of the input conditions is summarized as follows:

1.   Input A LOW and input B LOW:
     Q_1 is ON
     Q_2 is ON
     Q_3 is OFF
     Q_4 is OFF

The output is pulled HIGH through the ON resistance of Q_1 and Q_2 in parallel.

2. Input A LOW and input B HIGH:

Q_1 is ON
Q_2 is OFF
Q_3 is OFF
Q_4 is ON

The output is pulled HIGH through the ON resistance of Q_1.

3. Input A HIGH and input B LOW:

Q_1 is OFF
Q_2 is ON
Q_3 is ON
Q_4 is OFF

The output is pulled HIGH through the ON resistance of Q_2.

4. Input A HIGH and input B HIGH:

Q_1 is OFF
Q_2 is OFF
Q_3 is ON
Q_4 is ON

The output is pulled LOW through the ON resistance of Q_3 and Q_4 connected in series to ground.

Figure 14–31(b) shows a two-input CMOS NOR gate, which operates as follows for each of the possible input logic level conditions:

1. Input A LOW and input B LOW:

Q_1 is ON
Q_2 is ON
Q_3 is OFF
Q_4 is OFF

The output is pulled HIGH through the ON resistance of Q_1 and Q_2 in parallel.

2. Input A LOW and input B HIGH:

Q_1 is ON
Q_2 is OFF
Q_3 is ON
Q_4 is OFF

The output is pulled LOW through the ON resistance of Q_3 connected to ground.

3. Input A HIGH and input B LOW:

Q_1 is OFF
Q_2 is ON
Q_3 is OFF
Q_4 is ON

The output is pulled LOW through the ON resistance of Q_4 connected to ground.

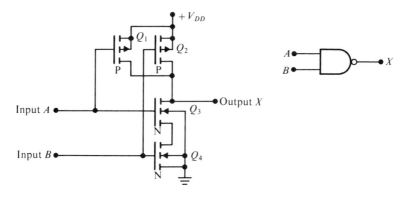

(a) CMOS NAND Gate

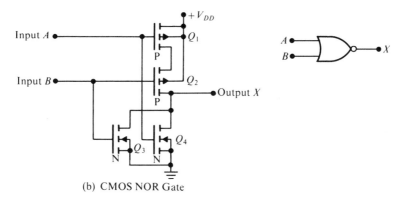

(b) CMOS NOR Gate

Figure 14–31.

4. Input *A* HIGH and input *B* HIGH:
    Q_1 is OFF
    Q_2 is OFF
    Q_3 is ON
    Q_4 is ON
    The output is pulled LOW through the ON resistance of Q_3 and Q_4 connected in parallel to ground.

PROBLEMS

14–1 Explain the basic difference between an integrated circuit and a discrete component circuit.

14–2 List three categories of ICs in the order of their complexity.

14–3 What is the basic difference between bipolar ICs and MOS ICs?

14–4 Draw a DTL NAND gate circuit with eight inputs.

14–5   A wire-OR connection of DTL gates is shown in Figure 14–32. Write the logic expression for the output and explain the operation.

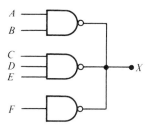

Figure 14–32.

14–6   Show how the wire-OR connection of Problem 14–5 can also be used as a wire-AND connection.

14–7   Explain why no connection (open) on a DTL gate input acts as a HIGH level.

14–8   The same basic explanation of Problem 14–7 applies also to TTL gates. Why?

14–9   Transistor Q_2 in the TTL circuit of Figure 14–16 is sometimes called a *phase splitter.* Explain.

14–10 A typical TTL input produces approximately 1.6 mA of current to the driving source in the LOW state. The Q_4 output transistor (Figure 14–16) can accept or "sink" approximately 16 mA without pulling out of saturation. Show how these characteristics determine the fan-out of a gate.

CHAPTER 15

Introduction to Microprocessors and Computer Organization

A microprocessor is an LSI digital system that can be programmed for processing digital data. In its physical form the microprocessor (μP) typically consists of a single IC chip (or a set of IC chips, depending on the type and manufacturer). A microprocessor differs from a digital system designed for a specific purpose in that it contains general logic units such as registers, counters, control logic, and arithmetic logic that are used to perform many different tasks, depending on the particular requirements of the system operation and how the system is programmed. The microprocessor is the processing portion of a computer in IC form, and a good way to begin understanding it is to look at basic computer operation and organization.

15–1   The Basic Computer

The digital computer, a combination of electronic circuitry organized and interconnected to follow a prescribed sequence of operations (program), solves problems using *data storage, data transfer,* and *data processing* techniques.

A computer system can be divided into five basic functional units or subsystems:

1. Memory subsystem.
2. Arithmetic logic unit (ALU).
3. Control subsystem.
4. Input subsystem.
5. Output subsystem.

A typical block diagram showing the interconnection of the basic subsystems forming a computer system is shown in Figure 15–1.

475

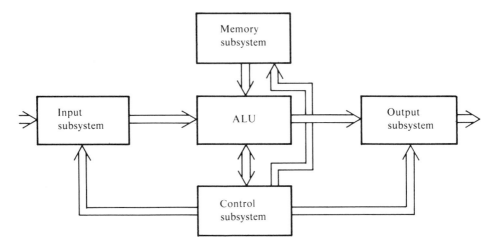

Figure 15-1. Typical Computer Block Diagram Showing the Five Basic Subsystems.

In many systems, the ALU and the control are organized into a subsystem called the *central processing unit* (CPU), shown in Figure 15-2, that is the heart of any processing system, whether it is a large, general-purpose computer or a microprocessor system dedicated to a specific task. In addition to the ALU and the control, it is convenient to include in the CPU certain basic registers and counters that are important in processor operation; these include the *program counter* (PC), the *instruction register* (IR), *index registers,* and *stack registers.* Each will be discussed in detail later.

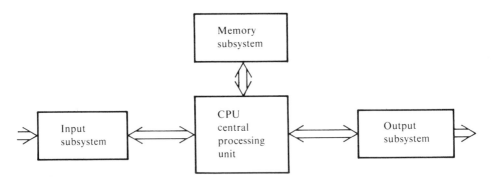

Figure 15-2. Computer Block Diagram with the ALU and the Control Subsystem Consolidated into the CPU.

15-2 Basic Computer Operation

The input subsystem allows external data to be fed into the computer for processing from *peripheral* devices such as keyboards, card readers, paper tape, magnetic tape, and data modems. In other words, the input subsystem is the interface between the processor and the outside world. Likewise, processed data from within the computer can be sent to the outside world via the output subsystem; this subsystem provides an interface with peripheral devices such as CRT displays, printers, storage units, or control of external systems, thus allowing the computer to communicate the results of its processing efforts.

The memory subsystem stores binary information in two basic forms, *program* and *data*. The program is the list of instructions that tell the system what tasks to perform. The data (sometimes called operands) are the numbers that are operated upon during the execution of a program. In a general-purpose computer system, the two types of information are usually stored in separate portions of the same main memory. In a microprocessor system, the program is sometimes stored in a separate read only memory (ROM), and a random access memory (RAM) is used for the data.

The control subsystem, which is usually considered a part of the CPU, "oversees" the entire system operation. It controls the timing for the input and output of information in the computer, and provides for obtaining information from the memory, for the proper execution of operations as directed by the program, and for returning information to the memory. In short, the control subsystem directs the operation of each of the other subsystems to assure that each task required by the program is carried out properly.

The arithmetic logic unit (ALU), which is also normally considered a part of the CPU, is the unit that performs the required operations on the data, such as add, subtract, multiply, divide, and other related operations, as directed by the control subsystem.

15–3   Basic Information Bus

Figure 15–3 shows a general data-busing arrangement for a processing system. The simplified illustration shows the basic busing concept and how each of the subsystems can be linked together over a common bus. Not all systems utilize this particular arrangement, but it serves to illustrate the meaning of busing in processor systems. (Data buses were also discussed in Chapter 13.)

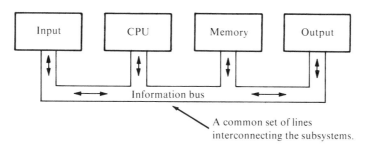

Figure 15–3.   Processor System Interconnected with a Bus Arrangement.

Let us assume that a processing system handles binary information (both program and data) in groups of eight bits. An eight-bit group is typically called a *byte* in computer terminology. In this case, the information bus consists of eight parallel lines that allow eight bits of information (byte) to be transferred from one subsystem to another simultaneously. A simple, everyday analogy is shown in Figure 15–4, where several houses on a block are linked by a single street. A vehicle going from house *A* to house *D* uses the same street as a vehicle going from house *C* to house *B*; there are no separate streets connecting each house. The only restriction is that two vehicles on the street cannot be at exactly the same location at the same time. In the example the street is analogous to an information bus, the houses to the various computer subsystems, and the vehicles to the groups of binary information or bytes (with each person in a vehicle being a bit of information).

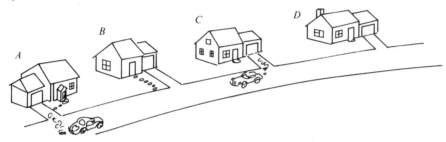

Figure 15–4.

15–4   Information Structure

Binary information used within a processing system is either program information or data to be operated on. The basic unit of information is the *word;* the *command* or *instruction word* carries program information, and the *data word* carries the *operands* (pieces of binary information such as numbers that are to be processed).

The instruction word is typically divided into several parts, each of which

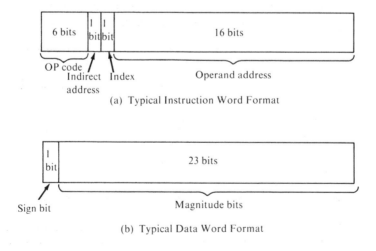

Figure 15–5.   Basic Information Word.

conveys certain information regarding the program instruction—a description of the task to be performed, an address in the memory of the data referred to, type of addressing to be used, and other supplementary information. For example, let us assume a particular system uses a 24-bit word length; Figure 15-5(a) shows a hypothetical word format for this system. Sixteen bits are allotted for the address information, six bits for the operation code (op code) that tells the CPU which operation to perform on the referenced data (add, subtract, etc.), and the remaining two bits for index and indirect address information (which we will discuss later).

The data word contains the magnitude and sign of a numerical quantity, as indicated in Figure 15-5(b). It should be realized that word formats will vary from one system to another, but all will contain the essential information just discussed.

15-5   The Memory Subsystem

In general, the memory subsystem is composed of a *memory cell bank,* a *memory address register* (MAR), a *memory data register* (MDR), and associated controls as shown in Figure 15-6. The purpose of the cell bank is storage of program and data information. As mentioned previously, the program and data can be stored in allotted sections of the same memory bank, or program memory and data memory can be physically separated (as is common in microprocessing systems). The MAR temporarily holds the *address* of the information that is to be read from or written into the memory bank, and the MDR temporarily holds the *information* itself when it is read out or written in. The associated controls essentially provide the proper timing for a memory read/write cycle or, in the case of a ROM, a read action.

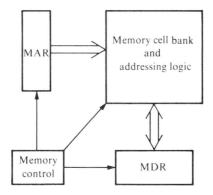

Figure 15-6.   Memory Subsystem Block Diagram.

An example of the sequence of events occurring during a read and a write operation will help demonstrate the basic function of the subsystem. Refer to Figure 15-7 for the following sequence of events for a read operation:

1.   The *address* of the information to be taken from the memory is loaded into the MAR from the CPU.

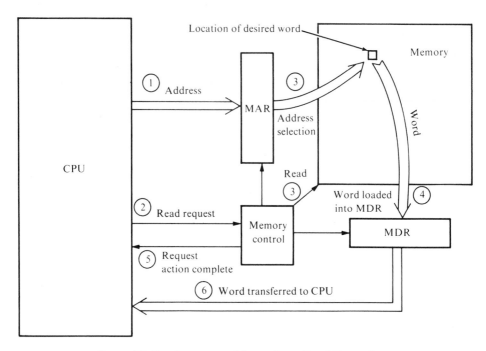

Figure 15–7.   Sequence of Events for a Read Operation.

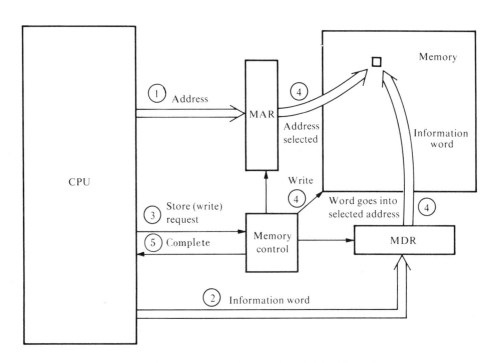

Figure 15–8.   Sequence of Events for a Write Operation.

2.   The CPU tells the memory control that it wants the word contained in the memory location specified by the contents of the MAR.
3.   The memory control initiates a read operation.
4.   The word contained in the specified address is read into the MDR.
5.   The memory control tells the CPU that the information requested is now contained in the MDR.
6.   The CPU takes the information from the MDR.

Now refer to Figure 15–8 for the following sequence of events for a write operation:

1.   The *address* of the information to be put into the memory is loaded into the MAR from the CPU.
2.   The information itself is loaded into the MDR.
3.   The CPU tells the memory control to store the word in the specified address.
4.   The memory control initiates a write operation.
5.   When the word has been stored, the memory control issues a *complete* signal.

These descriptions represent the basic ingredients of a memory subsystem; the details of organization and implementation will vary from system to system.

15–6   The Central Processing Unit (CPU)

As mentioned before, the CPU contains an *arithmetic logic unit,* a *control subsystem,* and an assortment of *registers.* In this section, we will look at the basic operation of these interrelated units to get an overall understanding of the CPU and its relation to other subsystems.

Figure 15–9 shows the basic components of the CPU. Again, details will vary from one system to another, but the basic functions appear in all processor systems in one form or another.

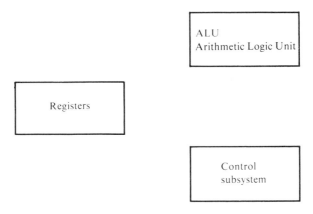

Figure 15–9.   CPU Components.

Arithmetic Logic Unit (ALU)

The basic function of this subsystem is to perform the arithmetic operations of add, subtract, multiply, and divide. In addition to these operations, most ALUs also have capabilities for Boolean operations, shift operations, and other editing functions. Some ALUs perform all of the arithmetic operations with an adder, and others have separate logic for subtraction, multiplication, and division. In general, a typical ALU contains three basic registers, the operational logic, and associated control logic; these basic ALU components are indicated in Figure 15–10. The basic functions of each of the three registers are itemized below:

1. The *A* register (accumulator) stores the sum in an addition process, the difference in a subtraction process, the multiplicand in a multiplication process, the divisor in a division operation, and one of the operands in a Boolean operation.

2. The *D* register stores one of the operands in an addition process, one of the operands in a subtraction process, the partial products and final product in a multiplication process, the dividend and partial remainders in a division process, and one of the operands in a Boolean operation.

3. The *M/Q* register stores the multiplier in a multiplication process and the quotient in a division process.

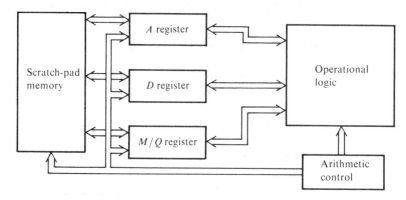

Figure 15–10.   Typical ALU Block Diagram.

In addition to the basic registers mentioned, sometimes an ALU will contain several other registers that comprise what is called a *scratch-pad* memory. Rather than putting intermediate results of a long calculation back into the main memory, they are put into the scratch-pad registers because they are needed only temporarily until the final result is achieved. A simple analogy to the scratch-pad concept in processors is a case of a student taking an exam where only the final answers are required on the exam paper. The student will normally use a piece of "scratch" paper to do all of the calculations required to arrive at a final answer, record only the final answer on the exam paper, and throw the scratch sheet away. In scratch-pad registers the stored data are thrown away once the final result is determined, and the final result is then put into the main memory.

To illustrate the fundamental ALU operation, let us take the problem

$$(1)(2) + (3)(4) + (5)(6) = ?$$

Refer to Figure 15–11 as we go through each step the ALU takes in calculating the result.

1. The first operand (1) is transferred from the memory subsystem to the A register under direction of the control subsystem.
2. The second operand (2) is transferred from the memory subsystem to the M/Q register under direction of the control subsystem.
3. The control subsystem tells the ALU to multiply the two numbers.
4. The ALU performs the multiplication operation and puts the product in the D register.
5. The ALU tells the control subsystem that the multiplication is complete.
6. The contents of the D register (the product of 1 and 2) are transferred to scratch-pad register.
7. The steps are repeated for the third and fourth operands (numbers 3 and 4).
8. Steps 1 through 5 are repeated for the fifth and sixth operands (numbers 5 and 6).
9. The first two products are now stored in the scratch-pad registers, and the last product is still in the D register.
10. The control subsystem transfers the contents of the second scratch-pad register into the A register.
11. The control subsystem tells the ALU to add the contents of the A register to the contents of the D register. The product $(5)(6)$ is now in the D register, and the product $(3)(4)$ is in the A register.
12. The ALU performs the addition operation and puts the sum in the A register.
13. The control subsystem transfers the contents of the first scratch-pad register into the D register.
14. The control subsystem tells the ALU to add the contents of the A register to the contents of the D register. The product $(1)(2)$ is now in the D register, and the sum of the two products (42) in step 11 is in the A register.
15. The ALU performs the addition operation and puts the sum in the A register. This is the final result.
16. The control subsystem transfers the contents of the A register to the memory, thus completing the entire operation.

The Control Subsystem and Its Registers

Everything that a processor does is determined directly or indirectly by a list of instructions (called the *program*) stored in the memory. The basic function of the control subsystem is to examine each instruction stored in the memory in the

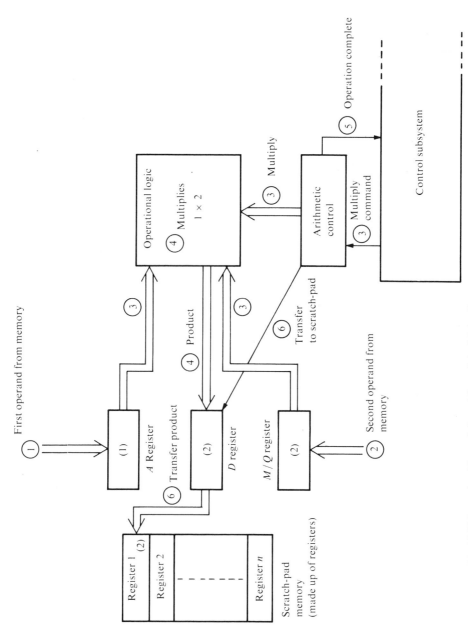

(a) ALU Operation Through Step 6. Steps 7 And 8 Are Repeats Except For Different Scratch-Pad Registers.

484

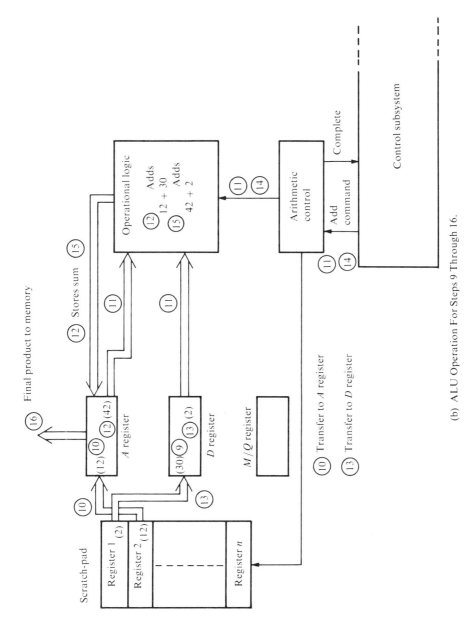

(b) ALU Operation For Steps 9 Through 16.

Figure 15–11.

485

proper order, and to determine the appropriate action required to carry out the instruction; once this is done, the control subsystem executes the instruction by setting up and controlling the proper sequence of operations.

To carry out its basic functions, the control subsystem has two phases or cycles, *fetch* and *execute,* and always alternates between them. In the fetch cycle, the control subsystem gets the next instruction from the memory. The following basic operation occurs during the fetch cycle; refer to Figure 15–12 as we go.

1.   The *program counter* (PC) contains the *address* of the next instruction in the program.
2.   The contents of the program counter are transferred to the memory address register (MAR).
3.   The CPU tells the memory subsystem that it wants the word located at the address specified by the contents of the MAR. A read action is initiated.
4.   The instruction word is read from the memory into the MDR.
5.   The instruction word is transferred from the MDR to the CPU's *instruction register* (IR).
6.   Once the instruction word is in the IR, the fetch cycle is complete and the control subsystem goes into the execute cycle.

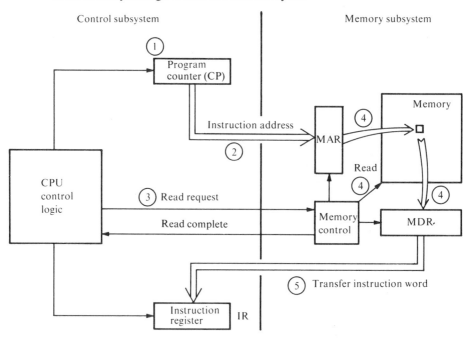

Figure 15–12.   Illustration of the Fetch Cycle.

In the *execute cycle,* the control subsystem performs the instruction it has fetched from the memory. The following basic operation occurs during the execute cycle; refer to Figure 15–13 as we go.

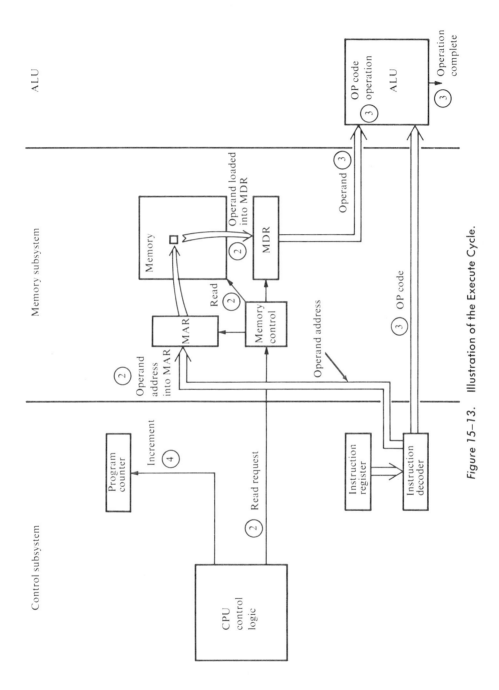

Figure 15-13. Illustration of the Execute Cycle.

1. The control subsystem *decodes* the instruction word contained in the instruction register to determine what is to be done.
2. The control subsystem gets the operand (data to be processed) from the memory as specified by the instruction word.
3. The operation specified in the instruction word (op code) is carried out by the processor.
4. The content of the program counter is increased by 1 (incremented); this now specifies the address of the next instruction in the program.
5. The execute cycle is complete and the control subsystem initiates another fetch cycle.

In this section we have covered the essential elements of a control subsystem's operation. Many processor systems contain additional functions that are of significant importance; we will discuss these additional CPU functions in the following sections.

15-7 Indexing

As mentioned earlier, an instruction word contains, among other things, the address of an operand (the number to be processed). Indexing provides a way of modifying the address automatically. To understand why this might be required, let us take a simple programming example.

Suppose that 20 numbers are contained at consecutive addresses in the memory, say, from address 501 to address 520, and we wish to add them. To perform the addition we must write a program for the processor system to follow, using the instruction set available for our particular system (any processor has a certain number of instructions that it can perform). To write the program we will use *transfer* instructions and an *add* instruction.

The program might typically be written as follows, using mnemonic statements to describe the instructions:

1.	XFERMA 501:	Transfer the contents of memory address 501 to the *A* register.
2.	ADD 502:	Add the contents of memory address 502 to the contents of the *A* register and store the result in the *A* register.
3.	ADD 503:	Add the contents of memory address 503 to the contents of the *A* register and store the result in the *A* register.
⋮	⋯⋯	⋯⋯⋯⋯⋯⋯⋯⋯⋯⋯⋯⋯⋯⋯⋯⋯⋯⋯⋯
20.	ADD 520:	Add the contents of memory address 520 to the contents of the *A* register and store the result in the *A* register.
21.	XFERAM 500:	Transfer the contents of the *A* register to memory address 500.
22.	HALT:	Terminate the program.

Steps 4 through 19 are in the program (but not shown for simplicity), and are the same as the other ADD instructions with the address increased by 1 in each successive step. The important thing to notice here is that it takes 20 ADD instructions to add the 20 numbers. This has several disadvantages. First, each instruction takes one address in the memory and therefore, if large quantities of numbers are to be added, a large amount of the memory is used to store the program. Second, if the list of numbers to be added is increased or decreased, instructions must be added or deleted from the program. Third, the more instructions in a program, the longer a programmer takes to write it.

In our example, the 20 instructions required to add the 20 numbers can be avoided by a modification of the address, called *indexing*, of the operand contained in the instruction word. Indexing is accomplished using *index registers*. An instruction is indexed by adding the contents of the index register to its operand address, and the index bit or *tag* in the instruction word tells the control subsystem whether or not to index the address. Refer to Figure 15–14 as we go through the basic steps in indexing.

1. The control subsystem examines the index bit or tag.
2. If the tag indicates an index, the control subsystem causes the operand address to be added to the number contained in the index register.
3. The modified address is then used to fetch the next operand.
4. The index register is increased by 1. Steps 3 and 4 are repeated until the indexed portion of the program is complete.

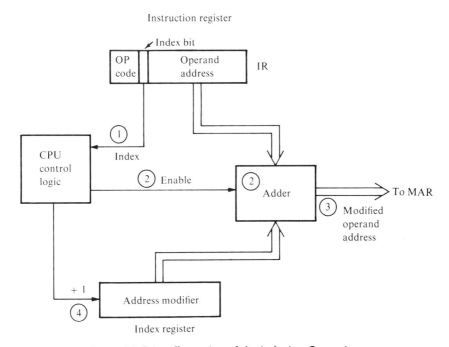

Figure 15–14. Illustration of the Indexing Operation.

To illustrate the usefulness of indexing, let us use our previous example of the addition of 20 numbers. If the ADD instruction is indexed, only one (rather than 20) ADD instruction is required in the program. The control subsystem will continue to update the instruction address with the index register until 20 additions have been completed; an example of this program follows. Some new instructions are used for setting up the index register and "looping" the program (repeating the ADD instruction until the twenty numbers have been added).

1. CLNR: Clear the index register to 0.
2. XFER 501: Transfer the contents of memory address 501 to the *A* register.
3. ADDN 501: Add the contents of the memory address specified (beginning with 501 and increased by the value of the index register each time) to the contents of the *A* register. This is an indexed instruction.
4. INNR 001: Increase by 1 the contents of the index register.
5. CPNR 020: Compare the contents of the index register to 20. This determines when the twentieth number has been added.
6. JMPL 003: "Jump" back to program step 3 if the content of the index register is less than 20. If content of the index register is equal to 20, continue to step 7.
7. XFER 500: Transfer the contents of the *A* register to memory address 500.
8. HALT: Terminate the program.

As you can see, the 22 steps of the original program have been reduced to eight steps by using indexing. This coverage of indexing is meant to familiarize you with the basic principle and is not intended as a thorough coverage of every aspect of the indexing function.

15–8   Stack Registers and Subroutines

In Section 15-7, you saw that indexing is a useful programming tool accomplished with the use of an index register. In this section we are going to see that the *subroutine,* also a common programming tool, is made possible by *stack registers,* which are used in what is commonly called a "push down-pop up" stack or "last in-first out" memory (LIFO).

A subroutine is essentially a program within a program. The subroutine is usually a set of instructions that perform a calculation required repeatedly during the course of the main program. Good examples of calculations sometimes incorporated as subroutines are exponentials, roots, and trigonometric functions.

The relation of the main program to a subroutine is very important. A special instruction in the main program tells the control subsystem to jump to the subroutine; when this happens, the control subsystem must "mark" its place in the main program so that it can pick up the main program at the correct point when the subroutine is complete. To mark its place, the control subsystem increments the program counter so that it contains the next instruction address in the main program. The contents of the program counter are then stored in a stack register for safe keeping. Next, the control subsystem loads the address (of the first subroutine instruction) contained in the jump-to-subroutine instruction into the program counter. The processor proceeds through the subroutine, and when it encounters an instruction telling it to go back to the main program, the control subsystem loads the stack register contents back into the program counter and proceeds with the main program. The essential purpose of the stack register is to remember the processor's place in the main program.

Subroutines within subroutines are often used, since at some point in a given subroutine another subroutine may be called for; this is referred to as *nesting*. The nesting concept requires that a "return" address must be stored each time the system jumps to another subroutine, so a stack register must be provided for each level of subroutine nesting. Refer to Figure 15–15 as we describe the steps involved in a nested subroutine. At this point we do not care about the details of the main program and subroutines.

1. Instruction 5 in the main program tells the processor to go to subroutine *A*, beginning at address 100.

2. The program counter is incremented to 6, which is the address of the next instruction in the main program. For simplicity, we are saying that instruction 5 is at address 5.

3. The program counter content is put into stack register 1.

4. Address 100 is loaded into the program counter.

5. The processor branches to address 100, fetches the first instruction of subroutine *A*, and proceeds to execute each succeeding instruction in the subroutine.

6. Instruction 8 of subroutine *A* (located at memory address 108) tells the processor to go to subroutine *B*, beginning at address 200.

7. The program counter is incremented to 109, which is the address of the next instruction in subroutine *A*.

8. The content of stack register 1 is "pushed down" into stack register 2, and the program counter content is put into stack register 1.

9. Address 200 is loaded into the program counter.

10. The processor branches to address 200, fetches the first instruction of subroutine *B*, and proceeds to execute each succeeding instruction in the subroutine.

11. At the end of subroutine *B*, a "branch-back" instruction is encountered, which causes the processor to go back and pick up where it left off in subroutine *A*.

12. To pick up in subroutine *A* at the proper point, the content of stack register 1 is loaded back into the program counter (address 109), and the processor resumes executing the remaining instructions in subroutine *A*. When the return address for subroutine *A* transfers from stack register 1 to the PC, the return address for the main program stored in stack register 2 "pops" back up into stack register 1.

13. At the end of subroutine A, a "branch-back" instruction is encountered, which causes the processor to go back and pick up where it left off in the main program.

14. To pick up in the main program at the proper point, the content of stack register 1 is loaded back into the program counter (address 6), and the processor resumes executing the remaining instructions in the main program.

This example illustrates two levels of subroutine nesting. Many systems can accommodate many more levels but require an additional stack register for each level accommodated.

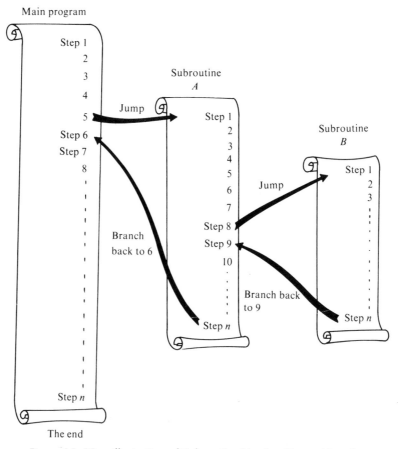

Figure 15–15.   Illustration of Subroutine Nesting (Second Level).

15–9   Interrupts

An interrupt is a *request* from an *external* device or peripheral unit for information or servicing by the computer. One example of this concept occurs when a CRT display unit requests updated information. The CRT must be refreshed periodically to maintain the display, so it must interrupt the computer operation and request updating at fixed intervals.

An interrupt will normally activate a "flag" bit, which is examined by the CPU after each instruction in the operating program is completed. When the flag bit indicates an interrupt request, the processor will branch to a routine designed to service the interrupting device. The same events occur as previously discussed in relation to a subroutine. The processor must leave the main program or subroutine it is executing when the interrupt occurs; to do this, it makes use of the stack registers to store the return address so that it can return to the proper place in the program after the interrupt has been serviced. The basic difference between an interrupt action and a subroutine action is that the interrupt is *externally* generated and the subroutine is *internally* generated.

Many processor systems are capable of handling more than one interrupt on a priority basis.

15–10   Input/Output (I/O)

As mentioned before, the input and output subsystems are the means by which the computer communicates with external devices or peripherals; they can provide single or multiple input and output ports. A port is actually the "pipeline" into or out of the system that provides for the transfer of information. Multiple ports provide almost simultaneous communication with several peripherals.

Two basic types of I/O techniques are utilized in processor systems, the *programmed I/O* and the *direct memory access* (DMA). In the programmed I/O operation, a specific program instruction is required to transfer data into or out of the system. The DMA method allows data to be transferred directly into the memory from peripheral devices, independent of the control subsystem and the operating program.

PROBLEMS

15–1   List the basic computer subsystems.

15–2   Sketch the basic block diagram for a computer.

15–3   Define a microprocessor.

15–4   Discuss the basic functions of the following units:
      (a)   input and output subsystems   (b)   ALU
      (c)   memory subsystem   (d)   control subsystem

15–5   What is the CPU?

15–6   What is meant by the term *peripheral* in a computer system? Give some examples.

15–7   Discuss the difference between program information and data.

15–8   Define the term *data bus*.

15–9   What does the term *word* mean in computer terminology?

15-10 Sketch the basic structure of a computer word and define each part. What is the difference between an instruction word and a data word?

15-11 Define MAR and explain its purpose.

15-12 Define MDR and explain its purpose.

15-13 Discuss what happens in terms of the sequence of events that occurs when a word is stored in the memory.

15-14 Discuss what happens in a sequence of events when a word is retrieved (read) from the memory.

15-15 Discuss the sequence of events for the ALU when processing the following calculations:

    (a) $3 + 2 + 7$ (b) 5×8 (c) $3 \times 2 + 9 \times 4$ (d) $\dfrac{6}{3} + 5$

15-16 What is the purpose of the scratch-pad memory?

15-17 Define the fetch cycle.

15-18 Define the execute cycle.

15-19 What subsystem initiates the fetch and execute operations?

15-20 Discuss the purpose of the program counter.

15-21 Discuss the purpose of the instruction register.

15-22 Outline the basic steps of the control subsystem during the fetch cycle.

15-23 Outline the basic steps of the control subsystem during the execute cycle.

15-24 Define indexing, and explain how the index register is used.

15-25 What is a subroutine?

15-26 Discuss the purpose of a stack register.

15-27 What is the significance of the term *push down-pop up?*

15-28 Define LIFO.

15-29 What is an interrupt, and how is it normally handled by a processing system?

15-30 Explain the basic difference between an interrupt and a subroutine.

15-31 What does *DMA* mean?

Glossary

Access time   The time required to get a piece of information from a memory.

Accumulator   The register in a binary adder in which the sum is stored.

A/D conversion   The process of converting an analog quantity or signal into digital form.

Addend   In addition, the number being added to another number called the **augend.**

Adder   A digital circuit that can perform addition of numbers

Address   The location of a given storage cell in a memory.

Adjacent   In reference to the Karnaugh map, adjacent cells are those that differ in their position by only one bit.

Alphanumeric   A system consisting of both numerals and alphabetic characters.

ALU   Arithmetic logic unit, generally a part of the central processing unit in computers and microprocessors.

Amplitude   In terms of pulse waveforms, the height or maximum value of the pulse.

Analog   Being continuous or having a continuous range of values, as opposed to a discrete set of values.

AND gate   A digital logic circuit in which a HIGH output occurs only when all of the inputs are HIGH. A coincident circuit.

Arithmetic   The four operations of add, subtract, multiply, and divide.

Astable   Having no stable state. A type of multivibrator that oscillates between two quasistable states.

Asynchronous   Having no fixed time relationship.

Augend   In addition, the number to which the addend is added.

Bar   In Boolean algebra, the symbol over a variable indicating the complement.

Base   One of the three regions in a bipolar transistor. Also, the number of symbols in a number system. The decimal system has a base of ten because there are ten digits.

BCD   Binary coded decimal, a digital code.

Binary   Having two values or states. The binary number system has two digits.

Binary fractional notation   A method of binary notation where the magnitude of a number is represented by the bits to the right of the binary point.

Bipolar   Referring to a junction type of semiconductor device. A PNP or an NPN transistor.

Bistable   Having two stable states. A type of multivibrator commonly known as a *flip-flop.*

Bit   Binary digit. A 1 or a 0.

Boolean algebra   A mathematics of logic.

Byte   A group of bits, usually eight.

Carry save   A method of binary addition where the carries are "saved" until the total sum is formed and then added to the sum.

Cascade   A configuration in which one element follows another.

Cell   A single storage element in a memory.

Character   A symbol, letter, or number.

Circuit   A combination of electrical and/or electronic components connected together in such a way as to perform a specified function.

Clear   To reset, as in the case of a flip-flop, register, or counter.

Clock   The basic timing signal in a digital system.

CMOS   Complementary metal oxide semiconductor.

Code   A combination of binary digits that represents information such as numbers, letters, and other symbols.

Code converter   An electronic digital circuit that converts one type of coded information into another form.

Collector   One of the three regions in a bipolar transistor.

Combinational logic   A combination of gate networks used to generate a specified function. Sometimes called *combinatorial logic*.

Comparator   A digital device that compares the magnitudes of two or more quantities and produces an output indicating the relationship of the quantities.

Complement   In Boolean algebra, the inverse function. The complement of a 1 is a 0, and vice versa.

Computer   A digital system that can be programmed to perform various tasks, such as mathematical operations, at extremely high speed.

Core   A magnetic memory element.

Counter   A digital circuit capable of counting electronic events, such as pulses, by progressing through a sequence of binary states.

CPU   Central processing unit.

D/A conversion   The process whereby information in digital form is converted into analog form.

Data   Information in numeric, alphabetic, or other form.

D flip-flop   A type of flip-flop in which the state of the device follows the *D* input.

Decade counter   A digital counter having ten states.

Decode   To determine the meaning of coded information.

Decoder   A digital circuit that converts coded information into a familiar form.

Delay   The time interval between the occurrence of an event at one point and its occurrence at a second point.

DeMorgan's theorems   (1) The complement of a product of terms is equal to the sum of the complements of each term. (2) The complement of a sum of terms is equal to the product of the complements of each term.

Difference   The result of a subtraction.

Digit   A symbol representing a given quantity in a number system.

Digital   Related to digits or discrete quantities.

Diode   An electronic device that conducts current in only one direction.

DIP   Dual in-line package.

Dividend   In a division operation, the quantity that is being divided.

Divisor   In a division operation, the quantity that is divided into the dividend.

DMA   Direct memory access.

Don't care   A condition in a logic network in which the output is independent of the state of a given input.

Down count   A counter sequence in which each successive state is less than the previous state.

Drain   One of the three terminals of an FET.

DRO   Destructive readout.

DTL   Diode-transistor logic.

Duplex   Bidirectional transmission of data along a transmission line.

Dynamic memory   A memory having cells that tend to lose stored information over a period of time and must be "refreshed." Usually capacitive type of storage.

ECL   Emitter-coupled logic.

Edge-triggered flip-flop   A type of flip-flop in which the input data are entered into the device and appear on the output on the same edge of the clock pulse.

Emitter   One of the three regions of a bipolar transistor.

Enable   To activate.

Encode   To convert information into coded form.

Encoder   A digital circuit that converts information into coded form.

End-around carry   The final carry that is added to the result in a 9's complement or 1's complement addition.

Error correction   The process of correcting errors occurring in a digital code.

Error detection   The process of detecting errors occurring in a digital code.

Even parity   Referring to a group of binary digits having an even total number of 1s.

Excess-3 code   A digital code where each of the decimal digits is represented by a four-bit binary code found by adding 3 to each of the digits. A type of binary coded decimal code.

Exclusive-OR   A logic function that is true if one but not both of the variables is true.

Execute   One of the two main cycles of a CPU.

Fall time   The time interval between the 10% point to the 90% point on the negative-going edge of a pulse.

Fan-in   The number of inputs to a logic gate.

Fan-out   The number of equivalent gate inputs that a logic gate can drive.

FET   Field-effect transistor.

Fetch   One of the two main cycles of a CPU.

FIFO   First-in–first-out memory.

Fixed point   A binary point having a fixed location in a binary number.

Floating point   A binary point having a variable location in a binary number.

Frequency   The number of pulses or cycles in one second for a periodic waveform. Expressed in PPS or hertz.

Full-adder   A digital circuit that adds two binary digits and an input carry digit and produces a sum digit and an output carry digit.

Full-duplex   Simultaneous bidirectional transmission of data on a transmission line.

Full-select current   In a coincident-current core memory, the amount of current required to switch fully a particular core.

Full-subtractor   A digital circuit that subtracts two binary digits and produces a difference digit and a borrow output digit. It can handle an input borrow digit.

Gate   A logic circuit that performs a specified logical operation such as AND, OR, NAND, or NOR.

Gray code   A type of digital code characterized by a single bit change from one code word to the next.

Half-adder   A digital circuit that adds two binary digits and produces a sum digit and an output carry digit. It cannot handle input carries.

Half-select current   In a coincident current core memory, the current in one of the select lines which is one-half the value required to switch fully the core.

Half-subtractor   A digital circuit that subtracts two binary digits and produces a difference digit and a borrow output digit. It cannot handle input borrows.

Hamming code   A type of error detection and correction code.

Hexadecimal number   A number with a base or radix of sixteen.

Hold time   The time interval required for the control levels to remain on the inputs to a flip-flop after the triggering edge of the clock pulse in order to clock reliably the flip-flop.

Hysteresis   A characteristic of a magnetic core or a threshold triggered circuit.

IC   Integrated circuit.

I²L   Integrated injection logic.

Increment   To increase the contents of a register or counter by 1.

Indexing   The modification of the address of the operand contained in the instruction word.

Index register   A register used for indexing.

Indirect address   The address of the address of an operand.

Information   In a digital system, intelligence represented in binary form.

Inhibit winding   In a core memory, the winding that is used to prevent the switching of the core to the 1 state.

Initialize   To put a logic circuit in a beginning state, such as to clear a register.

Input   The signal or line going into a circuit. A signal that controls the operation of a circuit.

Instruction   In a processor or computer system, the information that tells the machine what to do. Program information.

Instruction register   A register in the CPU that stores the instruction word when it is read from the memory.

Instruction word   The combination of a fixed number of binary bits that carry the program information.

Interrupt   The process of stopping the normal execution of a program in order to handle a higher-priority task.

Inversion   Converting a high level to a low level or vice versa.

Inverter   The digital circuit that performs inversion.

JK flip-flop   A type of flip-flop that can operate in the set, reset, no change, and toggle modes.

Johnson counter   A type of digital counter characterized by unique sequences of states.

Junction   The boundary between an N region and a P region in a semiconductor device.

Karnaugh map   An arrangement of cells representing the combinations of variables in a Boolean expression and used for a systematic simplification of the expression.

Leading edge   The first edge to occur on a pulse.

LED   Light-emitting diode.

LIFO   Last-in–first-out memory.

Linear-select memory   A type of memory having a single select line for each word.

Logic   In digital electronics, the decision-making capability of gate circuits in terms of yes/no or on/off type of operation.

Look-ahead-carry   A method of binary addition whereby carries from preceding stages are anticipated, thus avoiding carry propagation delays.

LSB   Least significant bit.

Magnetic core   A memory element made of magnetic materials and capable of existing in either of two states.

Magnitude   Size or value.

Master reset   Normally, the input to a counter, register, or other digital storage device that completely resets or clears the device.

Master-slave flip-flop   A type of flip-flop in which the input data are entered into the device on the leading edge of the clock and appear on the output on the trailing edge.

Memory address   The location of a cell in a memory array.

Memory array   An arrangement of memory cells.

Memory cell   An individual storage element in a memory.

Memory cycle   The read/write operation.

Microprocessor   A single integrated circuit or set of integrated circuits that can be programmed to process data.

Minuend   The number being subtracted from in a subtraction operation.

Modified counter   A counter that does not sequence through all of its "natural" states.

Mod-2 addition   Exclusive-OR addition. A sum of two bits with the carry being dropped.

Modulus   The maximum number of states in a counter sequence.

Monostable   Having only one stable state. A multivibrator characterized by one stable state commonly called a **one-shot.**

MOS   Metal oxide semiconductor.

MSI   Medium-scale integration.

Multiplex   To put information from several sources onto a single line or transmission path.

Multiplexer   A digital circuit capable of multiplexing digital data.

Multiplicand   The number being multiplied.

Multiplier   The number used to multiply the multiplicand.

NAND gate   A logic gate that performs a not-AND operation.

Natural count   The maximum modulus of a counter.

NDRO   Nondestructive readout.

Negative logic   The system of logic where a LOW represents a 1 and a HIGH represents a 0.

Nesting   Referring to the arrangement of subroutines where the program exits one subroutine and goes to another, eventually returning to the first. Several levels of this can occur.

Nixie tube   A vacuum tube used for digital readouts.

NMOS   N-channel metal oxide semiconductor.

Noise immunity   The ability of a circuit to reject unwanted signals.

Noise margin   The difference between the maximum low output of a gate and the maximum acceptable low level input of an equivalent gate. Also, the difference between the minimum high output of a gate and the minimum acceptable high level input of an equivalent gate.

NOR gate   A logic gate that performs a not-OR operation.

NOT circuit   An inverter.

NPN   Referring to the junction structure of a bipolar transistor.
Numeric   Related to numbers.

Octal   A number system having a base or radix of 8.
Odd parity   Referring to a group of binary digits having an odd total number of 1s.
One-shot   A monostable multivibrator.
Op code   Operation code. The part of a computer instruction word that designates the task to be performed.
Operand   A quantitity being operated on in a processing system.
OR gate   A logic gate that produces a HIGH output when any one or both of its inputs are HIGH.
Oscillator   An electronic circuit that switches back and forth between two unstable states.

Parallel   Lying side by side but never meeting.
Parity   Referring to the oddness or evenness of the number of 1s or 0s contained in a code word.
Period   The time required for a periodic waveform to repeat itself.
Periodic   Repeating at fixed intervals.
PMOS   P-channel metal oxide semiconductor.
PNP   Referring to the junction structure of a bipolar transistor.
Positive logic   The system of logic where a HIGH represents a 1 and a LOW represents a 0.
PPS   Pulses per second. A measure of the frequency of a pulse waveform.
Preset   To initialize a digital circuit to a predetermined state.
PRF   Pulse repetition frequency.
Priority encoder   A digital logic circuit that produces a coded output corresponding to the highest valued input.
Processor   A digital system that performs specified operations on digital data and produces a desired result.
Product   The result of a multiplication.
Product-of-sums   A form of Boolean expression that is the ANDing of ORed terms.
Program   A list of tasks that are arranged in a specified order and that control the operation of a processor system. A list of instructions that tell the system what to do and when to do it.
Program counter   A counter in the CPU of a computer system that keeps up with the place in the program. It acts as a "bookmark" that tells us where we are in the program at any given time.
Propagation delay   The time interval between the occurrence of an input transition and the corresponding output transistion.
Pulse   A sudden change from one level to another followed by a sudden change back to the original level.
Pulse duration   The time interval that a pulse remains at its high level (positive-going pulse) or at its low level (negative-going pulse). Typically measured between the 50% points on the leading and trailing edges of the pulse.
Pulse width   Pulse duration.

Quad   A group of four.
Quotient   The result of a division.

Race   A condition in a logic network where the differences in propagation times through two or more signal paths in the network can produce an erroneous output.

Radix   The base of a number system. The number of digits in a given system.

RAM   Random-access memory.

Read   The process of retrieving information from a memory.

Recirculating shift register   A shift register in which the serial output is connected back to the input so that the contents of the register are never lost.

Refresh   The process of renewing the contents of a dynamic storage element.

Regenerative   Having feedback so that an initiated change is automatically continued, such as when a multivibrator switches from one state to the other.

Register   A digital circuit capable of storing and moving (shifting) binary information.

Reset   The state of a flip-flop storing a binary 0. Equivalent to the clear function.

Ring counter   A digital circuit made up of a series of flip-flops in which the contents are continuously recirculated.

Ringing   A damped sinusoidal oscillation.

Ripple Counter   A digital counter in which each flip-flop is triggered with the output of the previous stage.

Rise time   The time required for the positive-going edge of a pulse to go from 10% of its full value to 90% of its full value.

ROM   Read-only memory.

RS flip-flop   A reset-set flip-flop.

Select line   A line in a linear-select memory that, when activated, addresses a given word in that memory. A line in a coincident-select memory that is used to select the X or Y portion of the memory address.

Semiconductor   A material used to construct electronic devices such as integrated circuits, transistors, and diodes.

Sequential logic   A broad category of digital circuits whose logic states depend on a specified time sequence.

Serial   An in-line arrangement where one element follows another, such as in a serial shift register. Also, the occurrence of events, such as pulses, in a time sequence rather than simultaneously.

Set   The state of a flip-flop storing a binary 1.

Set-up time   The time interval required for the control levels to be on the inputs to a digital circuit, such as a flip-flop, prior to the triggering edge of the clock pulse.

Shift   To move binary data within a shift register.

Shift register   A digital circuit capable of storing and shifting binary data.

Silicon   A semiconductor material.

Simplex   A mode of data transmission whereby the data can be sent in only one direction.

SSI   Small-scale integration.

Stack register   A processor register used to store the return location in a main program or subroutine when the system branches from the main program to a subroutine or from one subroutine to another.

Stage   One flip-flop in a counter or register.

Static memory   A memory composed of storage elements such as magnetic cores or flip-flops that are capable of retaining information indefinitely.

Storage   The memory capability of a digital device.

Store   The process of memorizing data.

Strobe   A pulse used to sample the occurrence of an event at a specified point in time in relation to the event.

Subroutine   A program that is normally used to perform specialized or repetitive operations during the course of a main program.

Subtractor   One of the operands in a subtraction.

Subtrahend   The other operand in a subtraction.

Sum   The result of an addition.

Sum-of-products   A form of Boolean expression that is the ORing of ANDed terms.

Synchronous   Having a fixed time relationship.

Terminal count   The final state of a counter sequence.

T flip-flop   A type of flip-flop that toggles or changes state on each clock pulse.

Three-state logic   A type of logic circuit having the normal two-state (1, 0) output and, in addition, an open state in which it is disconnected from its load.

Toggle flip-flop   T flip-flop.

Trailing edge   The second transition of a pulse.

Transistor   A semiconductor device exhibiting gain characteristics. When it is used as a switching device, it can approximate an open or a closed switch.

Transition   A change from one level to another.

Trigger   A pulse used to initiate a change in the state of a logic circuit.

UHF   Ultrahigh frequency.

Up-count   A counter sequence in which each binary state has a successively higher value.

Up-down counter   A counter capable of sequencing through an up-count or a down-count.

Variable modulus counter   A counter in which the maximum number of states can be changed.

VHF   Very high frequency.

Weight   The value of a digit in a number based on its position in the number.

Weighted code   A digital code that utilizes weighted numbers as the individual code words.

Wire-AND   An arrangement of logic circuits in which the gate outputs are hand-wired together to form an "implied" AND function.

Word   A group of bits representing a complete piece of digital information.

Write   The process of storing information in a memory.

APPENDIX A

Powers of Two

2^n	n	2^{-n}
1	0	1.0
2	1	0.5
4	2	0.25
8	3	0.125
16	4	0.062 5
32	5	0.031 25
64	6	0.015 625
128	7	0.007 812 5
256	8	0.003 906 25
512	9	0.001 953 125
1 024	10	0.000 976 562 5
2 048	11	0.000 488 281 25
4 096	12	0.000 244 140 625
8 192	13	0.000 122 070 312 5
16 384	14	0.000 061 035 156 25
32 768	15	0.000 030 517 578 125
65 536	16	0.000 015 258 789 062 5
131 072	17	0.000 007 629 394 531 25
262 144	18	0.000 003 814 697 265 625
524 288	19	0.000 001 907 348 632 812 5
1 048 576	20	0.000 000 953 674 316 406 25
2 097 152	21	0.000 000 476 837 158 203 125
4 194 304	22	0.000 000 238 418 579 101 562 5
8 388 608	23	0.000 000 119 209 289 550 781 25
16 777 216	24	0.000 000 059 604 644 775 390 625
33 554 432	25	0.000 000 029 802 322 387 695 312 5
67 108 864	26	0.000 000 014 901 161 193 847 656 25
134 217 728	27	0.000 000 007 450 580 596 923 828 125
268 435 456	28	0.000 000 003 725 290 298 461 914 062 5
536 870 912	29	0.000 000 001 862 645 149 230 957 031 25
1 073 741 824	30	0.000 000 000 931 322 574 615 478 515 625
2 147 483 648	31	0.000 000 000 465 661 287 307 739 257 812 5
4 294 967 296	32	0.000 000 000 232 830 643 653 869 628 906 25
8 589 934 592	33	0.000 000 000 116 415 321 826 934 814 453 125
17 179 869 184	34	0.000 000 000 058 207 660 913 467 407 226 562 5
34 359 738 368	35	0.000 000 000 029 103 830 456 733 703 613 281 25
68 719 476 736	36	0.000 000 000 014 551 915 228 366 851 806 640 625
137 438 953 472	37	0.000 000 000 007 275 957 614 183 425 903 320 312 5
274 877 906 944	38	0.000 000 000 003 637 978 807 091 712 951 660 156 25
549 755 813 888	39	0.000 000 000 001 818 989 403 545 856 475 830 078 125
1 099 511 627 776	40	0.000 000 000 000 909 494 701 772 928 237 915 039 062 5
2 199 023 255 552	41	0.000 000 000 000 454 747 350 886 464 118 957 519 531 25
4 398 046 511 104	42	0.000 000 000 000 227 373 675 443 232 059 478 759 765 625
8 796 093 022 208	43	0.000 000 000 000 113 686 837 721 616 029 739 379 882 812 5
17 592 186 044 416	44	0.000 000 000 000 056 843 418 860 808 014 869 689 941 406 25
35 184 372 088 832	45	0.000 000 000 000 028 421 709 430 404 007 434 844 970 703 125
70 368 744 177 664	46	0.000 000 000 000 014 210 854 715 202 003 717 422 485 351 562 5
140 737 488 355 328	47	0.000 000 000 000 007 105 427 357 601 001 858 711 242 675 781 25
281 474 976 710 656	48	0.000 000 000 000 003 552 713 678 800 500 929 355 621 337 890 625
562 949 953 421 312	49	0.000 000 000 000 001 776 356 839 400 250 464 677 810 668 945 312 5
1 125 899 906 842 624	50	0.000 000 000 000 000 888 178 419 700 125 232 338 905 334 472 656 25
2 251 799 813 685 248	51	0.000 000 000 000 000 444 089 209 850 062 616 169 452 667 236 328 125
4 503 599 627 370 496	52	0.000 000 000 000 000 222 044 604 925 031 308 084 726 333 618 164 062 5
9 007 199 254 740 992	53	0.000 000 000 000 000 111 022 302 462 515 654 042 363 166 809 082 031 25
18 014 398 509 481 984	54	0.000 000 000 000 000 055 511 151 231 257 827 021 181 583 404 541 015 625
36 028 797 018 963 968	55	0.000 000 000 000 000 027 755 575 615 628 913 510 590 791 702 270 507 812 5
72 057 594 037 927 936	56	0.000 000 000 000 000 013 877 787 807 814 456 755 295 395 851 135 253 906 25
144 115 188 075 855 872	57	0.000 000 000 000 000 006 938 893 903 907 228 377 647 697 925 567 626 953 125
288 230 376 151 711 744	58	0.000 000 000 000 000 003 469 446 951 953 614 188 823 848 962 783 813 476 562 5
576 460 752 303 423 488	59	0.000 000 000 000 000 001 734 723 475 976 807 094 411 924 481 391 906 738 281 25
1 152 921 504 606 846 976	60	0.000 000 000 000 000 000 867 361 737 988 403 547 205 962 240 695 953 369 140 625
2 305 843 009 213 693 952	61	0.000 000 000 000 000 000 433 680 868 994 201 773 602 981 120 347 976 684 570 312 5
4 611 686 018 427 387 904	62	0.000 000 000 000 000 000 216 840 434 497 100 886 801 490 560 173 988 342 285 156 25
9 223 372 036 854 775 808	63	0.000 000 000 000 000 000 108 420 217 248 550 443 400 745 280 086 994 171 142 578 125
18 446 744 073 709 551 616	64	0.000 000 000 000 000 000 054 210 108 624 275 221 700 372 640 043 497 085 571 289 062 5
36 893 488 147 419 103 232	65	0.000 000 000 000 000 000 027 105 054 312 137 610 850 186 320 021 748 542 785 644 531 25
73 786 976 294 838 206 464	66	0.000 000 000 000 000 000 013 552 527 156 088 805 425 093 160 010 874 271 392 822 265 625
147 573 952 589 676 412 928	67	0.000 000 000 000 000 000 006 776 263 578 034 402 712 546 580 005 437 135 696 411 132 812 5
295 147 905 179 352 825 856	68	0.000 000 000 000 000 000 003 388 131 789 017 201 356 273 290 002 718 567 848 205 566 406 25
590 295 810 358 705 651 712	69	0.000 000 000 000 000 000 001 694 065 894 508 600 678 136 645 001 359 283 924 102 783 203 125
1 180 591 620 717 411 303 424	70	0.000 000 000 000 000 000 000 847 032 947 254 300 339 068 322 500 679 641 962 051 391 601 562 5
2 361 183 241 434 822 606 848	71	0.000 000 000 000 000 000 000 423 516 473 627 150 169 534 161 250 339 820 981 025 695 800 781 25
4 722 366 482 869 645 213 696	72	0.000 000 000 000 000 000 000 211 758 236 813 575 084 767 080 625 169 910 490 512 847 900 390 625

APPENDIX B

Conversions

Decimal	BCD (8421)	Octal	Binary	Decimal	BCD (8421)	Octal	Binary
0	0000	0	0	28	00101000	34	11100
1	0001	1	1	29	00101001	35	11101
2	0010	2	10	30	00110000	36	11110
3	0011	3	11	31	00110001	37	11111
4	0100	4	100	32	00110010	40	100000
5	0101	5	101	33	00110011	41	100001
6	0110	6	110	34	00110100	42	100010
7	0111	7	111	35	00110101	43	100011
8	1000	10	1000	36	00110110	44	100100
9	1001	11	1001	37	00110111	45	100101
10	00010000	12	1010	38	00111000	46	100110
11	00010001	13	1011	39	00111001	47	100111
12	00010010	14	1100	40	01000000	50	101000
13	00010011	15	1101	41	01000001	51	101001
14	00010100	16	1110	42	01000010	52	101010
15	00010101	17	1111	43	01000011	53	101011
16	00010110	20	10000	44	01000100	54	101100
17	00010111	21	10001	45	01000101	55	101101
18	00011000	22	10010	46	01000110	56	101110
19	00011001	23	10011	47	01000111	57	101111
20	00100000	24	10100	48	01001000	60	110000
21	00100001	25	10101	49	01001001	61	110001
22	00100010	26	10110	50	01010000	62	110010
23	00100011	27	10111	51	01010001	63	110011
24	00100100	30	11000	52	01010010	64	110100
25	00100101	31	11001	53	01010011	65	110101
26	00100110	32	11010	54	01010100	66	110110
27	00100111	33	11011	55	01010101	67	110111

Decimal	BCD (8421)	Octal	Binary	Decimal	BCD (8421)	Octal	Binary
56	01010110	70	111000	78	01111000	116	1001110
57	01010111	71	111001	79	01111001	117	1001111
58	01011000	72	111010	80	10000000	120	1010000
59	01011001	73	111011	81	10000001	121	1010001
60	01100000	74	111100	82	10000010	122	1010010
61	01100001	75	111101	83	10000011	123	1010011
62	01100010	76	111110	84	10000100	124	1010100
63	01100011	77	111111	85	10000101	125	1010101
64	01100100	100	1000000	86	10000110	126	1010110
65	01100101	101	1000001	87	10000111	127	1010111
66	01100110	102	1000010	88	10001000	130	1011000
67	01100111	103	1000011	89	10001001	131	1011001
68	01101000	104	1000100	90	10010000	132	1011010
69	01101001	105	1000101	91	10010001	133	1011011
70	01110000	106	1000110	92	10010010	134	1011100
71	01110001	107	1000111	93	10010011	135	1011101
72	01110010	110	1001000	94	10010100	136	1011110
73	01110011	111	1001001	95	10010101	137	1011111
74	01110100	112	1001010	96	10010110	140	1100000
75	01110101	113	1001011	97	10010111	141	1100001
76	01110110	114	1001100	98	10011000	142	1100010
77	01110111	115	1001101	99	10011001	143	1100011

Answers to Odd-Numbered Problems

Chapter 1

1–1   *Bit* is a contraction of *binary digit.*
Two bits.

1–3   0.5 mA

1–5   (a)   0.6 microsecond (μs)   (b)   0.4 μs   (c)   2.8 μs   (d)   10 V

1–7   250 Hz

1–9   The flip-flop has storage or memory capability; gates do not.

1–11   AND gate

Chapter 2

2–1   (a)   6   (b)   4   (c)   1   (d)   87   (e)   82   (f)   74
(g)   50   (h)   13   (i)   872   (j)   618   (k)   309   (l)   8645

2–3   (a)   23   (b)   34   (c)   98   (d)   118   (e)   137   (f)   187

2–5   (a)   3   (b)   4   (c)   7   (d)   8   (e)   9   (f)   12
(g)   11   (h)   15

2–7   (a)   51.75   (b)   42.25   (c)   65.875   (d)   120.625
(e)   92.65625   (f)   113.0625   (g)   90.625   (h)   127.96875

2–9   (a)   5 bits   (b)   6 bits   (c)   6 bits   (d)   7 bits   (e)   7 bits
(f)   7 bits   (g)   8 bits   (h)   8 bits   (i)   9 bits

2–11   (a)   100   (b)   100   (c)   1000   (d)   1101   (e)   1110
(f)   11000

2–13   (a)   1001   (b)   1000   (c)   100011   (d)   110110
(e)   10101001   (f)   10110110

2–15   (a)   010   (b)   001   (c)   0101   (d)   00101000   (e)   0001010
(f)   11110

2–17   (a)   10   (b)   001   (c)   0111   (d)   0011   (e)   00100
(f)   01101

2–19   (a)   1010   (b)   10001   (c)   11000   (d)   110000
(e)   111101   (f)   1011101   (g)   1111101   (h)   10111010
(i)   100101010

2-21 (a) 10 (b) 23 (c) 46 (d) 52 (e) 67 (f) 367
(g) 115 (h) 532 (i) 4085

2-23 (a) 001011 (b) 101111 (c) 001000001
(d) 011010001 (e) 101100000 (f) 100110101011
(g) 001011010111001 (h) 100101110000000 (i) 001000000010001011

2-25 (a) 00111000 (b) 01011001 (c) 101000010100
(d) 010111001000 (e) 0100000100000000 (f) 1111101100010111

Chapter 3

3-1 (a) 00010000 (b) 00010011 (c) 00011000
(d) 00100001 (e) 00100101 (f) 00110110
(g) 01000100 (h) 01010111 (i) 01101001
(j) 10011000 (k) 000100100101 (l) 000101010110

3-3 (a) 000100000100 (b) 000100101000 (c) 000100110010
(d) 000101010000 (e) 000110000110 (f) 001000010000
(g) 001101011001 (h) 010101000111 (i) 0001000001010001
(j) 0010010101100011

3-5 (a) 80 (b) 237 (c) 346 (d) 421 (e) 754
(f) 800 (g) 978 (h) 1683 (i) 9018 (j) 6667

3-7 (a) 00010100 (b) 00010010 (c) 00010111
(d) 00010110 (e) 01010010 (f) 000100001001
(g) 000110010101 (h) 0001001001101001

3-9 (a) 0100 (b) 0110 (c) 1001
(d) 01000011 (e) 01001011 (f) 01011100
(g) 10001001 (h) 10101000 (i) 010000111010
(j) 010001111100 (k) 010101100100 (l) 100000110011
(m) 0100010110000100 (n) 0101011010101100 (o) 1001101101110100

3-11 (a) 0011 (b) 0110 (c) 1001
(d) 1100 (e) 01000110 (f) 01011000
(g) 01111010 (h) 10100101 (i) 010001101001
(j) 010010101011

3-13 (a) 1001 (b) 1100 (c) 100100
(d) 1010000 (e) 111101111 (f) 1000100010
(g) 1010101010 (h) 110101001010 (i) 111111100000
(j) 11001100110011

3-17 The right-most bit is the parity bit.
(a) 11011 (b) 10010 (c) 1010110
(d) 1110001 (e) 10101111 (f) 101110011011
(g) 101110011000110 (h) 101111101010101 (i) 101110101111111
(j) 1110110110010000

3-19 (a) No error (b) No error (c) No error (d) Error
(e) No error (f) Error (g) Error (h) No error

3-21 (a) 1011010 (b) 0100110 (c) 0010110
(d) 1110000 (e) 111001111 (f) 001011000

(g)   011000011   (h)   1011100011   (i)   0110000110
(j)   0010110000

3–23 (a) No error   (b) Corrected code: 1000011
      (c) Corrected code: 0110011   (d) Corrected code: 1011010
      (e) No error   (f) No error
      (g) Corrected code: 1001100   (h) Corrected code: 0000000
      (i) No error   (j) Corrected code: 1011001

3–25 See Figure A–1.

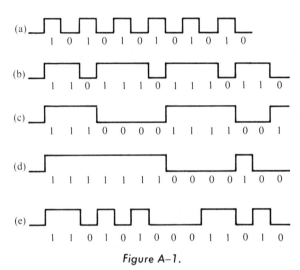

Figure A–1.

Chapter 4

4–1   See Figure A–2.

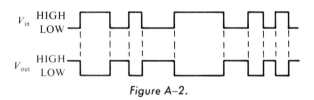

Figure A–2.

4–3   See Figure A–3.

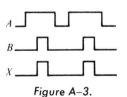

Figure A–3.

4–5   See Figure A–4.

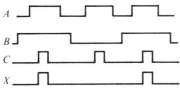

Figure A–4.

4–7   See Figure A–5.

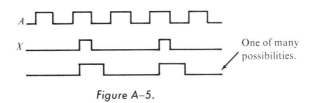

One of many possibilities.

Figure A–5.

4–9   See Figure A–6.

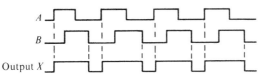

Figure A–6.

4–11   See Figure A–7.

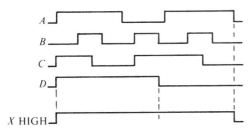

Figure A–7.

4–13   See Figure A–8.

A	B	C	X
0	0	0	1
0	0	1	1
0	1	0	1
0	1	1	1
1	0	0	1
1	0	1	1
1	1	0	1
1	1	1	0

Figure A–8.

4–15   See Figure A–9.

A	B	C	X
0	0	0	1
0	0	1	0
0	1	0	0
0	1	1	0
1	0	0	0
1	0	1	0
1	1	0	0
1	1	1	0

Figure A–9.

4–17   See Figure A–10.

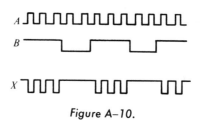

Figure A–10.

4–19   See Figure A–11.

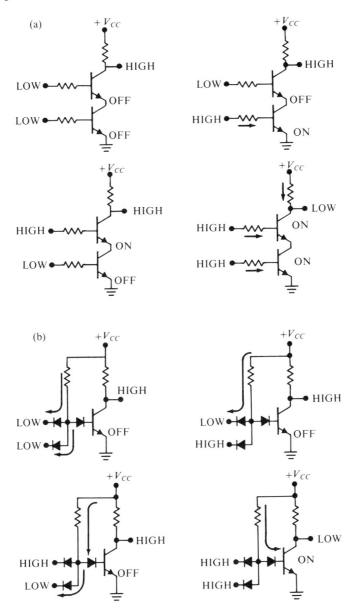

Figure A–11.

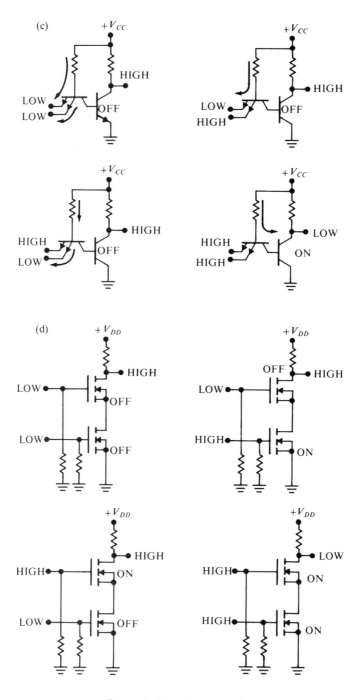

Figure A-11.   (Continued)

4-21   See Figure A–12.

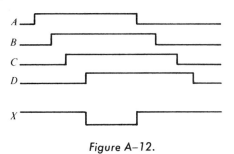

Figure A–12.

4-23   See Figure A–13.

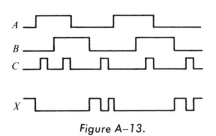

Figure A–13.

4-25   Two HIGHS produce a LOW for the NAND gate. A LOW produces a HIGH for the negative OR gate. The result is the same either way we look at the gate function. See Figure A–14.

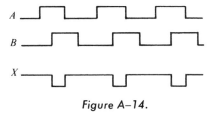

Figure A–14.

4-27   32 mW

4-29   0.3 V

4-31   3.3 V

4-33   Fan-in = 5

Chapter 5

5–1

(a)

A	B	X
0	0	0
0	1	0
1	0	0
1	1	1

(b)

A	B	C	X
0	0	0	0
0	0	1	0
0	1	0	0
0	1	1	0
1	0	0	0
1	0	1	0
1	1	0	0
1	1	1	1

(c)

A	B	X
0	0	0
0	1	1
1	0	1
1	1	1

(d)

A	B	C	X
0	0	0	0
0	0	1	1
0	1	0	1
0	1	1	1
1	0	0	1
1	0	1	1
1	1	0	1
1	1	1	1

(e)

A	B	C	X
0	0	0	0
0	0	1	1
0	1	0	0
0	1	1	1
1	0	0	0
1	0	1	1
1	1	0	1
1	1	1	1

(f)

A	B	X
0	0	1
0	1	1
1	0	0
1	1	1

(g)

A	B	C	X
0	0	0	0
0	0	1	0
0	1	0	0
0	1	1	0
1	0	0	1
1	0	1	0
1	1	0	0
1	1	1	0

(h)

A	B	C	X
0	0	0	0
0	0	1	1
0	1	0	0
0	1	1	1
1	0	0	0
1	0	1	0
1	1	0	1
1	1	1	1

(i)

A	B	C	X
0	0	0	0
0	0	1	0
0	1	0	0
0	1	1	0
1	0	0	0
1	0	1	1
1	1	0	1
1	1	1	1

(j)

A	B	C	X
0	0	0	1
0	0	1	1
0	1	0	1
0	1	1	0
1	0	0	0
1	0	1	0
1	1	0	0
1	1	1	0

5-3   (a) $X = AB$ (b) $X = \overline{A}$ (c) $X = A + B$ (d) $X = A + B + C$

5-5   (a)

A	B	X
0	0	0
0	1	1
1	0	1
1	1	1

(b)

A	B	X
0	0	0
0	1	0
1	0	0
1	1	1

(c)

A	B	C	X
0	0	0	0
0	0	1	0
0	1	0	0
0	1	1	1
1	0	0	0
1	0	1	0
1	1	0	1
1	1	1	1

(d)

A	B	C	X
0	0	0	0
0	0	1	0
0	1	0	0
0	1	1	1
1	0	0	0
1	0	1	1
1	1	0	0
1	1	1	1

(e)

A	B	C	X
0	0	0	0
0	0	1	0
0	1	0	0
0	1	1	1
1	0	0	1
1	0	1	1
1	1	0	0
1	1	1	1

5-7   (a) $\overline{A} + B + \overline{C}D$ (b) $\overline{A} + \overline{B} + (\overline{C} + \overline{D})(\overline{E} + \overline{F})$
        (c) $\overline{A}B\overline{C}D + \overline{A} + \overline{B} + \overline{C} + D$ (d) $\overline{A} + B + C + D + A\overline{B}\overline{C}D$
        (e) $AB + (\overline{C} + \overline{D})(E + \overline{F}) + ABCD$

5-9   (a) $AC + AD + BC + BD$ (b) $AD + \overline{B}CD$ (c) $ABC + ACD$

5-11   (a)   Sum-of-products   (b)   Sum-of-products   (c)   Product-of-sums
        (d)   Product-of-sums   (e)   Sum-of-products   (f)   Sum-of-products
        (g)   Sum-of-products   (h)   Product-of-sums

5-13   (a) A (b) AB (c) C (d) A (e) $\overline{A}C + \overline{B}C$

5-15   (a) $BD + BE + \overline{D}F$ (b) $\overline{A}\overline{B}C + \overline{A}\overline{B}D$ (c) B
        (d) $AB + \overline{A}CD + \overline{B}CD$ (e) ABC

5-17 $X = ABC + \overline{A}BC + A\overline{B}C + AB\overline{C}$

5-19   See Figure A–15.

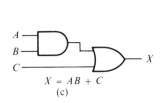

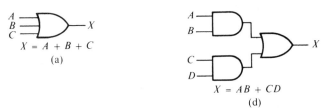

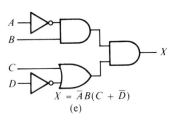

Figure A–15.

5-21   (a)   4   (b)   8   (c)   16   (d)   32

5-23   (a) $X = \overline{A}\overline{B} + \overline{B}C$ (b) $X = AC$ (c) $X = B$ (d) $X = \overline{C}$
          (e) $X = A + \overline{B}C$. No simplification possible.

5-25   (a) $X = \overline{B}(\overline{A} + C)$ (b) $X = AC$ (c) $X = B$ (d) $X = \overline{C}$
          (e) $X = (A + \overline{B})(A + C)$

5-27   (a) $X = A + B\overline{C} + CD$. No simplification possible.
          (b) $X = \overline{A}\overline{B}\overline{C} + ABC$ (c) $X = B\overline{C} + A\overline{C}D$ (d) $X = \overline{B}C$
          (e) $X = \overline{B} + \overline{D}$

Chapter 6

6-1   (a)

A	B	X
0	0	0
0	1	0
1	0	0
1	1	1

(b)

A	B	X
0	0	0
0	1	1
1	0	0
1	1	1

(c)

A	B	X
0	0	1
0	1	1
1	0	0
1	1	1

(d)

A	B	X
0	0	0
0	1	1
1	0	1
1	1	1

6–3

A	B	X
0	0	0
0	1	1
1	0	1
1	1	0

6–5 (a) $X = (AB + C)D + E$ (b) $X = \overline{(\overline{A} + B)\overline{BC}} + D$

(c) $X = (AB + \overline{C})D + \overline{E}$ (d) $X = \overline{(AB + CD)(EF + GH)}$

(e) $X = [(\overline{ABC})D][\overline{EFG}]$

6–7

A	B	C	D	X
0	0	0	0	0
0	0	0	1	0
0	0	1	0	0
0	0	1	1	0
0	1	0	0	0
0	1	0	1	0
0	1	1	0	1
0	1	1	1	1

A	B	C	D	X
1	0	0	0	0
1	0	0	1	0
1	0	1	0	1
1	0	1	1	0
1	1	0	0	0
1	1	0	1	0
1	1	1	0	1
1	1	1	1	0

6–9 See Figure A–16.

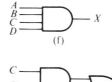

$$X = AB$$

Figure A–16.

6–11 See Figure A–17.

(a)

(b)

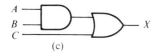
(c)

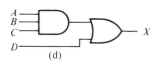
(d)

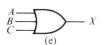
(e)

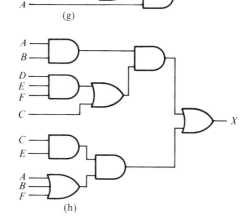

Figure A–17.

6–13 See Figure A–18.

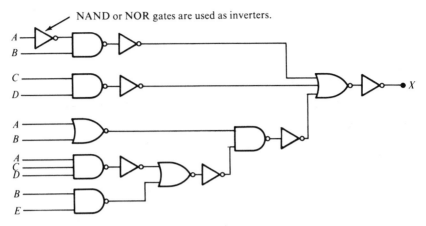

(a) This is an implementation of the expression
as is without modification or simplification.
It can be done in a simpler way.

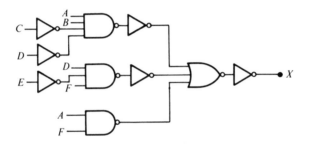

(b) Again, this is a straightforward implementation
of the expression as it "reads."

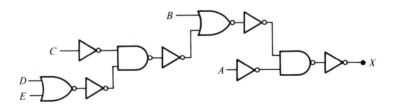

(c) Again, not the simplest—
but the most straightforward.

Figure A–18.

6–15   In each circuit in Figure A–19, an inverter symbol is shown for a NOR gate used as an inverter by connecting all inputs together.

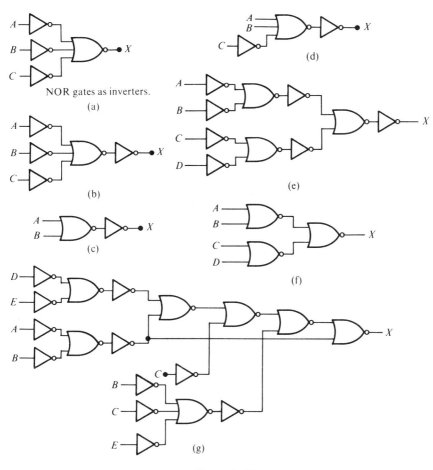

NOR gates as inverters.

(a)

(b)

(c)

(d)

(e)

(f)

(g)

Figure A–19.

6–17   See Figure A–20.

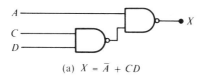

(a) $X = \bar{A} + CD$

Figure A–20.

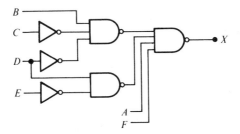

(b) $X = \overline{A} + \overline{F} + D\overline{E} + B\overline{C}\overline{D}$

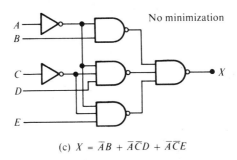

(c) $X = \overline{A}B + \overline{A}\overline{C}D + \overline{A}\overline{C}E$

Figure A–20. (Continued)

6–19 See Figure A–21.

$$X = \overline{A}C + A\overline{C}\overline{D} + BCD$$

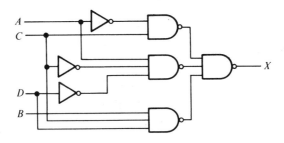

Inverter symbols represent NAND gates connected
as inverters (all inputs connected).

Figure A–21.

6–21   See Figure A–22.

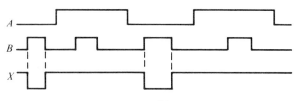

Figure A–22.

6–23   See Figure A–23.

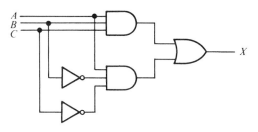

Figure A–23.

6–25   See Figure A–24.

$X \equiv$ light ON.

$A \equiv$ front door switch ON.

$B \equiv$ back door switch ON.

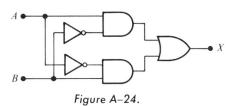

Figure A–24.

6–27   The output waveform is incorrect.

6–29   Gate G_2 failed with output high.

6–31   See Figure A–25. Pin numbers are circled.

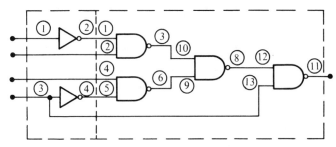

Figure A–25.

Chapter 7

7–1   Expression for both circuits: $X = A\bar{B} + \bar{A}B$. Exclusive-OR.

7–3   (a) G_1 output = 0   (b) G_1 output = 1   (c) G_1 output = 1
          G_2 output = 0 G_2 output = 1 G_2 output = 0
          G_3 output = 0 G_3 output = 0 G_3 output = 1
          G_4 output = 0 G_4 output = 0 G_4 output = 0
          G_5 output = 0 G_5 output = 0 G_5 output = 1

7–5   Sum = 1100

7–7   (a) 1   (b) 1   (c) 1   (d) 0   (e) 0

7–9   0001

7–11 See Figure A–26.

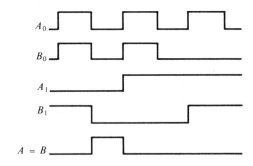

Figure A–26.

7–13   (a)   (b)   (c)
          G_1 output = 1 G_1 output = 1 G_1 output = 1
          G_2 output = 0 G_2 output = 1 G_2 output = 1
          G_3 output = 1 G_3 output = 0 G_3 output = 1
          G_4 output = 0 G_4 output = 0 G_4 output = 1
          G_5 output = 0 G_5 output = 0 G_5 output = 1
          G_6 output = 0 G_6 output = 0 G_6 output = 0
          G_7 output = 1 G_7 output = 0 G_7 output = 0
          G_8 output = 0 G_8 output = 0 G_8 output = 0
          G_9 output = 0 G_9 output = 0 G_9 output = 0
          G_{10} output = 1 G_{10} output = 0 G_{10} output = 0

(a)   (b)   (c)

G_{11} output = 0 G_{11} output = 0 G_{11} output = 0
G_{12} output = 0 G_{12} output = 0 G_{12} output = 0
G_{13} output = 0 G_{13} output = 1 G_{13} output = 0
G_{14} output = 0 G_{14} output = 0 G_{14} output = 0
G_{15} output = 0 G_{15} output = 1 G_{15} output = 0
    $A > B$ $A < B$ $A = B$

7-15   See Figure A–27.

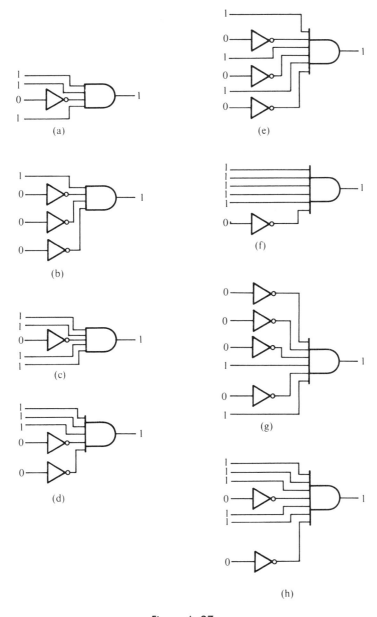

Figure A–27.

7–17 See Figure A–28.

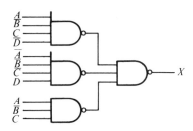

Figure A–28.

7–19 See Figure A–29.

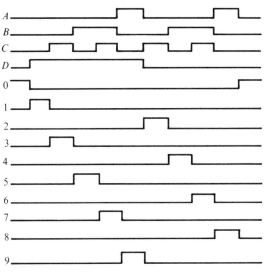

Figure A–29.

7–21 Output code:

A	B	C	D
1	0	1	1

Not a valid BCD code word. This circuit requires that a *single* input be active at any given time.

7–23

A	B	C	D
1	0	0	1

7–25 All 0s on the inputs to G_6 produce a 0 on its output.
All 1s on the inputs to G_2 produce a 1 on the G_7 output.
All 1s on the inputs to G_4 produce a 1 on the G_8 output.

7–27 Input D_1 is selected; therefore the output is a 1.

7-29 See Figure A-30.

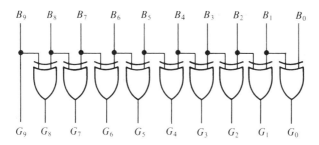

B_9 B_8 B_7 B_6 B_5 B_4 B_3 B_2 B_1 B_0

G_9 G_8 G_7 G_6 G_5 G_4 G_3 G_2 G_1 G_0

Figure A-30.

(a) 1 0 1 0 1 0 1 0 1 0 Binary
     1 1 1 1 1 1 1 1 1 1 Gray

(b) 1 1 1 1 1 0 0 0 0 0 Binary
     1 0 0 0 0 1 0 0 0 0 Gray

(c) 0 0 0 0 0 0 1 1 1 0 Binary
     0 0 0 0 0 0 1 0 0 1 Gray

(d) 1 1 1 1 1 1 1 1 1 1 Binary
     1 0 0 0 0 0 0 0 0 0 Gray

7-31
(a) BCD: 0010   (b) BCD: 1000   (c) BCD: 00010011
     Binary: 0010   Binary: 1000   Binary: 1101
(d) BCD: 00100110   (e) BCD: 00110011   (f) BCD: 01000101
     Binary: 11010   Binary: 100001   Binary: 101101
(g) BCD: 01100001   (h) BCD: 01110000   (i) BCD: 10000100
     Binary: 111101   Binary: 1000110   Binary: 1010100
(j) BCD: 10011001
     Binary: 1100011

7-33
(a) 0010   (b) 1010   (c) 1011
(d) 100100   (e) 101101   (f) 10000

Chapter 8

8-1 Assuming Q starts LOW, the waveform is as shown in Figure A-31.

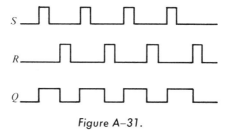

S

R

Q

Figure A-31.

8-3 See Figure A-32.

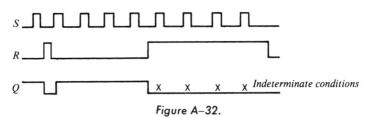

Figure A-32.

8-5 See Figure A-33.

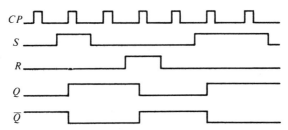

Figure A-33.

8-7 See Figure A-34.

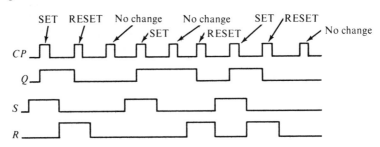

The *S* and *R* inputs can vary as long as
the proper conditions exist when the *CP*s occur.

Figure A-34.

8-9 See Figure A-35.

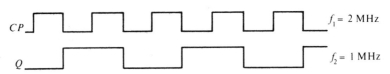

The flip-flop is negative edge triggered.

Figure A-35.

8–11 See Figure A–36.

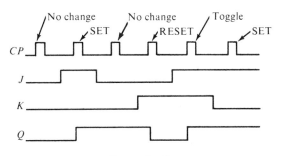

Figure A–36.

8–13 See Figure A–37.

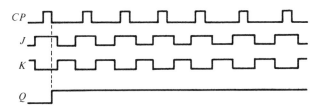

Figure A–37.

8–15 See Figure A–38.

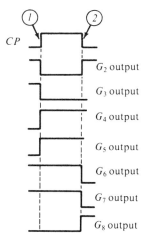

Figure A–38.

8–17   See Figure A–39.

Changes in *J* and *K* occur slightly after the positive-going clock edge.

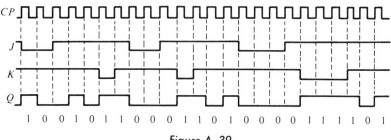

1 0 0 1 0 1 1 0 0 0 1 1 0 1 0 0 0 0 1 1 1 1 0 1

Figure A–39.

8–19   See Figure A–40.

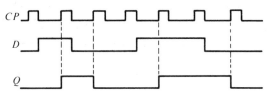

Figure A–40.

8–21 I_{tot} = 150 mA
P_{tot} = 750 mW

8–23   0011000

8–25   See Figure A–41.

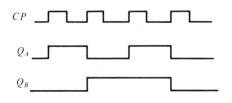

Figure A–41.

Chapter 9

9–1   (a) Reset   (b) Set   (c) Reset   (d) Set

9–3   (a) 2   (b) 2   (c) 3   (d) 4   (e) 4   (f) 5
(g) 6   (h) 7

9-5   See Figure A–42.

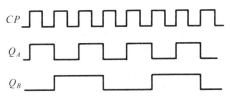

Figure A–42.

9-7   See Figure A–43.

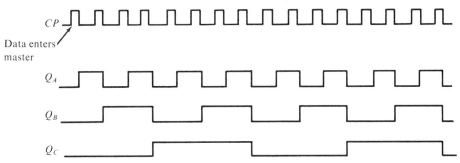

Figure A–43.

9-9   12 ns

9-11   The counter goes through the sequence 0000 to 1001. An example analysis for the first clock pulse is as follows:

$$J_A = K_A = 1. \quad \text{Therefore } Q_A \text{ goes to a 1.}$$
$$J_B = K_B = 0. \quad \text{Therefore } Q_B \text{ remains a 0.}$$
$$J_C = K_C = 0. \quad \text{Therefore } Q_C \text{ remains a 0.}$$
$$J_D = K_D = 0. \quad \text{Therefore } Q_D \text{ remains a 0.}$$

A similar procedure can be used for the remaining clock pulses.

9-13   1010, 1011, 1100, 1101, 1110, 1111

9-15   (a)   A single toggle flip-flop   (b)   Two flip-flops
          (c)   A modulus-5 counter   (d)   A decade counter
          (e)   A decade counter followed by one toggle flip-flop
          (f)   A decade counter followed by two toggle flip-flops
          (g)   A modulus-5 counter followed by five toggle flip-flops
          (h)   A decade counter followed by two modulus-5 counters
          (i)   Three decade counters   (j)   Four decade counters

9–17 See Figure A–44.

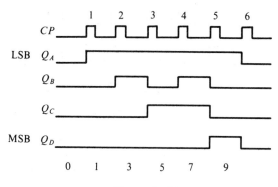

Figure A–44.

9–19 See Figure A–45.

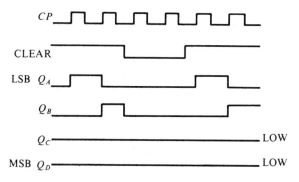

Figure A–45.

9–21 See Figure A–46.

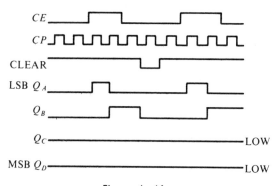

Figure A–46.

9-23 See Figure A–47.

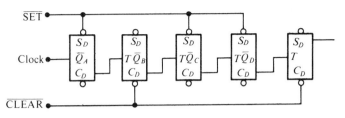

Figure A–47.

9-25 0000, 1111, 1110, 1101, 1010, 0101, 1010, 0101, 1010, . . .
The counter will "lock up" in the 0101 and 1010 states.

9-27 See Figure A–48.

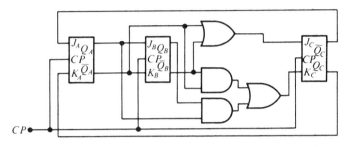

Figure A–48.

9-29 See Figure A–49.

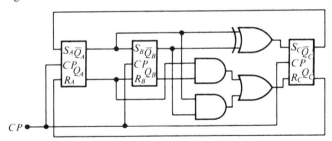

Figure A–49.

9–31 000000000
          100000000
          110000000
          111000000
          111100000
          111110000
          111111000
          111111100
          111111110
          111111111
          011111111
          001111111
          000111111
          000011111
          000001111
          000000111
          000000011
          000000001
          000000000
          See Figure A–50.

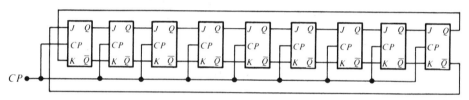

Figure A–50.

9–33 A 15-stage ring counter with the following stages SET: 8, 4, and 13.

9–35 See Figure A–51.

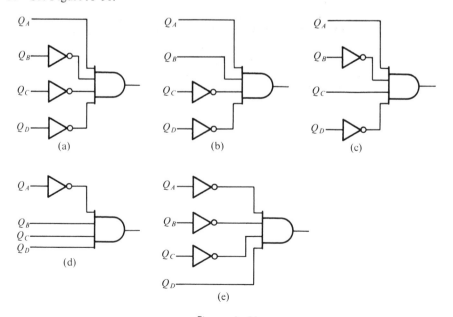

Figure A–51.

Chapter 10

10–1 When the flip-flop is SET, it is in one of its stable states which represents a binary 1. When it is RESET, it is in its other stable state which represents a binary 0. Since the flip-flop can remain indefinitely in either state, it acts as a memory or storage element.

10–3

After clock pulse	Q_A	Q_B	Q_C	Q_D
1	0	1	1	1
2	1	0	1	1
3	1	1	0	1
4	0	1	1	0
5	1	0	1	1
6	0	1	0	1
7	1	0	1	0
8	0	1	0	1
9	0	0	1	0

10–5 The output looks like the input delayed by 10 clock pulses.

10–7 See Figure A–52.

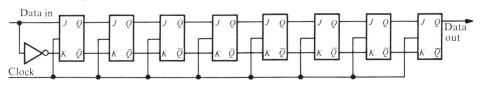

Figure A–52.

10–9

Initially	01001100
CP_1	10011000
CP_2	01001100
CP_3	00100110
CP_4	00010011
CP_5	00100110
CP_6	01001100
CP_7	00100110
CP_8	01001100
CP_9	00100110
CP_{10}	01001100
CP_{11}	10011000

10–11 See Figure A–53.

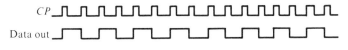

Figure A–53.

10–13 5816

10–15 See Figure A–54.

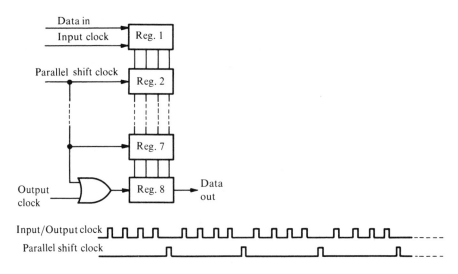

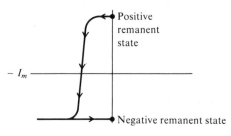

Figure A–54.

Chapter 11

11–1   See Figure A–55.

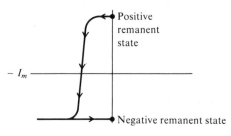

Figure A–55.

11–3   Because a core can reside in either of two magnetic states indefinitely, these states can be used to represent a binary 1 or 0.

11–5   It detects the occurrence of an induced voltage when the core is switched and thereby indicates the state of the core by the magnitude and duration of the induced voltage.

11–7   64 bits
One sense line

11–9   6 memory planes with 32 bits each

11-11 See Figure A-56.

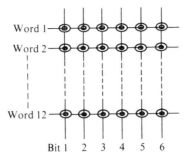

Figure A-56.

11-13 (a) 8   (b) 12   (c) 16   (d) 20

11-15 A static storage cell can retain information indefinitely as long as power requirements are met. A dynamic cell loses its stored information after a period of time due
to capacitive discharge and must be recharged periodically.

11-17 (a) 4   (b) 5   (c) 6   (d) 7   (e) 8   (f) 10

11-19 Information is not destroyed when it is read from memory.

11-21 (a) Coincident-select, linear-select   (b) RAM, associative
     (c) Static, core, dynamic, flip-flop   (d) RAM, ROM
     (e) FIFO

11-23 See Figure A-57.

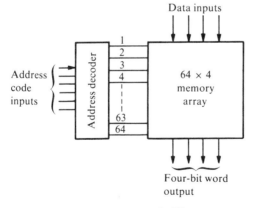

Figure A-57.

11-25

A_2	A_1	A_0	0_3	0_2	0_1	0_0
0	0	0	0	0	0	1
0	0	1	0	0	1	0
0	1	0	0	0	1	1
0	1	1	0	1	0	0
1	0	0	0	1	0	1
1	0	1	0	1	1	0
1	1	0	0	1	1	1
1	1	1	1	0	0	0

11-27   Remove all diodes except the following:
        63, 51, 41, 42, 44, 45, 30,
        21, 22, 23, 24, 25, 27, 11,
        12, 14, 15, 1, 3, 5, 7, 8, 9

11-29   See Figure A-58.

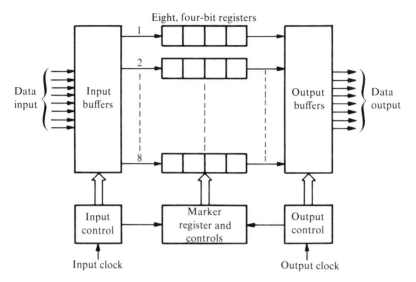

Figure A-58.

Chapter 12

12-1   Take the 9's or 10's complement of each negative number, and, using the appropriate method, add.

12-3   (a)   0.1001   (b)   0.1101   (c)   1.10111   (d)   1.101010
         (e)   0.1010111   (f)   1.1100011

12-5   (a)   0.0011   (b)   0.0011   (c)   1.0110   (d)   1.1110
         (e)   1.0001   (f)   0.0101

12-7   The serial adder using the 1's complement has a disadvantage over the 2's complement adder in that the final carry must be added to the accumulated sum, requiring an additional add cycle.

12-9

	Accumulator	Register B	FA Sum	FA C_o
Initially	1.1000	0.0110	0	0
After CP_1	0.1100	0.0011	1	0
After CP_2	1.0110	1.0001	1	0
After CP_3	1.1011	1.1000	1	0
After CP_4	1.1101	0.1100	1	0
After CP_5	1.1110	0.0110		

The 1's complement of the negative number is used for addition. The result is in 2's complement form.

12–11 (a) 01001100   (b) 00010000   (c) 01101101

These are the final results. The step-by-step procedure is left for the reader.

12–13 To determine the sign of the larger number

12–15 Overflow can occur only when both signs of the original numbers are the same and the sign of the sum differs from the original sign. For example, see Figure A–59.

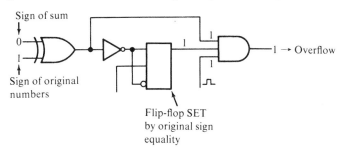

Figure A–59.

12–17 See Figure A–60.

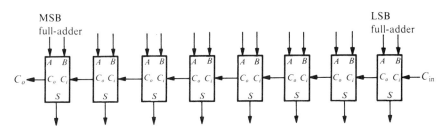

Figure A–60.

12–19 (a) 10111   (b) 1000010   (c) 1001001

The proper result occurs in each case.

12–21 (a) 1000   (b) 1001   (c) 000100010   (d) 00010010
     (e) 00010011   (f) 00011000

12–23 (a) Original sum $= 0101$ (b) Original sum $= 1101$
         Corrected sum $= 1011$ Corrected sum $= 0011$
         Final sum $= 0101$ Final sum $= 0011$
         Carry out $=\ \ \ 0$ Carry out $=\ \ \ 1$

     (c) Original sum $= 0010$ (d) Original sum $= 1001$
         Corrected sum $= 1000$ Corrected sum $= 1111$
         Final sum $= 1000$ Final sum $= 1001$
         Carry out $=\ \ \ 1$ Carry out $=\ \ \ 0$

     (e) Original sum $= 0001$ (f) Original sum $= 1010$
         Corrected sum $= 0111$ Corrected sum $= 0000$
         Final sum $= 0111$ Final sum $= 0000$
         Carry out $=\ \ \ 1$ Carry out $=\ \ \ 1$

Other points within the adder are left to the reader.

12–25 See Figure A–61.

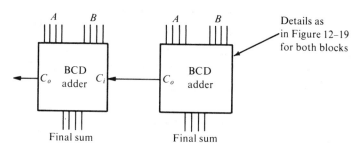

Figure A–61.

12–27 See Figure A–62.

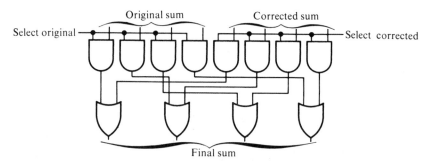

Figure A–62.

12–29 The final product is 1.110001. See Figure A–63.

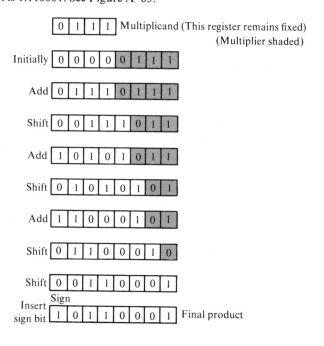

12–31 One possibility is shown in Figure A–64.

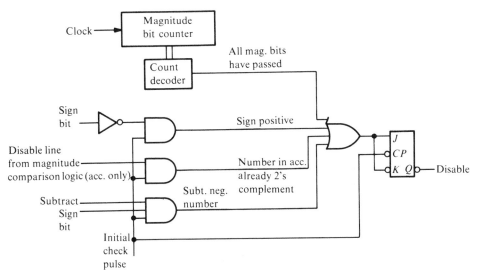

Figure A–64.

Chapter 13

13–1 $350\,\mu s$

13–3 See Figure A–65. A delay circuit.

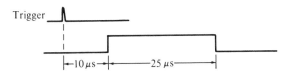

Figure A–65.

13–5 We get a square wave output with a period of 20 μs. See Figure A–66.

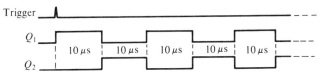

Figure A–66.

13–7

(a) Decrease frequency	(b) Increase frequency	(c) Increase frequency
Increase period	Decrease period	Decrease period
Decrease duty cycle	Increase duty cycle	Decrease duty cycle
No change in recovery time	Decrease recovery time	No change in recovery time

(d) Decrease frequency
     Increase period
     Increase duty cycle
     Increase recovery time

(e) No change in frequency
     No change in period
     No change in duty cycle
     Increase recovery time of Q_1

13–9   See Figure A–67.

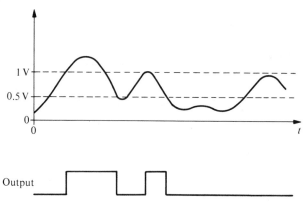

Figure A–67.

13–11   See Figure A–68.

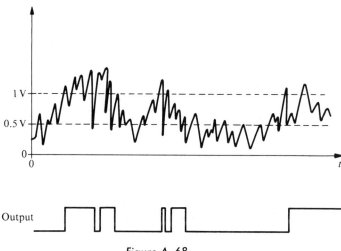

Figure A–68.

13–13   The cable, wire, etc., over which data are transmitted from one point to another

13–15   The single-ended uses a single signal-carrying conductor with a common ground return. The differential uses a pair of signal-carrying conductors, one for the signal and the other for the complement of the signal.

13–17   See Figure A–69.

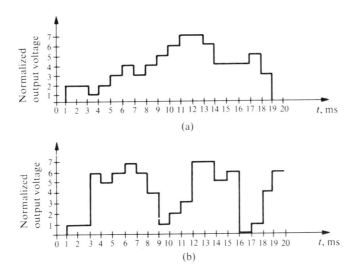

Figure A-69.

13-19 See Figure A-70.

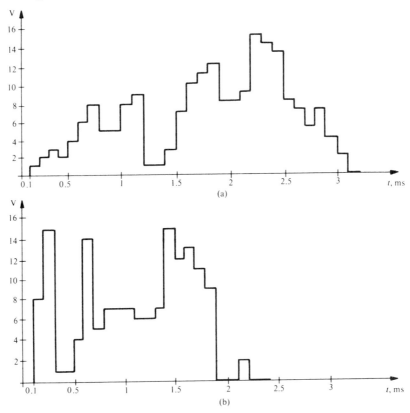

Figure A-70.

13–21 This can be shown by repeated application of the voltage divider rule starting at the 2^0 input.

13–23 001, 010, 100, 101, 110, 111, 111, 111, 111, 101,
101, 110, 110, 110, 110, 101, 100, 011, 010, . . .

13–25 See Figure A–71.

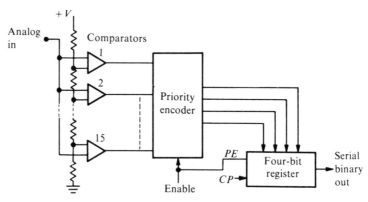

Figure A–71.

13–27 100, 110, 101, 100, 011, 010, 010, 010, 011, 100, 101, 101, 100.

Chapter 14

14–1 In an integrated circuit, all components are built-up on a *single* piece of semiconductor material, and all interconnections are formed on the single chip. A discrete component circuit consists of individual or separate components interconnected by wiring or other means, such as etched connections on a printed circuit board.

14–3 Bipolar ICs consist of junction transistors, whereas MOS ICs consist of FETs. There are, of course, many differences in the electrical characteristics and fabrication techniques.

14–5 $X = \overline{AB + CDE + F}$

It is called a wire-OR connection because each term is ORed together and complemented (actually NORed).

14–7 The input diode cannot be forward biased with an open input and is therefore like a reverse biased junction that occurs when a HIGH is applied.

14–9 The reason is that the collector signal is used to drive one of the output transistors and the emitter signal is used to drive the other. These two signals are inversions of each other, which means they are 180° out of phase.

Chapter 15

15–1 Input, output, memory, arithmetic logic unit, and control

15–3 A microprocessor is an LSI digital system that can be programmed for processing digital data.

15–5 The CPU is the central processing unit of a computer or microprocessor. It consists of an ALU, control logic, and several registers.

15–7 Program information contains instructions that tell the processing system what tasks to perform. *Data* refers to the information to be operated on, such as numbers.

15–9 The basic unit of digital information consisting of a group of bits.

15–11 Memory address register. It temporarily holds the address of information that is to be read from or written into the memory.

15–13 1. The address of the information to be put into the memory is loaded into the MAR from the CPU.

    2. The information itself is loaded into the MDR.

    3. The CPU tells the memory control to store the word in the specified address.

    4. The memory control initiates a write operation.

    5. When the word has been stored, the memory control issues a complete signal.

15–15 (a) 1. The number 3 is transferred from memory to the A register.

         2. The number 2 is transferred from memory to the D register.

         3. The control subsystem tells the ALU to ADD these two operands.

         4. The ALU performs the addition and puts the sum into the A register.

         5. The number 7 is transferred from memory to the D register.

         6. The control subsystem tells the ALU to ADD the contents of the A register to the contents of the D register.

         7. The ALU performs the addition and puts the final sum into the A register. The sum is 12.

         8. The control subsystem transfers the final sum in the A register to memory, thus completing the entire operation.

  (b) 1. The number 8 (multiplicand) is transferred from memory to the A register.

         2. The number 5 (multiplier) is transferred from memory to the M/Q Register.

         3. The control subsystem tells the ALU to MULTIPLY these two operands.

         4. The ALU performs the multiplication and puts the product into the D register. The product is 40.

         5. The control subsystem transfers the product to memory, thus completing the entire operation.

  (c) 1. The number 3 is transferred from memory to the A register.

         2. The number 2 is transferred from memory to the M/Q register.

         3. The control subsystem initiates a MULTIPLY operation.

         4. The ALU performs the multiplication of the two operands and puts the product in the D register. The product is 6.

         5. The contents of the D register are transferred to "scratch-pad" memory.

         6. Steps 1 through 4 are repeated for the operands 9 and 4. The product is 36.

         7. The first product now in the "scratch pad" is transferred back into the A register. The second product is in the D register.

         8. The control subsystem tells the ALU to ADD the contents of the A register to the contents of the D register.

      9.   The ALU performs the addition and puts the sum into the A register. The sum is 42.

     10.   The control subsystem transfers the final result in the A register to memory, thus completing the entire operation.

(d)   1.   The number 6 (dividend) is transferred from memory to the D register.

      2.   The number 3 (divisor) is transferred from memory to the A register.

      3.   The control subsystem tells the ALU to DIVIDE these two operands.

      4.   The ALU performs the division and puts the quotient into the M/Q register.

      5.   The contents of the M/Q register are transferred into the D register.

      6.   The number 5 is transferred from memory to the A register.

      7.   The control subsystem tells the ALU to ADD the contents of the A and D registers.

      8.   The ALU performs the addition and puts the sum into the A register.

      9.   The control subsystem transfers the sum to memory.

15–17 One of the two main cycles of the control subsystem in which the next instruction to be executed is taken from the memory.

15–19 Control subsystem

15–21 It stores the instruction word fetched from memory.

15–23 1. The control subsystem decodes the instruction word contained in the IR.

       2. The control subsystem gets the operand from the memory as specified by the instruction word.

       3. The operation specified by the op code portion of the instruction word is carried out by the processor.

       4. The program counter is incremented.

15–25 A program that is normally used to perform specialized or repetitive operations during the course of the execution of the main program. A program within a program.

15–27 Data entered into this type of register arrangement can be thought of as being pushed further down into the stack. As data are taken out, each successive stored word moves back toward the top of the stack so that the last word entered is the first to be taken out.

15–29 An interrupt is a request from an external device or peripheral unit for information or servicing by the computer. The processor normally branches to a subroutine for executing the interrupt request.

15–31 Direct memory access

Index